Taxation and Mining

Taxation and Mining

Nonfuel Minerals in Bolivia and Other Countries

Malcolm Gillis
Institute Fellow, Harvard Institute
for International Development,
and Lecturer in Economics,
Harvard University

and

Meyer W. Bucovetsky	Associate Professor of Economics, Institute for Policy Analysis, University of Toronto
Glenn P. Jenkins	Lecturer in Economics, Harvard University, and Institute Associate, Harvard Institute for International Development
Ulrich Petersen	Professor of Mining Geology, Harvard University
Louis T. Wells	Professor of Business Administration, Graduate School of Business Administration, Harvard University
Brian D. Wright	Assistant Professor of Economics, Yale University

Ballinger Publishing Company • Cambridge, Massachusetts
A Subsidiary of J.B. Lippincott Company

This book is printed on recycled paper.

International Standard Book Number: 0-88410-458-3

Library of Congress Catalog Card Number: 77-23806

Printed in the United States of America

Library of Congress Cataloging in Publication Data

Gillis, Malcolm.
Taxation and mining.

Includes bibliographical references.
1. Mineral industries—Taxation—Bolivia. 2. Underdeveloped areas—Mineral industries—Taxation. I. Title.
HD9506.B62G5 336.2'78'33820984 77-23806
ISBN 088410-458-3

For the memory of
Coyle McKinnon Gillis

Contents

List of Figures

List of Tables

Foreword

The Harvard Institute for International Development (HIID) is a multi-disciplinary center with links to all parts of Harvard University. The Institute offers a variety of programs related to the developing countries. These programs are focused not only upon teaching and research, but technical assistance and training of officials from developing countries. Major fields of interest at HIID include economic policy analysis, health, education, and rural development, and the analytic techniques required for effective and innovative work in these fields.

This book illustrates many aspects of the orientation of HIID programs. First, much of the methodology and conclusions reported herein is the product of stimulating interaction between practitioners of different disciplines: in this case economics, geology, and business administration. Second, as in so many undertakings in which HIID is involved, contributors were from a variety of national backgrounds: two Americans, two Canadians, one Australian and one Peruvian. Two of the contributors, including the principal author, are members of the HIID staff in Cambridge. Two other contributors, both of whom have been involved in several other HIID projects elsewhere, were drawn from the Harvard community, while two others hailed from other prestigious universities in the United States and Canada. Third, all contributors have been actively involved not only in teaching and research, but in technical assistance to policy makers in a diverse group of developed and developing nations, ranging from Asia, Africa and Latin America to North America, in a variety of fields. All contributors shared at least two traits: strong interest in the analysis and design of public policy toward nonrenewable resources and a conviction that the complexities of the subject are so great as to preclude effective and relevant research by any single discipline such as economics.

The contributors have produced a result that is also typical of HIID's work. It is, first of all, based on sound, up-to-date technical analysis in each of the disciplines involved. However, the essential quality of the work is that this technical expertise is brought to bear on practical policy issues. Indeed, the work is an outgrowth of a request by the Bolivian government for advice and is responsive to that need. Because the issues addressed are problems common to many countries with mineral resources, the volume should appeal to a broad audience encompassing several types of specialists in economics, natural science, and business administration, as well as public officials involved in the design and administration of government policy toward the mining sector.

Lester E. Gordon
Director
Harvard Institute
for International Development

Acknowledgments

A volume of this scope could not have been successfully completed without active and energetic support of a large group of people in Bolivia and elsewhere. Therefore, the authors wish to acknowledge the contributions made by various persons in Bolivia and the United States. However, such recognition does not in any way implicate such individuals, or their institutions, or for that matter, Harvard University, in the opinions expressed in this volume. We particularly wish to acknowledge the cooperation provided in Bolivia by a number of persons in the Ministry of Mining and Metallurgy, the Ministry of Planning and Coordination, and the Ministry of Finance.

Special gratitude is due to the following persons in the Ministry of Mining and Metallurgy: Minister Gen. José Antonio Zelaya Salinas, Undersecretary Armando de Urioste Sánchez Bustamante, Director of Mines Guillermo Murgia Sánchez, Director of Metallurgy Gonzalo Gutierrez Moscoso, Director of Planning Alvaro Ugalde Canedo, and Head of the Department of Economic Studies Rolando Jordán Pozo.

In the Ministry of Planning, we are indebted to Gaston Villa, Undersecretary; David Vargas Chacón; and Jorge Montalvo.

Various persons in the Ministry of Finance were helpful, including David Blanco Blandes, Bernardo Carvajal, Carlos Otálora, and Oscar Saavedra.

In other institutions, we are particularly thankful to Dr. Enrique García of the Industrial Park, Jorge Felix Ballivian of the National Statistical Institute, Antonio Sanchez de Lozada of the private mining sector, and Dr. Fernando Urquidi of the Bolivian Geological Survey.

Colleagues at Harvard and other institutions read and provided critical—and uniformly useful—comments and suggestions for revision of the manuscript. These included Michael Roemer and Joseph Stern, Institute Fellows at HIID; Professor Richard Musgrave, Department of Economics at Harvard; Professor Ralph Beals of Amherst College; and Professor Carl Shoup, emeritus, Colombia University. Their contributions are also acknowledged in specific chapters.

Special recognition and thanks are due to Mary-Anita Browne, our secretary in the Harvard Institute for International Development, for her high-level efficiency and extraordinary energies in typing various drafts of the chapters and the final versions as well. We also wish to thank Elvira Sánchez-Bustamante de Morató for typing various early drafts of some sections of this volume.

Chapter One

Taxation, Mining, and Economic Development

Malcolm Gillis

GENERAL INTRODUCTION

Much has been written over the past decade, particularly since 1973, concerning taxation of hydrocarbon resources, from the point of view of both exporting and consuming nations. But there is a relative scarcity of published material dealing with a broad range of issues in the taxation of nonfuel minerals, and most of what is available is focused upon issues specific to the developed countries, particularly the United States.[1] There exist several valuable and comprehensive volumes dealing with mineral concessions policy in which substantial attention is devoted to tax considerations associated with foreign investment in the mining sectors of less developed countries.[2] But there are virtually no studies which examine the complex of micro and macro issues involved in governmental tax policy toward both foreign and domestic investors in nonfuel minerals, and almost no published work which has as a major component the discussion of key issues in taxation of state-owned mineral enterprises. This latter omission is particularly unfortunate, given the sizeable and growing share of state-owned enterprises in world mineral production.

This volume seeks to develop a basis for a clearer understanding of the complicated issues involved in the design and implementation of taxes on nonfuel minerals, particularly—but not exclusively—in less developed countries (LDC's) where mining[a] furnishes a substantial share of total export earnings, taxes and domestic value-added, especially where state enterprises are a significant feature of the mining sector. The case study upon which much of the discussion is based is that of the Bolivian mining sector, but the authors believe the spectrum of topics chosen and the approach taken in this volume are such that the implications derived from the Bolivian and related cases are likely to be of general interest to a broad audience of economists, geologists and mining officials in the public and private sectors of developed and developing countries alike.

MINERAL RESOURCES AND ECONOMIC STRUCTURE: SOME INTERNATIONAL COMPARISONS

A number of nations around the world rely heavily upon mining as a source of national income, foreign exchange and fiscal resources. While Bolivia is not nearly so dependent upon her mining sector as Liberia, Gabon, Zambia, Zaire or Papua New Guinea (see Table 1-1), and although Bolivian mining production per capita (at U.S. $70 in 1974) is low compared to such countries as Canada or Chile (U.S. $284 and $175 respectively in 1974) the Bolivian economy remains highly dependent upon mining in spite of recent rapid growth in both the hydrocarbon and commercial agriculture sectors.[3] Indeed, Bolivia shares many features with sister nations that rely heavily upon the mining sector as a means of national livelihood. Like Chile, Gabon, Jamaica, Sierra Leone and Zambia, the mining sector in Bolivia accounts for more than 10 percent of Gross Domestic Product (GDP). Seldom does mining account for more than 60 percent of total exports, as in Bolivia, Liberia, Jamaica, Sierra Leone and Zaire. And rarely does the mining sector provide more than 25 percent of a country's total tax revenue, but this is the case for Bolivia, New Caledonia, Gabon, Jamaica, Papua New Guinea and a handful of other countries.

Minerals taxation is a sufficiently complicated subject even in nations where the mining sector is of limited relative significance. But in those countries where mining is responsible for an inordinately high proportion of national income, export earnings and taxes, the formulation, evaluation and reform of minerals tax policy consistent with national development objectives becomes an even more difficult

[a]Throughout this book, the terms "mining" and "mining sector" refer exclusively to the nonfuel minerals sector. This convention was chosen not only to simplify exposition, but because in many less-developed countries, including Bolivia, traditional usage of the label "mining sector" excludes energy resources.

Table 1-1. Importance of the Mining Sector, Several Countries (nonfuel minerals only)[a]

Share of Mining in GDP, Exports and Taxes	*Country*
A. *Mining Sector as Percent of GDP*	
0-2.4%	Brazil, Colombia, Ecuador, Eire, Finland, Honduras, Mexico, Pakistan, Philippines, Thailand
2.5-4.9%	Indonesia, Peru
5-9.9%	Canada, Malaysia
10-14.9%	Bolivia, Chile
15-19.9%	Gabon, Sierra Leone
20-29.9%	Jamaica
30% and above	Liberia, New Caledonia, Papua New Guinea, Zaire, Zambia
B. *Mining Sector Exports as Percent of Total Exports*	
0-9.9%	Colombia, Ecuador, Eire, Ghana, Honduras, Indonesia, Mexico, Pakistan, South Korea
10-19.9%	Brazil, Canada, Gabon, Mexico, Philippines, Thailand
20-59.9%	Malaysia, Morocco, Peru
60-79.9%	Bolivia, Jamaica, Liberia
80% and above	Chile, New Caledonia, Papua New Guinea, Sierra Leone, Zaire, Zambia
C. *Mining Sector Taxes as Percent of Total Central Government Tax Revenue*	
0-9.9%	Colombia, Ecuador, Eire, Honduras, Indonesia
10-24.9%	Chile, Malaysia, Thailand
25% and above	Bolivia, Gabon, Jamaica, Liberia, New Caledonia, Papua New Guinea, Zaire, Zambia

[a]Except for Sierra Leone (1968-69), data refer to the years 1971-1974. Information upon which this table is based was derived from a wide variety of sources detailed at the end of Chapter Five, Table 5-3.

task, and requires a thorough grasp of the nature of the interdependence between the mining sector and the national and international economy. In such a context, the study of mining taxes cannot be usefully done in isolation from the national and international setting in which mining takes place. Rather, such an undertaking necessarily becomes a study of the mining sector in its technical, economic and social aspects. This latter theme recurs repeatedly throughout this volume.

The need for a broad perspective on minerals tax policy is particularly critical in circumstances where state-owned enterprises loom large in the national mining picture, given that such enterprises are often expected to pursue objectives other than, or in addition to, those commonly set for privately owned mineral enterprises. Such objectives often include employment, reduction of regional economic imbalances, reduction of dependence on foreign capital, and a variety of "social" functions not commonly imposed on private sector firms.[4] The extent of public (i.e., state-owned) enterprise participation in minerals production around the world has not been extensively documented. Whereas publicly owned enterprises in the natural resources sectors have been exceedingly rare in recent years in North America[b] and Australia, and almost nonexistent in nonfuel minerals, the role of state-owned firms in nonfuel minerals exploitation is substantial among the developing countries (see Table 1-2), and from all indications appears to be steadily growing in relative importance.[5]

Table 1-2. Some Indicators of the Relative Role of State Enterprises in Mining, Selected Countries (nonfuel minerals only)

Country	*Percent Share of State-Owned Mining Firms*
A. Brazil	• 62% of net assets in the mining sector (1972)
	• CVRD accounted for 80% of iron and steel exports (1972)
B. Bolivia	• 63% of mineral exports (1974)
	• 33% of total government tax revenues (1974)
	• 30% of total employment in mining (1974)
C. India	• 24% of value-added in the mining sector
D. Indonesia	• 95% of tin exports (1975)
	• 37% of other nonfuel mineral exports (1975)
	• 75% of total employment in mining (1974)
E. Korea	• 33% of value-added in the mining sector
F. Mexico	• 37% of mining production (1976)
G. Zambia	• 98% of mineral exports and value-added in the mining sector

Sources: For Bolivia, Indonesia and Zambia, Chapter Five, Table 5-3. For others, see Malcolm Gillis, "Taxation, Mining and Public Ownership" in University of Arizona School of Mines, *Taxation of Nonrenewable Resources in the Western States* (Lincoln Institute Monograph No. 77-2, Cambridge, Mass., 1977).

[b]Four important exceptions in Canada in recent years include: SOQUEM (the Quebec Mining Exploration Company); Eldorado Nuclear Limited (a federal crown corporation); Syncrude (a joint venture for exploiting Alberta's oil sands, involving the federal government and the provincial governments of Alberta and Ontario), and Sidbec-Normines (a large iron-ore project at Fire Lake, 51 percent owned by provincial governments).

In some countries, state-owned firms have been responsible for well over half of nonfuel mineral exports since 1970. These include not only Bolivia, but Brazil, Chile, and Indonesia. The case of Brazil is particularly instructive. There, state-owned mining firms control over 62 percent of net assets in the sector, and the public enterprise CVRD (Cia Vale do Rio Doce) accounts for over 80 percent of iron ore mining and exports. In Bolivia, COMIBOL (Corporacion Minero de Boliva) is typically responsible for nearly two-thirds of mine production, one-third of mining sector employment and between 15 and 20 percent of total tax collections. In other countries, state-owned firms were responsible for less than half, but still a significant share of overall mining activity. For example, in 1971-72, state enterprises accounted for about one-third and one-quarter of value-added in the South Korean and Indian mining sectors, respectively. In a very few countries, state-owned firms account for virtually all mineral production. This is the situation for copper in Zambia, and for the socialist countries of Burma and Congo-Brazzaville. But there remain a substantial number of developing nations in which private domestic and foreign firms continue to dominate the nonfuel minerals sectors. Such is the case for the Dominican Republic, Honduras, Jamaica, Malaysia, Papua New Guinea, Philippines, and Thailand. But if there is a discernible trend in any of this latter group of nations, it is toward increased involvement (whether alone or in joint venture arrangements) of public sector entities, particularly in Jamaica, Malaysia, and Thailand.[6]

PERSPECTIVES ON COSTS AND BENEFITS OF INVESTMENT IN MINERAL RESOURCES IN DEVELOPING COUNTRIES

From a broad social perspective, a nation's stock of mineral resources may be viewed as essentially no different from unfinished capital goods, or alternatively, a ready-made stock of natural capital that can be utilized to shortcut the normally slow and painful process of accumulation of human capital by usual means.[7] In order to tap this natural capital, investment of human and physical capital in exploration, exploitation and processing is required. But unless the view is taken that investment in additional capacity in mining is inherently desirable,[c] there are strong reasons for believing that investment in the mining sector should be subject to the same criteria that would normally be applicable to investment in any type of capital good: whether the prospective investment will pay. Put another way,

[c]See the next section of Chapter One for a digression on this point.

investment in mining, as in any other sector, can be justified only if it serves to produce benefits commensurate with the costs involved, or if there are strong social policy or strategic reasons for ignoring these costs. Therefore, formulation of effective public policy toward investment in mineral resources involves a thorough examination of all prospective benefits and costs.

Among the benefits most often cited for new investments, not only by mining industry leaders but government officials in less developed nations as well are:

1. increased labor income arising from expanded employment in the mining sector;
2. increased capital income for domestic shareholders in mineral projects;
3. increased labor and capital income for domestic firms who furnish supplies and materials to mineral projects;
4. increased access to, and transfer of, knowledge about technology, markets and managerial knowhow;
5. promotion of the development of lagging or isolated regions within a country;
6. a widening of national foundations for resource-based industrialization (RBI);
7. expansion in net foreign exchange earnings;
8. expansion of fiscal resources, in the form of government tax and other revenues.

Without much doubt, most of these benefits are indeed present, to a greater or lesser degree, in any significant mineral investment project. But the magnitude of many of them can easily be overstated, particularly those such as numbers 4-6, which in addition, cannot be easily measured. In any case, experience in a variety of nations indicates some have been more significant than others in a less-developed country setting.

Because modern mineral projects typically involve very substantial amounts of capital[d] and normally require access to sophisticated technology, many developing nations with important mineral endowments have found it essential to turn to large foreign enterprises for development of their respective mining sectors. In other cases, such countries have chosen to meet large-scale capital requirements

[d]One fairly large sample of investment plans of 368 nonfuel minerals projects in 91 countries indicates that the *average* size of such investment will be U.S. $168.5 million over the period 1976-1982. See Gillis, "Taxation, Mining and Public Ownership," p. 12.

in mining through government investment in state-owned mineral enterprises. In both cases, the first three types of benefits have in the past not proven to be particularly substantial.

Expanded labor income for domestic workers directly employed in minerals projects has, by and large, not constituted a significant benefit, both because of the highly capital-intensive nature of modern mining operations, and (in many cases) the location of mineral deposits in remote areas of the country, away from major population centers. Even in nations such as Bolivia, where mining has dominated economic life for decades, employment in the mining sector has rarely exceeded three percent of the labor force. And in the typical case, the wage bill is often no more than three to six percent of gross mining revenues.[8] Nor have domestic benefits in the form of capital income for domestic shareholders typically been sizable. The basic reason is that modern mining has, over the past half century, been dominated by foreign firms having the requisite capital, technology, and market access for successful exploitation of large mineral deposits.[9] And although one of the chief aims of the joint venture movement in less developed countries has been that of increasing the share of domestic investors (type 2 benefits) in overall capital returns from minerals projects, the discernible results of joint venture arrangements in this respect cannot be adjudged to have been very substantial by the mid-seventies.[10] With regard to type 3 benefits, a common expectation is that new investment in mining will generate additional income for domestic laborers and shareholders in industries linked to the mineral project, such as domestic suppliers of materials. But often this expectation is not fully realized, for a variety of reasons including the marked import intensity[11] of much mineral investment and subsequent operation, the remoteness of many minerals projects in areas well away from accessible domestic suppliers (thereby often making it easier to import materials from abroad rather than procuring them domestically), and in many cases, generous import duty exemptions on materials and supplies granted by governments in the past to foreign minerals firms.[12]

Type 4 benefits, those arising from transfer to the national economy of mining and metallurgical technology—particularly from foreign firms—can potentially be significant over the longer term. But in the short term, domestic capacity for absorbing new mineral technology may be and often is limited by a lack of indigenous geologists, mining engineers and metallurgists. But in general, nearer term benefits from transfer of technology have not been large in spite of LDC efforts to promote technological transfer through insistence on joint ventures. Indeed, there are indications that, at least in early stages of a foreign investment project, parent firms treat joint ventures *less* favorably than wholly owned subsidiaries and in particular tend to be *less* inclined to transfer technology to a joint venture without payment.[13]

Investments in mineral projects—whether by foreign or state-owned enterprises—are commonly expected to play a major role in the development of lagging or backward regions in LDC's. Unfortunately, the relative significance of this type of benefit has not been systematically documented, much less reliably established. The magnitude of this type of benefit can be, and often is, easily overstated, given that the external benefits created by mining firms (infrastructure such as roads, schools, other community services) are often double counted,[14] whereas pecuniary benefits in the form of expanded labor and capital income to linked industries are already included in type three benefits discussed above. In addition, much of the infrastructure built in the course of a mineral project is so highly specific to mining (rail spurs from mine sites, mining camps, access roads, etc.) so as to be of very limited general utility to activities other than mining.

It is sometimes argued that an important benefit arising from investment in mineral resource projects is that it furnishes a broader basis for mounting a natural strategy of resource-based industrialization (RBI). Under such a strategy, investment in the extractive sector is viewed as a superior means of achieving more rapid industrialization through expanded domestic processing geared toward either foreign markets (outward-looking RBI) or self-sufficiency in domestic markets (inward-looking RBI). But it is not obvious that RBI is better suited to achieve national development goals than other potential development strategies, and the record of experience with RBI cannot be said to have been unambiguously advantageous for countries that have followed it, particularly the inward-looking variant.[15]

The last two types of benefits that may flow from investments in mineral projects are considerably easier to assess and subject to fewer ambiguities of interpretation, insofar as their contribution to national development is concerned, particularly in the short to medium term. Benefits from both expanded foreign exchange earnings and fiscal revenues can be quantified rather easily in ex-post terms and, within limits, can be estimated with a reasonable degree of confidence in ex-ante terms. Additional net foreign exchange earnings from mining allow a country greater command over external resources required in the development process, while expanded fiscal revenues confer greater capacity for transferring both domestic and foreign resources to the public sector. But foreign exchange benefits do not accrue automatically to an LDC minerals producing country; explicit provision must be made for policies that will help insure that expected foreign exchange benefits actually materialize. Where mining projects are undertaken by foreign enterprise, capture of foreign exchange benefits[e] for an LDC is heavily dependent (especially in the early stages of operation) upon tax policy to insure that the projects generate "retained value"[f] for the national economy. The concept of retained value as used by such authors as Mikesell,[16] measures the overall foreign exchange contribution of a minerals project to the economy in which the project is located. The notion of retained value is similar to, but narrower than, measures such as the "terms of investment," that are often employed to assess the distribution of gains from investment between an investor and the host economy.[17] More specifically, for any given minerals project, retained value (RV) constitutes the sum of all mining revenues retained in a producing country, and includes labor income of domestic workers (W_d), that

---

[e]Of course, when LDC minerals projects are geared largely toward satisfaction of demand for the domestic market, foreign exchange benefits would accrue in the form of foreign exchange saved rather than in direct generation of exchange proceeds. But with some important exceptions (Brazil and India) the overwhelming proportion of mineral production in LDC's is exported, typically well over 95 percent (see Malcolm Gillis et al., *Taxation and the Mining Sector in Bolivia,* (Report prepared for the Ministry of Mining and Metallurgy, Bolivia; Harvard Institute for International Development, December 1975), Chapter Two.

[f]Following Mikesell, the breakdown of gross revenues (R) of a mining project between retained value (RV) and other components may be expressed as follows:

$$1)\ R = RV + M + I + L + P + W_f + U$$

where: M = imports; I = interest paid on external loans and credits; L = loan repayments; P = profits transferred abroad; W_f = salaries of expatriates accruing abroad; U = unidentified items.

Given the same tax treatment for both domestic and foreign firms, the percentage of RV to R will vary according to a variety of factors, including nature of deposit, nature of financing arrangement and the stage of the project, but will clearly be different for foreign than domestic investment projects. In the adaptation of Mikesell's retained value concept as used in this chapter, the concept may be symbolically expressed as:

$$2)\ RV = W_d + C_d + DP\,(1-Z) + K_d + T_d + Q$$

where: W_d = labor income and fringe benefits for domestic workers; C_d = expenditures by expatriate employees in the domestic economy; DP = domestic purchases of goods and services; Z = import content (in %) of DP; K_d = capital payments to domestic shareholders; T_d = taxes and duties paid to the domestic treasury, Q = other payments.

portion of payments to expatriate workers spent locally (C_d), domestic purchases of goods and services (DP) net of their import content, capital payments to domestic shareholders (K_d), plus taxes and duties paid to the domestic treasury (T_d) and other, miscellaneous payments (Q). For mineral projects in LDC's retained value is typically less than half of gross revenues (R) generated by the project when foreign firms are involved, and may vary between 50 and 80 percent when domestic firms undertake projects.

In Peru over the period 1966-72, taxes and quasi-taxes constituted slightly more than 60 percent of retained value generated by the large Toquepala copper mine.[g] Other components of retained value were relatively small, in spite of concerted governmental efforts to increase local purchasing of materials and employment of Peruvian nationals in the Toquepala project. Wages and salaries were but 8.1 percent of gross revenue and 16.6 percent of value retained for the Peruvian economy. The experience of Sierra Leone also underscores the importance of taxation in generating retained value for the domestic economy. There, retained value in mining in 1963-65 was but 43 percent of gross revenues, and taxes accounted for 47 percent of total retained value.[18] Where state-owned mining enterprises are concerned, tax policy may or may not be the principal means of ensuring substantial foreign exchange benefits. The experience of some countries, including Indonesia,[19] suggests that failure to fully subject the income of state-owned extractive firms to prevailing tax laws may result in significant, and permanent, losses in foreign exchange benefits (and therefore retained value) that would otherwise be expected from natural resource exploitation.

In summary, while it is important to identify, and to the extent possible quantify, all significant benefits that may accrue to an LDC from exploitation of its mineral resource endowments, it seems prudent to conclude that tax policy must be the backbone of public policy toward investment in mineral resources that is intended to result in an equitable sharing of benefits (in the case of foreign investment) or to safeguard broader national interests from narrower, sectoral interests (where state-owned enterprises are involved).

If not all of the benefits claimed for mineral resource development can be quantified with any degree of precision, most of the potential societal costs often thought to be associated with mining investment are even less subject to systematic identification and measurement. Among the costs most often cited are:

---

[g]Raymond F. Mikesell, *op. cit.*, p. 69. In turn, retained value (total revenue from the project minus external payments) was 49 percent of gross revenue over the period 1966-1972. In the earlier years of the project (1960-65) retained value was but 30 percent of gross revenue.

1. Increased vulnerability to shocks generated in world markets,
2. Increased national dependence (in the broader sense) upon the outside world in general or upon particular developed countries,
3. Unwanted environmental consequences, including erosion and air and water pollution,
4. Potential effects of inframarginal mineral projects upon non-mineral exports and import-substituting industry.

A frequently voiced claim is that heavy reliance upon mineral exports, particularly single commodities, involves substantial costs in the form of heightened economic instability for minerals-producing countries. The argument has its roots in the observation that, owing to low price elasticities of demand and supply for minerals on world markets, world mineral markets tend to be highly volatile, characterized by sharp cyclical (and in 1974, abrupt) changes in world demand and supply. Countries such as Bolivia, Chile, Peru and Zambia, where foreign exchange earnings and government revenue are highly dependent upon mineral exports, are therefore particularly vulnerable to external shocks over which they have no control and which are not easily anticipated. In fact, it is clear, for countries such as Bolivia, that extreme variability in mineral export earnings has been an important source of inflationary pressures, not only in boom periods but often in years of depressed world market conditions.[20]

But marked economic instability is not inherent in heavy dependence upon mineral export earnings. Rather, such instability might just as plausibly be attributed to inadequate government policy for dealing with an abrupt surge in export earnings and tax receipts during boom conditions. If then the government, and its chief constituencies, become accustomed to markedly higher levels of public spending in the course of the boom, institutional rigidities (and political realities) tend to preclude the type of government policy responses required to cope with subsequent sharp declines in mineral export values. In the first instance, policies for partial sterilization[h] of a sudden increase in foreign exchange reserves could materially reduce inflationary pressures, as in Bolivia in most recent years except 1973 and 1974. Given failure to sterilize exchange reserves accruing in the boom phase and the near universal impossibility (and often undesirability) of reducing government expenditures in response to a precipitous decline in government revenues (even when such expenditures have only recently been swollen

[h]In this context, "sterilization" refers to the insulation of the domestic monetary base from sudden changes in foreign exchange reserves. Typically, sterilization involves measures to prevent increased foreign exchange reserves from accumulating too rapidly as reserves of the Central Bank (see Chapter Three).

during the boom), the inflationary impact of deficit finance implicit in sharp cutbacks in mineral export earnings cannot be so easily handled.

Although many LDC leaders in international forums have cited the costs arising from high dependence upon unstable export earnings as a strong reason for curbing the growth of mineral investment in their countries, it does seem clear that much of the perceived costs of this instability can be reduced by deft management of foreign exchange reserves. Finally, it might be noted that the consequences of export instability for income growth in LDC's are not unambiguously negative. According to Caves and Jones, statistical analysis has failed to show that countries that experience marked fluctuations in export earnings grow slower, as a group, than those with modest fluctuations.[21]

In recent years, a radically different type of "dependence" that is claimed to arise from participation in world trade, and by extension, from reliance upon mineral resource exports, has received a substantial amount of attention, particularly from neo-Marxists. Whereas the notion of dependency discussed above refers to stabilization problems associated with vulnerability of economies that rely heavily on mining for foreign exchange and taxes, the newer group of "dependency" theorists consider a much wider range of dimensions of "dependency," including not only trade dependence, but financial, cultural, technological, market and managerial dependence as well.[22] The thrust of many of these recent writings on "dependency" is nothing less than an explanation for underdevelopment based upon the dependence of third world countries upon the capitalist economies of the industrial West and Japan; in short, the claim is made that dependence causes underdevelopment. Some adherents to the neo-Marxist school of dependency view its costs as great enough to justify a complete withdrawal of developing countries from world markets for mineral products, coupled with pursuit of economic autonomy through measures to insulate themselves from the workings of capitalistically organized world markets.[23] In effect, then, this branch of the dependency school categorically denies the benefits from international trade.

A heightening of dependency of this type would not necessarily be counted as a cost by all, or even most, governments, or by the groups within LDC's who stand to benefit most from development of their mineral resources. But in some countries, including Guinea and Burma, political leaders have for years been disposed to view mineral resource development from one or another version of the dependency perspective, and in such disparate countries as Jamaica and

Papua New Guinea one may discern in the mid-seventies a growing inclination to employ the rhetoric, and often the outlook, of the dependency school, in spite of the fact that many of the basic propositions of dependency theory remain decidedly ambiguous. In fact, even some sympathetic reviewers of dependency theory conclude that advocates tend to overstate the effects of dependency.[24]

The wave of concern over the environment that crested in the United States in the early seventies led to legislation such as the Water Pollution Control Act of 1972 and the Clean Air Act Amendment of 1970, both of which have served to limit mineral resource development, particularly on the public lands in the western states. While it would be inaccurate to say that important less developed mining countries have tended to overlook potential environmental costs from resource exploitation, it is nevertheless true that control of the pollution and erosion which may accompany any mining operation has not been of overriding concern in most mineral concessions agreements signed in the LDC's over the past two decades. This is not to say that the issue has been ignored; in fact, the mining codes of several LDC's, including Bolivia, contain explicit and lengthy sections dealing with penalties for damage of private and governmental property rights where pollution or contamination of the environment occurs. And in Malaysia, strong measures have been taken in recent years to reduce erosion and pollution arising from hydraulic methods of mining eluvial tin deposits. Further, a growing awareness of, and disinclination to incur, environmental damage may be noted in a number of mineral concession agreements signed since 1970[25] and in model concession contracts drawn up by expert consultants.[26] The new provisions involve not only the prescription of penalties for pollution and erosion, but enjoin companies to take positive steps to prevent environmental damage. Indeed, in recent years many multinational mining concerns themselves appear to have become more sensitized to environmental issues, as in many cases special departments for environmental study and implementation have been established.[27]

Finally, large inframarginal minerals projects can involve significant short- to medium-term frictional costs in the form of reduced profitability for a country's nonmineral export sector and for import-substituting industry. But here again, government policy can be deployed to reduce the significance of this particular cost. This issue can perhaps best be presented first in terms of polar examples, then by moving to more common cases.

Nature has been particularly generous, in the form of natural resource endowments, to some nations with small populations. In energy resources, Kuwait is the prime example, while the island of Nauru in the South Pacific is an extreme illustration of nature's munificence in nonfuel minerals.[28] In such countries, per capita export earnings are large enough to provide every family with a high standard of living quite apart from income from labor or capital. In such extreme circumstances, a free market rate of exchange (relative to, say, the U.S. dollar) would tend to be so high that practically everything would be imported; neither export-substituting nor import-substituting industry would be attractive to prospective investors under such conditions. This phenomenon is often called the "Kuwait effect." Furthermore, the high value of the currency, by making imported capital goods far cheaper than would be the case without very large mineral exports, provides powerful incentives for capital intensity for those productive activities that do continue to be undertaken.

Few less developed nations have had the fortune to face the challenges of dealing with extreme cases of "Kuwait effects." But a number of countries in which the mining sector furnishes a very large share of total export proceeds have encountered milder but nevertheless troublesome forms of "Kuwait effects" over the past decade. Examples, other than in hydrocarbons, include Bolivia, Gabon, Zaire, Liberia and Zambia. In all of these countries, the mining sector of course contributes materially to overall economic welfare, not only as a source of income but through its effect in making possible lower domestic prices of imports.[29] But some argue that the blessings made possible by large mineral exports may not be entirely unmixed.[30]

The argument basically runs as follows: One effect of having minerals dominate the export picture is precisely that all foreign goods are cheaper than would be the case without exportation of substantial amounts of foreign capital deposited in past eons. This state of affairs would normally be viewed as representing an unambiguous gain in welfare for all interest groups, with perhaps two important exceptions: import-competing domestic producers and domestic exporters of nonmineral products, both of whom will experience reductions in income owing to the lower prices of competing imports (facing the former) and the lower prices for exports (facing the latter) in domestic currency. If governments are more sensitive to the complaints and political pressures of domestic producers than to the plaudits of domestic consumers, they will tend to react to the situation by imposing (or increasing) tariffs and other trade restrictions to provide protection against imported goods (or, in the case of damaged exporters, subsidies to exports). This scenario is a plausible one for at least two reasons. First, it may be argued that individuals tend to be much more aware of developments and policies that affect them as income earners than those that affect them as consumers.[31] Second, because producers of items affected by an appreciated exchange rate are far fewer in number than consumers of those items, the former may more easily organize into effective pressure groups than the amorphous mass of consumers, who may barely perceive the effects of tariffs. In any case, trade restrictions of this type would benefit a few people (producers) a great deal, and harm many people (consumers) a little, so that the producers have much stronger incentives for pressuring governments for further protection than have consumers for opposing such measures. If producers are initially successful in inducing the government to enact trade restrictions to counter the decline in their incomes caused by exchange rate appreciation arising from the advent of large mineral projects, the story will not normally end there. To take the case of import-competing industry, to the extent that higher tariffs or quotas do serve to curb imports, the exchange rate will appreciate further, necessitating still further increases in the tariff level in order to maintain protection at previous levels.

If, as a result of such policies, value added in all the production of all tradeable goods received uniform levels of effective protection, and if at the same time nonmineral exports were freed (through subsidies) from the burden of the exchange rate overvaluation that would tend to arise,[32] the above phenomenon would not normally represent a particularly serious concern. But in practice no tariff system in present use provides for uniform effective protection. Rather, most studies of effective protection show that tariff policy in developing countries tends to become convoluted toward protection of final consumer products, to the detriment of domestic intermediate and capital goods industries.[33] As a result, welfare costs from misallocation of resources, as well as government revenue losses from import restriction, can be substantial. As Due notes, this type of misdirected tariff policy is not unique to mineral-exporting countries—nor to developing countries in general—but the effects of disproportionate reliance upon mineral resource exports constitutes fertile ground for the rise of convoluted tariff policy.

Tax policy can be deployed to reduce some, but not all, of the costs cited above. Clearly a tax system designed to sterilize abrupt increases in export earnings arising from mineral booms can moderate the destabilizing pressures on the national economy that otherwise would tend to occur both during and after boom conditions have passed. In addition, tariff policies which avoid "cascading" of duty rates (low nominal rates of import duty on capital and intermediate goods and high rates of duty on finished goods) can help reduce some of the costs of convoluted trade policies which may thrive in the presence of mild or moderate "Kuwait effects" that may arise in countries where mineral exports do have significant effects on the economy. Tax policy can also be used to reinforce incentives and regulations for protecting the environment against damage from mining operations. But tax policy would seem to be ill suited as a device for reducing the tangible or, more importantly, intangible costs of "dependency" in those countries where such costs are perceived to have major importance.

As the next section indicates, there is a general expectation among Bolivian policy-makers that tax policy should play an important role in expanding the benefits from, and reducing the costs of, mineral resource development in that country.

GOALS AND INSTRUMENTS OF GOVERNMENT POLICY TOWARD THE MINING SECTOR IN BOLIVIA

Overall Objectives of Government Policy

Any study that is focused upon one sector of a national economy is vulnerable to a number of pitfalls, not the least of which is the possibility of biased conclusions that may arise from a myopic view of the workings of the economy, and of insufficient appreciation of the opportunities and limitations facing the economy as a whole. The Bolivian economy, like other national economies, is a general equilibrium system, in which priorities set for one sector cannot be fully divorced from those formulated for other sectors, much less from broader societal goals. Thus, any examination of development strategy for the mining sector, and of course, any assessment of the present and potential role of tax policy in this strategy, must be viewed in light of not only narrower sectoral objectives, but overall national economic and social goals.

The broader medium- and longer-term objectives of government economic and social policy have been explicitly restated in a series of directives of the National Planning Agency (CONEPLAN),[3 4] together with underlying priorities for productive sectors, infrastructure and rural development. Among other features, these documents recognize the possibilities for potential conflict among the array of basic objectives, particularly in the medium term, and provide qualitative comment upon the relative weights to be attached to certain medium-term objectives in the event of conflict.

For a longer term, basic objectives for national development as given by the Planning Agency are as follows:

1. Reduction in income inequality, both across households and between regions.
2. Rapid growth in per capita Gross Domestic Product (GDP).
3. Sustained growth in employment, leading ultimately to full employment.
4. Improvements in general living levels, through imported nutrition, housing and greater general accessibility of education and health services.
5. Maintenance of a reasonable degree of economic stability, both in terms of the price level and the balance of payments.

There are a number of medium-term objectives adjudged to be supportive of achievement of the longer-term goals, including the solution to Bolivia's century-old problem of access to the sea, steps to increase internal savings, attracting foreign savings on a selective basis, and the channeling of internal and external savings into activities with potential for a high economic rate of return from the national point of view. Finally, public policy pronouncements suggest that a high premium is to be placed upon the more complete integration of the indigenous *campesino* population into society, through expanded programs of rural development, particularly in the Altiplano, coupled with measures for promoting the dispersal of population toward thinly populated regions.

Economic growth is stated to be the overriding medium-term economic policy objective,[35] as the position has been taken that only the vigorous pursuit of economic growth will enable significant progress in attaining other objectives such as reduction of income inequality, regional balance, full employment and price stability. Some of the policy guidelines issued by the Planning Agency take explicit note of the possibility of conflict between economic growth and other economic and social objectives, particularly that of reducing income inequality. Indeed, the nature of the conflict between growth and income distribution goals, together with arguments for applying a heavier weight to either objective, has been a much-discussed topic in recent economic development literature. A number of authors have strongly argued in favor of assigning a much heavier weight to income distribution goals, even at the expense of growth, while others continue to emphasize the importance of economic growth. This literature is now quite extensive[36] and has not escaped the attention of Bolivian planners, who have nevertheless been explicit in stating that (at least for the present) in the event of conflict between growth and other objectives the former criterion is to be assigned primary (but not exclusive) importance.

Policy Objectives and Instruments for the Mining Sector

Although general national policy objectives for the economy as a whole have been explicitly delineated in CONEPLAN documents, that same degree of explicitness is not nearly so evident with regard to sectoral goals, particularly as they pertain to the mining sector. Although there exists a broad awareness both of what is expected of the mining sector and of the desired general pattern of government policy toward the sector, neither the objectives nor the means (instruments) have been clearly spelled out in the past. The following section presents the authors' interpretation of prevailing government policy objectives for the mining sector, distilled from a variety of government documents[37] and from discussions with government officials, and also attempts to identify those policy instruments that are widely viewed within the government as supportive of these goals.

Ultimate Goals

The nonrenewable mineral resources of Bolivia continue to represent a significant and valuable part of the national patrimony, and by law are the property, not of individuals or firms, but of the people at large.[i] The underlying, or ultimate objective of government policy toward the mining sector, then, is taken to be that of ensuring maximum benefits from mineral exploitation for both present and future generations of Bolivian citizens. Given this goal, policy should be directed toward the optimal use of the natural and human resources of the sector over time,[38] where the basic, but not sole, criterion for "optimal use" is prospective economic returns, properly discounted. In this way the sector may continue to provide, at lowest social cost, a sizable share of the nation's fiscal and foreign exchange resources over the relevant future, with a minimum wastage of both human talents and of dwindling resource endowments. Development

[i]In Bolivia, as in virtually all less-developed countries, property rights in minerals cannot be vested in individuals. Rather, mining rights are owned by the government. The United States is one of the few nations in which property rights in minerals can be vested in individuals, as in all states, title to the surface land rights generally conveys title to subsoil rights as well. In Canada, Australia and Eire, the overwhelming proportion of mineral rights are retained by the state, not private individuals. See Gillis, "Taxation, Mining and Public Ownership," pp. 4-7.

of the mining sector is not, however, viewed as an end in itself, but as a means for supporting the economic diversification of the nation, reducing national dependency and vulnerability to shocks generated in the world economy, as well as providing for more robust, but stable development of the nation's mining regions.

If these are indeed close approximations to the underlying or *ultimate* objectives of government policy toward the mining sector, then we may identify several *proximate* objectives that the government appears to view as supportive of the ultimate objectives. These proximate objectives are more concrete in nature and more amenable to the policy instruments (means) at the government's disposal.

Proximate Objectives and Means of Government Policy Toward the Sector

In order to avoid mixing of objectives and instruments, it is useful to identify an objective as a goal, the attainment of which may have a value in and of itself. Instruments, or *means*, then might be viewed as methods for achieving objectives. It is rarely the case that decision makers, whether in the public or private sector, habitually make sharp distinctions between objectives and means, and this is no less true for Bolivia than for other countries, developed and developing. Thus, the classification in Table 1-3 below should be viewed merely as a serious effort both to clarify key policy issues and to place them in as sharp a focus as possible, but not as the last word on the subject. It is to be noted that some of the instruments are intended to serve more than one objective. Also, although the potential for direct conflict between objectives at the sectoral level appears minimal, it is nevertheless true that many objectives will be competing for limited fiscal and financial resources.

The alternative tax reform "packages" presented in Chapter Eight are designed to be supportive, to the extent possible, of the objectives outlined in Table 1-3. But as may be concluded from inspection of this table, it is clear that tax policy is but one (albeit an important one) of the many instruments that can be brought to bear upon efforts to insure that the mining sector continues to contribute to the support of a broad variety of national objectives. The table also helps to make apparent the correctness of the view that mineral tax reform undertaken in isolation from reform of other public policies and institutions is not likely to materially further the broad aims of government policy toward mining. However, tax reform, because it has a bearing upon a number of the basic objectives, undoubtedly has a significant potential role to play.

Table 1-3. Apparent Government Objectives and Policies: Mining Sector

Proximate Objectives	*Instruments*
1) *Economic Diversification:* Diversification of Production and Reduction of Vulnerability to Instability in Export Markets	1) Provision for a resurgence in mineral development through tax reform, creation of an exploration fund and a stable climate of regulations.
	b) Fuller use of mining potential in that part of the mineral belt set aside as "fiscal reserves."
	c) Coherent programs for survey and inventory of mineralized areas, especially of the small mining sector.
	d) Active participation in international efforts to stabilize and raise prices of, and real returns from mineral exports.
2) *Fiscal and Foreign Exchange Resources:* Continued Generation of High Share of Fiscal Resources and Foreign Exchange from the Mining Sector	a) Promotion of expanded exploration and development, as in 1-a above.
	b) Tax reform geared to expenditure requirements of the central government.
	c) Fuller and more rational use of mineralized areas in the nation's fiscal reserves.
3) *Productivity:* Higher Productivity and Efficiency, especially in Nationalized Mines	a) Rationalization of public investment decisions in the mining sector and in complementary infrastructure, in the framework of sensible project analysis.
	b) Policies and measures to promote higher recovery of metallic content in mining, beneficiation, and smelting.
	c) Research and development activities to lay groundwork for improved methods of exploitation, beneficiation, and smelting.
	d) Tax reform to remove artificial incentives for waste and misallocation of resources.
	e) Strengthening of accounting, marketing and management methods and skills, particularly in state mineral enterprises.

Table 1-3. (cont.)

Proximate Objectives	*Instruments*
4) *Working Conditions:* Improvements in Conditions of Work, Safety and Retirement Prospects for Laborers.	a) Improvement in institutions and regulatory mechanisms affecting the mining sector.
	b) Better use of limited investment resources in social infrastructure in mining districts.
	c) Reform of social security system, especially health benefits.
	d) Tax policy which more fully recognizes certain outlays for improving health and safety conditions (ventilation, maintenance and repair of equipment, medical care, etc.).
5) *Overall National Returns:* Higher Value Added from minerals processing in Bolivia.	a) Promotion of investment in the national metallurgical industry in projects of demonstrated high prospective economic returns from the national point of view.
	b) Promotion of improved beneficiation methods, especially to improve recuperation.
	c) Improvements in physical infrastructure in the national mining belt.
	d) Tax reform and foreign investment policy geared to a higher degree of domestic processing.
	e) Improved marketing (commercialization) of mineral exports, to reduce losses from uninformed sales policy and wastes in transport.
6) *Private Sector:* Ensure Favorable Conditions for Development of a Viable Private Mining Sector Alongside a Strong State Enterprise Sector	a) Provision of stable regulatory and tax climate.
	b) Selection of foreign investment proposals complementary to aims and expectations for the private domestic mining sector.
	c) More efficient conduct of credit policy; improving access to credit for the small mining sector.

THE SCOPE OF THIS VOLUME

Chapter Two summarizes the salient structural and operating features of the Bolivian mining sector. Here, key nontax factors in sectoral development are identified and placed in perspective with the broad fiscal issues of relevance to mining in Bolivia. Chapter Three examines the macroeconomic impact of the mining sector and also provides an explicit explanation of the model of the sector's role in the economy. Prime focus here centers on the fiscal and monetary policy implications of two basic variables: changes in the volume of production and fluctuations in metal prices. Hopefully, as more adequate national accounts and mining statistics are developed, the model can be expanded and improved, so as to provide a firmer basis for analysis of the interplay between changes in the mining sector and global economic activity in Bolivia.

Chapter Four presents a detailed discussion of the basic considerations involved in the design of minerals tax policy in a variety of settings, including that of Bolivia. Chapter Five contains a general discussion of the evolution of minerals taxes around the world, together with summary discussions of mineral tax legislation in seventeen countries. Chapter Six involves a detailed scrutiny of the existing system of Bolivian mining taxes, and also compares the tax treatment of mining with other sectors of the economy. Chapter Seven includes a thorough evaluation of the impact of the existing tax system upon the mining sector, in terms of the system's effects on incentives for efficiency in exploitation and processing, investment in new capacity, government revenues and other important goals. Chapter Eight presents two alternative options for mining tax reform, both of which are intended to be supportive of basic public policy objectives set for this important sector. Finally, Chapter Nine discusses a variety of nontax issues thought to have an important bearing on the prospects for successful reform of mining taxes, in Bolivia and elsewhere.

NOTES TO CHAPTER ONE

1. See for example College of Mines, University of Arizona, and Lincoln Institute, *Nonrenewable Resource Taxation in the Western States* (Cambridge, Mass., Lincoln Institute Monograph No. 77-2).

2. Two of the best examples of such studies are David N. Smith and Louis T. Wells, Jr., *Negotiating Third World Mineral Agreements: Promises at Prologue* (Cambridge, Mass.: Ballinger Publishing Co., 1975); and Raymond F. Mikesell, *Foreign Investment in Copper Mining: Case Studies of Mines in Peru and Papua New Guinea* (Baltimore: Johns Hopkins Press, 1975).

3. For a comprehensive review of the economic structure in Bolivia in the late seventies, together with an exhaustive examination of the full range of economic policy issues facing Bolivia, see Richard A. Musgrave et al., *Fiscal Reform for Bolivia* (Cambridge, Mass., forthcoming, 1977).

4. For a listing of the variety of goals typically set for state-owned enterprises, see International Legal Center, *Law and Public Enterprise in Asia* (New York: Praeger Special Studies in Economic Development, 1976), p. 17.

5. For an assessment of the share of state-owned mineral enterprises in world mining investment planned over the period 1976-1982, see Malcolm Gillis, "Taxation, Mining and Public Ownership," in *Taxation of Nonrenewable Resources in the Western States.*

6. Gillis, *ibid.*, pp. 3-11.

7. For a similar view, see Orris C. Herfindahl, "Mineral Surveys and Economic Development" in David Brooks (ed.), *Resource Economics: Selected Works of Orris C. Herfindahl* (Baltimore: Johns Hopkins Press, 1974), pp. 236-251.

8. Even in the relatively labor-intensive Bolivian mining sector, the wage bill generally constitutes no more than six percent of total mine revenues. See Chapter Three.

9. There are indications, however, that the share of state-owned enterprises in world mining investment is increasing steadily. See Gillis, "Taxation, Mining and Public Ownership," pp. 3-10.

10. For a survey of the impact of the joint venture movement in LDC's, see Louis T. Wells, "Foreign Investment in Joint Ventures: Some Effects of Government Policies in Less-Developed Countries," Economic Development Report No. 167, Harvard Development Advisory Service, Cambridge, Mass., 1970).

11. For a Bolivian example of import intensity in mining, see Chapter Three. For a Peruvian case, see Mikesell, *op. cit.*, pp. 70-71.

12. Blanket import-duty exemptions for foreign-source materials and supplies were common in many mineral concessions agreements signed before 1970, as for example in Indonesia and Ghana. But in recent years, government pressures for local buying have begun to change this pattern and in many cases have actually led to the development of local businesses to supply the needed items. See Smith and Wells, *op. cit.*, pp. 104-105.

13. See Raymond Vernon, *The Economic and Political Consequences of Multinational Enterprise, An Anthology* (Boston: Harvard Graduate School of Business Administration, 1972), p. 35.

14. See Glenn P. Jenkins, "Evaluation of Social Externalities from Investments in Community Infrastructure by Mining Enterprises" (Harvard Institute for International Development, Cambridge, Sept. 1976).

15. See Michael Roemer, "Resource-Based Industrialization in the Developing Countries" (Harvard Institute of International Development, Development Discussion Paper No. 21, January 1, 1977).

16. See Mikesell, *op. cit.*, pp. 66-71.

17. The "terms of investment" refer to the ratio between the net social benefits accruing to the host country and the private returns accruing to

investors abroad. See Tony Killick, "The Benefits of Foreign Direct Investment and Its Alternatives," *Journal of Development Studies*, January 1973, pp. 301-315.

18. Computed from information presented in Table 1, Killick, *ibid.*

19. See Bruce Glassburner, "In the Wake of General Ibnu: Crisis in the Indonesian Oil Industry," *Asian Survey*, Vol. XVI, No. 12, December 1976, pp. 1099 *ff.*; and Malcolm Gillis and Charles E. McLure, Jr., "Tax Policy and Public Enterprises" (paper prepared for Musgrave Commission on Fiscal Reform for Bolivia, August 1976), Section II.

20. This argument is developed more fully in Chapter Three.

21. See Richard Caves and Ronald W. Jones, *World Trade and Payments: An Introduction* (Boston, Little, Brown and Company, 1976), p. 554.

22. For a critique of basic elements of "dependency theory" as applied to resource-based industrialization strategy, see Roemer, *op. cit.*, pp. 56-60. For a recent summary of various aspects of the radical perspective on development economics in a broader context, see E. Wayne Nafziger, "A Critique of Development Economics in the U.S.," *Journal of Development Studies*, Vol. 13, No. 1, Oct. 1976, pp. 18-33.

23. See, for example, Chiweizu, *Black Scholar*, forthcoming (July-August 1977); and Timothy Shaw, "Zambia: Dependence and Underdevelopment," *Canadian Journal of African Studies*, Vol. 10, No. 1, 1976, pp. 3-32.

24. See Thomas E. Weisskopf, *Dependence as an Explanation of Underdevelopment: A Critique*, Discussion Paper No. 66, Center for Research on Economic Development, February 1977.

25. See, for example, Article 2, Section d of the 1977 "third generation" concession agreement between the Government of Indonesia and Rio Tinto.

26. Smith and Wells, *op. cit.*, p. 219.

27. This has been the case for Utah International, for example.

28. See John F. Due, "Mineral Resources, Taxation and the Developing Economies" (manuscript, University of Illinois, Urbana, November 1976), pp. 1-2.

29. See Chapter Three of this volume for an estimate of the gain in economic welfare made possible by the existence of a mining sector in Bolivia.

30. John Due, *op. cit.*, pp. 3-4.

31. This argument has been developed by Anthony Downs in his *Economic Theory of Democracy* (Harper and Row, 1957), pp. 253-257.

32. See Malcolm Gillis and Charles E. McLure, Jr., "Coordination of Tariffs and Internal Indirect Taxes" in Richard A. Musgrave and Malcolm Gillis, *Fiscal Reform for Colombia* (Cambridge, Mass.: International Tax Program, 1971).

33. See, for example, Bela Belassa and Associates, *The Structure of Protection in Developing Countries* (Baltimore: Johns Hopkins Press, 1971).

34. CONEPLAN, "Directivas para la Conducción de la Política Económica y Social a Mediano Plazo," mimeographed, La Paz (August 28, 1975).

35. *Ibid.*, pp. 6-7.

36. See, for example, J. Penn, *Income Distribution* (New York: Praeger, 1971), Chapter Three; Hollis Chenery et al., *Redistribution with Growth* (London and New York: Oxford University Press, 1974), especially Chapter

Two; William R. Cline, *The Potential Effects of Income Redistribution on Economic Growth: Latin American Cases* (New York: Praeger Special Studies, 1972); and John Adler, "Development and Income Distribution," *Weltwirtschaftliches Archiv*, Heft 3, 1972.

37. These include: CONEPLAN, *op. cit.*, pp. 9-10; CONEPLAN, *Plan Operativo*, 1975, vol. I (La Paz, May 1975); Ministerio de Mineria y Metalurgia, *El Sector Mineria, 1971-1974* (mimeo., La Paz, August 1975), Section III.

38. For an eminently sensible view of what constitutes the proper rule of activity for natural resource industries, see Orris C. Herfindahl, "Natural Resource Information and Economic Development," in Raymond Mikesell (ed.), *Natural Resources in Economic Development* (New York: Praeger, 1968), pp. 7-9.

Chapter Two

Perspectives on the Bolivian Mining Sector

Malcolm Gillis and
Ulrich Petersen

PRINCIPAL FEATURES OF THE BOLIVIAN MINING SECTOR

Historical Background

Any interpretation of the economic history of Bolivia, from the early colonial era to the present, would necessarily be based upon one central theme: the extreme national dependency upon nonrenewable resources, particularly nonfuel minerals. Short periods of relaxation of this dependency, such as during a brief rubber-based boom in the late nineteenth and early twentieth century, are but variations on this central theme. For nearly four centuries, from the early mining of the once-rich silver deposits of Cerro Potosi under the Spanish Crown through the initial exploitation of the then rich, but narrow-veined tin deposits of Oruro and Potosi in the first quarter of the twentieth century, the nation's fortunes have been inextricably linked with world prices of first silver, then tin and lately both tin and petroleum. Although modern Bolivia enjoys only moderate mineral endowments relative to such countries as Canada or Chile (see Chapter One), the nation is by no means bereft of mineral reserves or potential mineral resources, remains highly dependent upon nonfuel mineral and hydrocarbon deposits both as a source of import finance and public sector revenues.

A history of dependence upon mineral resources is, of course, not in itself a cause for regret, as resource-poor nations such as India, Uruguay or Mali could attest. Indeed, the successful, diversified development of resource-rich countries such as Canada or Australia has been fueled in no small part by returns on mineral resource endowments. By and large, these two countries historically met with some success in both internalizing these returns and channeling them toward productive investment not only in the mining sector, but in productive capacity elsewhere in the economy.

The history of mineral development in Bolivia up to at least 1950 shows a significantly different pattern: for most of the past 400 years, the larger part of returns to the sector were not internalized, and that small part of the economic surplus generated in mining that was available as a national resource was largely misallocated. This was patently obvious throughout the colonial period, during which the Quinto Real from the fabled silver mines of Potosi contributed materially not only to the treasury of the Spanish Crown but to inflationary pressures in preindustrial Europe. It is only somewhat less obvious for the first half of this century, up to the nationalization of the large mining companies. During most of this period, national retained value (the sum of returns accruing to the Bolivian economy through taxes, factor payments, domestic material purchases, etc.)[1] from mineral exploitation was minimal. Except for the period 1936-1947, mineral taxes were typically quite small relative to both total export values and to profits in the sector.[2] Further, given the relatively high degree of capital intensity in the bulk of Bolivian mining activity, factor payments to labor (including expatriate labor) accounted for a relatively small share of total mine revenue throughout most of the period prior to nationalization.[3] Moreover, given the high degree of import-intensity of the mining sector's purchase of material inputs,[4] domestic purchases of materials never accounted for more than a small fraction of the value of mining output. Finally, in spite of the maze of taxes in the form of foreign exchange controls characteristic of the period 1930-1947, little reinvestment occurred during this period, as the largest mining firms engaged in large-scale repatriation of funds (as in the case of truly foreign-owned firms) or capital flight (in the case of the nominally Bolivian Patino interests).

Thus, there was minimal direct impact of the mining sector (via wages, domestic procurement, etc.) on the rest of the economy. The sector provided relatively small tax revenues and provided limited foreign exchange for meeting general import needs. Not only was reinvestment in the industry minimal after the mid-thirties,[5] but because of systematic capital flight, few of the resources generated by the sector found their way into increments to domestic productive capacity in other sectors of the national economy.

The situation did not improve significantly following the nationalization of the larger mines in 1952. Several years of labor unrest and hyperinflation, concurrent with simmering political discontent (which on occasion erupted in political upheaval) through 1971, took at least as heavy a toll on the mining sector as on any other sector of the economy. It was not until the late 1950s that the total physical volume of mine production recovered to prenationalization levels, and the value of mining exports, in *current* dollars, did not surpass 1951-52 levels until 1967. In the meantime, serious exploration virtually ceased in the metallic minerals sector, and the capital stock in mining continued to deteriorate steadily. Although the volume and value of certain metals (silver, zinc, bismuth) have increased significantly in percentage terms since 1968, this increase was from a very small base. But for the dominant export, tin, the physical volume of output in 1974 was almost the same as in 1968. With International Tin Council (ITC) quotas in effect throughout virtually all of 1975, exports of that mineral fell almost to levels reached in the troubled years between 1961 and 1963.

In spite of the virtual stagnation of the sector over the best part of the last two decades, the nation remains highly dependent on mining both as a source of foreign exchange and for fiscal resources. As noted in Chapter Three, the nonfuel minerals sector is only a minor source of direct employment, as the approximately 75,000 workers account for only about three percent of the labor force. But value-added of the sector accounted for 12.1 percent of gross domestic product in 1974 (see Table 2-1). Not only in this proportion for mining just slightly below that of agriculture (14.7 percent)

Table 2-1. Nonrenewable Natural Resources: Relative Share in Total Central Government Tax Revenues and Gross Domestic Product, 1972-1974

	1972		*1973*		*1974*	
	Percent Share in		*Percent Share in*		*Percent Share in*	
	GDP	*Total Taxes*	*GDP*	*Total Taxes*	*GDP*	*Total Taxes*
A. Mining sector (nonfuel minerals)	10.9	18.7	12.3	40.3	12.1	32.6
B. Hydrocarbons (oil and gas)	3.9	13.8	4.5	19.7	8.4	27.8
C. Total share (A + B)	14.8	32.5	16.8	60.0	20.5	60.4

Sources: Tax Data: Chapter Six, Tables 6-1 and 6-2. GDP Data: Instituto Nacional de Estadística (INE)

and commerce (15.1 percent), but the relative importance of mining in overall GDP in Bolivia is exceeded in only a few countries (see Table 1-1), most of which have much more primitive economic structures. But in terms of the generation of foreign exchange and fiscal resources, the nonrenewable natural resources sectors in general and the mining sector in particular continue to dominate the economic life of the nation, as Table 2-1 indicates. The extractive minerals sector as a whole accounted for 32.5 percent of total tax revenue in 1972 and for roughly 60 percent in 1973 and 1974. Similarly, the nonfuel minerals sector still accounted for nearly 19 percent of central government revenues in 1972, after two decades of stagnation. This share grew to over 40 percent in 1973, before falling to about 33 percent in 1974, as hydrocarbons revenues expanded significantly in the aftermath of the Mid-East war of October 1973. The nonfuel minerals sector continues to dominate the national export picture, accounting for an average of 81 percent of total export value in 1970-72 and 68 and 60 percent in 1973 and 1974 respectively. Mining and hydrocarbons combined were responsible for 88 percent and 90 percent of total exports in 1973 and 1974 (see Table 2-2). Together with the hydrocarbons sector, exploitation of mineral resources has been responsible for as little as 13.8 percent of GDP in 1970 and as much as 20.5 percent in 1974.[6]

Structure of the Mining Sector

Exploitation of Minerals. Mineral production is dominated by the large state enterprise, COMIBOL (Corporación Minera de Bolivia),

Table 2-2. Relative Share of Mining Sector and Hydrocarbon Exports in Total Exports,[a] 1970-74

	1970	*1971*	*1972*	*1973*	*1974*
1. Exports of nonrenewable resources as % of total exports	94.3	93.2	88.3	87.6	90.0
2. Mining sector exports as % of total exports	(88.6)	(81.9)	(71.3)	(67.6)	(60.0)
3. Hydrocarbon exports as % of total exports	(5.7)	(11.3)	(17.1)	(20.0)	(30.0)
4. Other exports	5.7	6.8	11.6	12.4	10.0
	100.0	100.0	100.0	100.0	100.0

Source: Ministry of Mines and National Planning Agency (CONEPLAN).

[a]Both mining sector exports and total exports expressed on a gross basis (i.e., inclusive of "realization" costs (freight, smelting charges, etc.).

formed in 1952 to take over the properties of the large mining companies expropriated in that year. COMIBOL itself is the second largest tin enterprise in the world, ranking just behind the Indonesian state-owned tin producer, P.T. Timah. Between 1965 and 1971, COMIBOL was responsible for not less than 47 percent and not more than 59 percent of mineral exports by value. Fully one-fourth of COMIBOL's total recorded production is purchased from cooperatives and private miners, plus repurchase of stolen ores. In more recent years the enterprise accounted for about 63 percent of the annual value of mineral exports (see Table 2-3).[a] COMIBOL is a medium-sized, labor-intensive mining operation by international standards, with total current assets of U.S. $226.4 million at end-1974 and total sales of nearly $250 million (see Table 2-3) and 24,556 employees in 1974. COMIBOL and the state petroleum

Table 2-3. Gross Value[a] and Percentage Share of Total Mineral Exports, 1965-1974: COMIBOL, Medium, Small Mining Sectors and Others[b] (including ENAFBOL purchases from each sector)

	U.S. $ millions–nominal values									
Year	*COMIBOL*		*Medium Mining*		*Small Mining*		*"Other Exporters"*		*Total*	
1965	$ 64.7	51.8%	$22.8	18.3%	$19.2	15.4%	$18.2	14.5%	$124.9	100.0%
1966	77.8	59.2	27.4	20.9	18.4	14.0	7.8	5.9	131.4	100.0
1967	74.2	56.5	30.3	23.1	18.7	14.2	8.1	6.1	131.3	100.0
1968	78.5	56.5	35.0	25.2	20.8	15.0	4.7	3.4	139.0	100.0
1969	87.9	52.6	40.0	23.9	26.8	16.0	12.5	7.5	167.2	100.0
1970	97.0	47.3	52.9	25.8	34.1	16.6	20.9	10.2	204.9	100.0
1971	91.3	52.7	32.8	18.9	22.5	13.0	26.7	15.4	173.3	100.0
1972	111.1	63.8	36.3	20.9	22.4	12.9	4.3	2.5	174.1	100.0
1973	140.9	62.4	54.2	24.0	19.3	8.5	11.5	5.1	225.9	100.0
1974	242.6	62.9	90.0	23.3	38.3	9.9	15.1	3.9	386.0	100.0

Source: For 1965-1971, Ministry of Mining and Metallurgy, *Bolivia: Minería en Números* (La Paz, 1975), pp. 48, 52, 56, 62. For 1972-1974, above source, plus supplementary tables, Ministry of Mining.

[a]Gross value of exports are defined inclusive of "realization costs" (freight, smelting charges, etc.). Such realization costs are typically 22-35 percent of gross export values, depending on the mineral in question.

[b]ENAF's share in total value of mineral exports averaged 14.3 percent from 1972 through 1974. Figures for each sector (COMIBOL, medium mines, etc.) include their sales to ENAF.

---

[a]Inasmuch as less than two percent of the value of any metallic mineral is sold for domestic consumption, exports may be taken as a useful surrogate for total production.

company (YPFB) are by far the two largest productive entities in Bolivia. COMIBOL alone accounted for nearly 40 percent of the total value of all Bolivian exports in 1974 and for 58 percent of total taxes paid to the central government. In 1973-74 the enterprise produced 67 percent of total tin, 60 percent of lead, 88 percent of zinc output, 91 percent of bismuth, but very little tungsten (wolfram) and antimony.

The Medium Mining Subsector. The medium mining subsector has consisted in any given recent year of between twenty-seven and twenty-nine private firms, of which two are foreign-controlled and four to five others have varying degrees of minority participation by foreign interests. In general, the medium mining sector is made up of private firms that do not meet the legal requirements for classification as "small miners." (The latter enjoy certain limited tax benefits and are subject to less stringent accounting requirements.) There are, however, between four and six "small mining" firms that are larger than some of the smaller medium mining enterprises, in terms of output tonnage, number of employees, and assets.

The medium mining subsector has typically accounted for between 20 and 25 percent of mining sector export values since 1965, but this has ranged from a low of 18.3 percent in 1965 to a high of 25.8 percent in 1970 (see Table 2-3). Medium mining firms are dominant in the production of tungsten and antimony, as they are normally responsible for over 50 percent of total exports of the former and nearly 70 percent of the latter. Identifiable taxes paid by the subsector in 1974 (those for which collections are discriminated among the three groups of firms in the sector) amounted to about 25 percent of total central government taxes on the mining sector, or about eight percent of total central government tax revenues. This subsector directly employed between 7,000 and 8,000 workers in 1974.

Results of a survey of the medium mining firms concerning average sales, profits and number of workers are summarized in Table 2-4. (The survey is more fully discussed in Chapter Three.)

The averages in Table 2-4, however, mask a number of traits of the subsector as a whole. As is evident from data presented in Appendix A, there are considerable differences in size between medium mining firms. In 1973, 14 of the 26 firms in the subsector had sales of less than U.S. $500 thousand, and in 1974, 15 of 28 had sales of less than U.S. $500 thousand. But in 1973 and 1974, at least two firms had sales in excess of U.S. $15 million. The majority of the firms had fewer than 300 employees, while five in 1973 and four in 1974 had between 500 and 1,000 employees.

Table 2-4. Indicators of Size, Medium Mining Firms, 1973 and 1974 (20 of 27 firms in 1973, 24 of 28 firms in 1974), dollar values in U.S.$ thousands

	1973	*1974*
1. Average sales value per firm	$2,515	$2,625
2. Average profits per firm	302	532
3. Average number of employees per firm	300	293

Source: Appendix A.

Small Mines and Cooperatives. The remaining enterprises in the mining sector are private firms belonging to the "small mining" subsector and the cooperatives, plus between 100 and 120 private firms (termed "other" producers) that have been able to escape classification as either "medium" or "small" mining firms. The small miners and cooperatives are considerably more labor-intensive than the medium mining firms, and together account for well over half of the total number of employees in the mining sector. The small mining firms, which number between 3,000 firms in poor years (e.g., 1975) and 4,500 in prosperous years (e.g., 1973), employ from 15,000 to 25,000 workers. The cooperatives account for another 20,000 workers. The scant data available for these concerns suggests that the great majority of production of the small mines comes from less than ten percent of the firms. For nearly all such enterprises, only a fraction of the work force is employed full time, the predominant pattern being employment during winter for subsistence farmers or part-time work for family members of regular full-time miners. Seasonal or part-time employment is also present in COMIBOL and the medium mines, but to a far lesser degree.

Prior to 1971, the small mining firms were typically responsible for 15 percent of total value of mineral exports. This share began to decline in 1971, and by 1974 had fallen to just under 10 percent (see Table 2-3). The proportion of total exports by "other" producers has followed no discernible trend since 1965, but rather has been subject to very wide variation. The share of these exporters appears to have hovered between 5 and 8 percent for most of the last 10 years, except for years of great domestic economic and/or political instability (1965 and 1971), when such producers accounted for about 15 percent of total export value, usually at the expense of the medium mining subsector.[b]

[b]The largest of the group of "other" producers are firms that would normally be classed as medium mining firms (three such firms joined the Medium Miners' Association in 1974) but somehow have managed to "declassify" themselves, perhaps in order to avoid the heavier taxes imposed upon the medium mining sector.

The mining cooperatives date from the 1930s, when cooperative units were formed in response to high unemployment in the mining districts, especially in Potosi Department. The cooperatives now account for but a small fraction of output, but for between one-fourth and one-fifth the total number of workers in the mining sector. Some cooperatives are quite small, with but five members; others, with up to 1,000 members, have as many workers as are employed by the largest of the medium mining firms. The smaller cooperatives are truly hand-to-mouth operations, employing very primitive mining and ore-dressing techniques, often under highly hazardous conditions. The largest cooperatives operate under conditions that are only slightly more favorable. The principal cooperatives are typically located on portions of COMIBOL property, ceded to the entities after the deposits became unworkable with COMIBOL methods, or for political reasons. Major cooperatives are found in the once rich Bolsa Negra (Black Pocket) tungsten district in the Department of La Paz, in gold mining along the Tipuani River, and in and around the city of Potosi, where cooperatives continue to exploit marginal deposits of tin and silver on Cerro Potosi. The internal structure of the cooperatives varies primarily with the number of members, who are typically drawn from the ranks of former COMIBOL employees, many of whom have become too infirm (because of silicosis or mining accidents) to continue to work underground, or from retired or surplus COMIBOL surface labor.

Metallurgy

General Considerations. Until the completion of construction of the ENAFBOL (Empresa Nacional de Fundiciones de Bolivia) tin smelter in Oruro in February 1971, there was no significant domestic capacity in smelting and refining. By law, all new investments in smelting and refining activities from 1970 onward must be carried out by the state. ENAFBOL is nominally in charge of developing and operating smelters and refineries, although, as noted below, COMIBOL has begun to move into some smelting activity (sometimes in partnership with ENAFBOL and in some cases on its own) and a few private smelters were built prior to 1970.

The ENAFBOL tin smelter was designed in 1967-1968 for an ultimate 20,000 tons of annual output, and production began in 1970. Only 6,000 tons of initial capacity was installed, but the electric power, buildings and other infrastructure for 20,000 fine tons had already been built. Because of this, and because ENAFBOL must use charcoal rather than coal (there are no significant accessible coal deposits in Bolivia; the charcoal comes from domestic timber) the smelter was a high-cost operation from the beginning. In September 1975 capacity rose to 11,000 fine tons, and by 1978 is expected to increase to 30,000 fine tons, sufficient to smelt all tin mined in Bolivia.

The ENAFBOL tin smelter is a medium- to high-grade smelter relative to, say, the Capper Pass operation in Britain. Whereas Capper Pass normally takes concentrates of 20 to 25 percent grade, ENAFBOL prefers concentrates of 30 to 40 percent grade. ENAFBOL generally markets Grade A tin (99.85 percent tin content) to the United States, where later refining takes the grade to AA (99.90 percent). Latin American and U.S.S.R. buyers accept only grade AA tin. The enterprise normally procures about 65 percent of its tin concentrates from COMIBOL, 30 percent from a few medium miners (35 percent in 1974) and 5 percent from the small mines (1 percent in 1974). Smelting and other handling charges are geared to fees charged by foreign smelters (Capper Pass in Britain for medium- and low-grade concentrates; Texas City in the U.S. for high-grade concentrates). That is, when a producer sells to ENAFBOL, the smelting and other charges are roughly equivalent to charges abroad, plus nearly all of the difference that the producer would save on freight and other transport costs. By July of 1975, tin recovery in ENAFBOL was about 93 percent (7 percent losses). By 1977 ENAFBOL hopes to reach 95 percent recovery (this was the guarantee of the constructing firm). Losses of this magnitude matter little for zinc, but much for high-valued tin.

Small-scale tin smelting activity is carried out by two firms in the private sector, but this accounts for but a small fraction of tin-metal exports. COMIBOL also plans to operate a few small tin volatilization plants (pyrometallurgical plants that convert low-grade and/or colloidal concentrates into more commercially suitable grades) and is building, with assistance from the Soviet Union, two large energy-intensive volatilization plants in Potosi and Oruro.

For other metals, COMIBOL operates a bismuth smelter at Telemayu and is planning a bismuth refinery at the same location. ENAFBOL is now building an antimony smelter, with design capacity of 6,000 tons. In addition, there are two small-scale private sector antimony smelters in Potosi. Plans for a large zinc smelter, to be operated jointly by ENAFBOL and COMIBOL, were in the advanced discussion stage in late 1975. Tentative plans for a copper smelting operation with 20,000 tons per year of capacity were also under discussion in 1975. Finally, there are three private sector lead or lead-antimony smelters in Cochabamba, La Paz, and Potosi.

Technical Considerations. In Bolivian tin operations there has been a sustained tendency toward lower metallic content (grade) and greater complexity of the ore treated. This phenomenon, resulting from insufficient exploration and inadequate reserve levels, has meant reduction in the recovery of metallic values from a level which was generally in excess of 60 percent during the 1940s, to a range most commonly 35-50 percent, with some notable exceptions. High-grade (60 percent tin or more) concentrates formerly predominated, but now the average concentrate grade is only somewhat above 35 percent tin with a much higher incidence of impurities than in the nineteen-forties.

Gravity concentration has been the predominant method used for tin and tungsten, with flotation, magnetic separation and sintering used to remove impurities. Almost no effort is made to separate byproducts. In the past, most research and development was undertaken outside Bolivia and had been for alluvial material (thus not necessarily applicable to low-grade complex hard rock minerals) or frequently not published or freely available, as it was controlled by the tin smelting cartel. However, during the last three decades better ore dressing techniques were developed at public institutions, and within the last 15 years metallurgical research has intensified both in and for Bolivia. Although the technical aspects are now well advanced for tin flotation and pyrometallurgical methods, economic considerations have precluded implementation of this improved technology. Various improvements in gravity methods which permit recoveries of up to 70 percent have been commercially proven for some time, but again the investment and management constraints already noted seem to be the key factors for their limited application in Bolivia. For other metals, metallurgical processes are well established and accessible. For example, antimony flotation is in commercial operation in Bolivia; and with smelting facilities, grade and impurities problems could be overcome if investment and management constraints are relaxed.

Other Features of the Mining Sector

The following sections describe other features of the mining sector that are pertinent to any review of the present and potential future role of taxation in the sector, except for the managerial and accounting infrastructure in mineral enterprises, which is discussed more fully in Chapter Nine.

Marketing Arrangements. Very diverse patterns of marketing arrangements are found in Bolivia, not only with respect to types of producers but according to mineral and to ultimate overseas markets. No short discussion can capture the full extent of this diversity, but an effort is made in this section to characterize the essential features.

Prior to 1975, about 24 percent of Bolivian production of tin has been channelled to the state smelter (ENAFBOL). In 1974, COMIBOL and the medium mining firms sold 27 percent and 48 percent respectively of their production to ENAFBOL. Because ENAFBOL is a high-grade smelter and because a large part of small miners' production is low-grade (less than 20 percent concentrate), the small mining sector channelled less than 3 percent of its total output to ENAFBOL. As noted earlier, ENAFBOL's structure and level of charges are modeled after those of foreign smelters, except that ENAFBOL does not levy surcharges for pollution and collects nearly all of the difference in freight savings arising from shipment of metals rather than concentrates. Bolivia exports tin directly to 23 countries, with Britain (27 percent), the United States (26 percent), West Germany (10 percent), Brazil (8 percent) and the Soviet Union accounting for over 70 percent of the 1974 total. For zinc, the United States (64 percent) and Japan (26 percent) together accounted for 90 percent of the total in 1974. The United States also purchases the majority of Bolivian silver, wolfram, antimony, lead and bismuth output, while Japan takes over 90 percent of Bolivian copper exports.

In the case of the medium miners, 70 percent of total sales are typically made directly to domestic and foreign smelters. Agents and ore merchants are not utilized in such arrangements. Other sales are channelled through selling agents that do not take title to the minerals, but look for buyer contracts. In such cases, the contract is between the mining firm and the smelter. In still other arrangements, the miners sell through international companies who execute their own contracts with the firms, for later sale to overseas buyers. Finally, some medium miners sell a portion of their output to BAMIN (Banco Minero), the Mining Bank. Terms vary almost as much as the structure of marketing arrangements, ranging from payment on arrival in foreign smelters, to 80 percent payment on proof of shipment from Bolivia, to payment within 10 days for deliveries to ENAFBOL. Sales prices for tin and gold are geared to L.M.E. (London Metal Exchange) prices, whereas the major Bolivian zinc exporters receive prevailing "producers" prices. Practices vary considerably for other minerals.

Small miners and "other" producers are required by law to market their minerals through BAMIN. However, exceptions to this requirement are made, particularly in the case of the larger firms which can often find superior terms than those offered by BAMIN. In any case, 90 percent of the output of the small mining firms is channelled through BAMIN. The remainder is sold largely through commercial agents (*comercializadores*) who represent foreign buyers, principally smelters, abroad. BAMIN pays on delivery, after making deductions for *regalías*, export and other taxes, smelting costs, transport and handling. The cooperatives are normally obligated to sell to either BAMIN or COMIBOL.

Technological, Health and Work Conditions. By international standards the Bolivian mines would be labeled as "small" or "medium-sized" and requiring "narrow-vein mining." This type of mining does not normally involve risking large sums of money to test and develop reserves before productive operations can begin, except when the possibility exists for at least a medium-sized operation. Because of the narrowness and variability of mineralization in Bolivia, diamond drilling is commonly ineffective for proving reserves. Narrow veins of variable quality also limit mechanization possibilities and imply a small tonnage developed per meter of tunneling, resulting in a cost per ton of proven ore substantially higher than for wide-vein, bedded or disseminated deposits. The limit on mechanization requires that development and exploitation be rather labor-intensive, reducing levels of operating productivity. Further, rock breakage cannot be held to mineralized portions only. This fact results in substantial dilution, i.e., reduction in the metal content of the material reaching the mill. Because of low operating efficiency and dilution, the grade of ore that can be economically mined (cutoff grade) must be considerably higher than for other types of deposits elsewhere in the world.

In spite of the size and productivity limitations inherent in narrow-vein mining, there is nonetheless considerable scope for increased volume and efficiency in many mines. This potential has not been realized due to the investment and managerial constraints noted later in this chapter. Even in the medium mining sector and COMIBOL, it is the exception to find more than about 18 months of measurable working reserves of a cutoff grade adequate for sustaining production during periods of relatively low metal prices. The great majority of small mines, cooperatives, some medium mines and sectors of COMIBOL mines operate on almost a hand-to-mouth basis insofar as reserves are concerned.

Notwithstanding the observations at the beginning of this section, COMIBOL's Catavi group of mines contains the largest underground tin mine in the world, which alone accounts for nearly 20 percent of

Bolivia's tin production, or about 6 percent of 1974 world production. But the once exceedingly rich veins at Catavi have been largely exhausted over the last half-century. A lack of adequate exploration and development has led to a decrease in high-grade ore reserves and has required extraction of lower-grade ore by block-caving methods. Thus, the ore grade dropped from about 3.0 percent in 1940 to 0.9 percent in 1970 and finally to a very marginal 0.49 percent in 1975. Only 30 percent of the ore is mined from veins; the remainder is obtained from disseminated ore worked by the block-caving method. Partly due to inadequate maintenance of the mostly half-century old infrastructure within the mine, and partly due to poor supervision and hesitancy to invest in a mining area that has long been the center of labor union militancy, the rate of fatality and dismemberment for COMIBOL laborers who work underground is far higher than for most of the larger medium mining firms.[7] Health and work conditions in this group of mines are in general poor. Lack of both ventilation and water for drilling, and the resultant high dust levels, contribute to high levels of silicosis. As a result, COMIBOL workers in Catavi may expect an average of from 5 to 7 years of underground work. This compares to an average underground work expectancy of 15-20 years in some of the better-ventilated private medium mines working vein deposits. Further, very high altitudes characteristic of almost all COMIBOL mines also contribute to health problems. Of the 58 separate mining districts operated by COMIBOL, only three are located at altitudes of less than 3,000 meters, thirteen are at between 3,000 and 4,000 meters, thirty-six between 4,000 and 5,000 meters, and three at above 5,000 meters.

A great proportion of the concentration processes now used by COMIBOL is geared to the technology of the 1920s to 1940s. In some locations such as Catavi, the concentrator (*ingenio*) was tailored in the early twenties to the very high ore grades (up to 20 percent Sn in some cases, but on the average about 6 percent) characteristic of the early years of the mine. Since nationalization in 1952, only the considerable ingenuity of the COMIBOL staff has prevented mineral recoveries from slipping below 50 percent for tin. Even so, a large share of potential COMIBOL tin output is shunted to the many small mountains of tailings found on major COMIBOL properties, awaiting the day when tin prices and/or recovery technology justifies their exploitation.

As noted in further detail in Chapter Three, COMIBOL's costs are, by any standards, high in all phases of exploitation and concentration, particularly costs above ground. For both 1972 and 1973, 47 percent of costs at the mine (*costo puesto mina*) were incurred in the mine, 25 percent in concentration and 28 percent in "other costs." In the larger medium mines, the cost breakdown normally runs 70 percent for mining, 15 percent for concentration and 15 percent for "other" costs.[8]

The Medium Mines. Technological, health and work conditions vary greatly among the 26-30 companies comprising this subsector at any given time. Except for two large tin dredges used by two of the firms, exploitation activity is underground, primarily from veins of 40 centimeters to one meter's width, at average altitudes only slightly lower than for COMIBOL's mines. Estimates of the vintage of technology and of capital equipment across the subsector are unavailable, but substantial investment programs in five or six of the mines in 1966-69 did apparently serve to decrease vintage to levels significantly below the average for COMIBOL properties. For the major private tin producers (six mines with 70 percent of medium mine tin output), antimony producers (two firms accounting for 60 percent of 1973 production), one firm that produces 88 percent of all tungsten produced by medium mines, and one or two zinc mines, managers and technical staff and mine workers tend to be conversant in all aspects of modern technology appropriate for medium-scale, narrow-vein and dredge deposits. Major items of capital equipment tend to be well maintained and used to near-maximum capacity.

For most of the medium mining firms, worker turnover is considerably lower than in COMIBOL, owing largely to better health and safety conditions in most, but not all, of the medium mines. Ventilation, at least in the larger mines, tends to be more satisfactory, and superior maintenance and repair reduces the incidence of maiming accidents.

Small Mines and Cooperatives. Small mines and cooperatives are, in the main, characterized by primitive technological methods (both in exploitation and in concentration), low productivity and generally abysmal health and safety standards. Commonly, small mines and cooperatives engage in reworking the tailings from the larger concentrators, particularly those of COMIBOL. Others direct their efforts to such hazardous operations as extraction of safety pillars and minor ore blocks left by previous operators when mines were closed down. Finally, other small mines and cooperatives operate extremely narrow and short veins which cannot be mined economically at medium or large scales of operation. These undertakings generally are economically worthwhile only if costs are kept low through the use of inferior safety, health or working conditions for staff and workers.

Relative Importance of Particular Metals

Since the beginning of the century, tin has dominated the Bolivian minerals sector. This dominance has continued during the nineteen-seventies, in spite of a decline of nearly 2 percent in physical production of tin from 1970 through 1974. This metal accounted for close to 60 percent of the value of total mineral exports (excluding hydrocarbons) from 1970 through 1974, ranging from a low of about 50 percent in 1970 to a high of 65 percent in 1972 (see Table 2-5).

Bolivia, with 16.2 percent of world production, is the world's second-leading producer of tin, behind Malaysia (37.4 percent), and just ahead of Indonesia (13.7 percent) and Thailand (11.3 percent) (see Table 2-6). Bolivian tin comes primarily from cassiterite in small veins, whereas most Malaysian tin is won from alluvial and eluvial placer deposits. Virtually all Indonesian tin and most Thai tin comes from placer deposits as well.

Antimony in Bolivia is found in veins and is mined for itself, in contrast to the Peruvian case, where antimony is primarily a smelter byproduct. At the beginning of the decade, antimony was Bolivia's second most important mineral export, accounting for nearly 15 percent of mineral export value. In recent years, however, antimony has been overtaken by zinc, which was 9.7 percent of total 1974 mineral export value, whereas antimony was 7.5 percent. The medium mining sector dominates antimony production. Physical volumes of production of both zinc and antimony increased slightly over the five-year period. Zinc output in 1974, primarily from COMIBOL and a few narrow-vein medium mines, was about 4.5 percent higher than 1970 levels, whereas antimony output in 1974 was almost 13 percent higher than the 1970 level.[c] Bolivia produces a negligible share of the world's zinc, and in 1973 was responsible for 25.4 percent of world antimony production, just behind South Africa (27.1 percent) and just ahead of China (20.4 percent). Together the three are responsible for nearly 3/4 of the world's production of antimony; however, Bolivia has but 7.5 percent of world antimony reserves.[9]

Wolfram (tungsten) has typically been responsible for between five and eight percent of total mineral export values, ranging from 5.4 percent in 1974 to 8.6 percent in 1970. The physical volume of wolfram production remained virtually constant from 1970 through

[c]It is worth noting that annual physical volumes for production of Bolivian metals are virtually identical with export volumes, *except* in years in which substantial stockpiling occurs, as in 1975.

Table 2-5. Mineral Exports, 1970-1975 (values in thousands of U.S. dollars at current prices)

	1970		*1971*		*1972*		*1973*		*1974*	
	Value	*%*	*Value*	*%*	*Value*	*%*	*Value*	*%*	*Value*	*%*
Tin	102,047.1	49.8	105,878.6	61.1	113,540.8	65.2	130,992.6	58.0	229,813.9	59.5
Concentrates	(102,024.1)	(49.7)	(82,013.3)	(47.3)	(88,926.6)	(51.1)	(98,667.0)	(43.7)	(174,320.7)	(45.2)
Metal	(23.0)	(0.1)	(23,865.3)	(13.8)	(24,614.2)	(14.1)	(32,325.6)	(14.3)	(55,493.2)	(14.3)
Antimony	30,926.4	14.8	9,040.2	5.2	9,081.1	5.2	17,352.7	7.8	29,062.0	7.5
Zinc	14,318.6	7.0	15,270.5	8.8	15,437.6	8.9	25,693.4	11.4	37,322.8	9.7
Wolfram	17,576.9	8.6	13,610.3	7.9	10,284.3	5.9	11,128.1	4.9	20,900.2	5.4
Copper	12,498.8	6.1	8,297.0	4.8	8,761.7	5.0	13,440.4	5.6	15,999.6	4.2
Bismuth	7,957.8	3.9	5,791.9	3.3	2,646.0	1.5	4,377.9	1.9	12,456.6	3.2
Lead	7,806.5	3.8	5,948.6	3.4	5,776.1	3.3	8,346.8	3.7	11,482.8	3.0
Silver	10,531.4	5.1	8,342.0	4.8	7,590.0	4.4	12,561.2	5.6	26,841.8	7.0
Others[a]	1,844.0	0.9	1,213.3	0.7	1,044.7	0.6	1,485.1	1.1	1,930.1	0.5
Totals:	204,917.1	100.0	173,325.6	100.0	174,133.8	100.0	225,919.9	100.0	386,015.0	100.0

Source: Ministry of Mining and Metallurgy, *Bolivia: Minería en Números* (La Paz, 1975).

[a]Gold, sulfur, manganese, iron, cadmium, marble, asbestos, etc.

[Numbers may not add to totals due to rounding.]

Table 2-6. Total World Production, Tin-in-Concentrates, 1974

Country	*Production (metric tons)*	*Percentage shares 1974*
Malaysia	68,122	37.4
Bolivia	29,553	16.2
Indonesia	25,023	13.7
Thailand	20,359	11.3
Australia	10,331	5.7
Nigeria	5,657	3.1
Zaire	4,380	2.4
Rest of World	18,975	11.2
	182,400	100.0

Source: Derived from International Tin Council, *Monthly Statistical Bulletin*, Vol. 19, #4 April 1975, Table A-1.

1974, while total export values increased by nearly 19 percent over the five-year period. Bolivia is the third largest producer of tungsten in the western world, ranking just behind the U.S. and South Korea. Three communist nations (China, U.S.S.R. and North Korea) account for 50 percent of the world's production and China has 50 percent of world reserves.[10]

Copper typically accounts for about five percent of total mineral export earnings. Export values of this mineral grew by 25 percent over the period 1970-74, in spite of a 12 percent decline in tonnage. Bolivia is responsible for an infinitesimally small share of world copper production (less than 1/100 of one percent). In many other mining countries, such as Peru, bismuth is a smelting byproduct, but most Bolivian bismuth comes from bismuth mines proper, primarily COMIBOL mines. This particular metal normally is responsible for between 1.5 and 4.0 percent of mineral export values. Bolivia is a major producer of bismuth, accounting for 16 percent of world production in 1973.[11]

Lead was responsible for around three percent of total mineral export values each year over the past three years. Bolivian lead production comes largely from lead-zinc deposits (COMIBOL's Matilde Mine) and from lead-silver and tin-lead complex ores. Physical production declined substantially, by nearly 25 percent, from 1970 through 1974, although total lead export values rose by 47 percent. Bolivia produces a negligible share (less than 1/10 of one percent) of the world's lead production (1973).[12]

Silver has, over the past five years, accounted for between 4.5 and 7.0 percent of mineral export values. Marginal silver operations around Potosi still produce appreciable volumes of silver from very low-grade mines in which the metal is won from tin-silver and lead-silver complex minerals. As was the case for lead and copper, the total tonnage of recorded Bolivian silver production declined over the period 1970-74. But given the high value-to-bulk ratio for silver and the 160 percent increase in world prices from 1970 to end-1974, it is entirely possible that the proportion of total silver production that is smuggled has risen considerably since 1970, and this would not of course be reflected in the official export statistics. In spite of Bolivia's past fame as a silver-producing nation, it now produces less than one third of one percent of total world silver output.

All other minerals, including gold, sulphur, asbestos, iron and cadmium, constituted slightly less than one percent of total export values over the 1970-1974 period. Total reported gold production rose substantially (52 percent) from 1970 to 1974. This represented the effects of much-reduced nominal taxes on gold export, with a consequent decline in export smuggling, as much as any significant producer response to the drastic increases in gold prices over the period. Sulphur exports also rose significantly from the very small (16,300 metric tons) base of 1970, rising by nearly two and one-half times over the five-year period.

Prices of all Bolivian mineral exports except antimony rose substantially from 1970 to the end of 1974. Table 2-7 below lists average "official" prices in U.S. dollars for each of Bolivia's major mineral exports.[d] The greatest percentage increase, in terms of U.S. dollars, occurred for silver (160 percent) followed by zinc (150 percent) and tin (145 percent). Lead prices rose by 95 percent, copper by 43 percent, bismuth by 41 percent and wolfram only by 8 percent. However, by mid-1975 prices of nearly all these metals except wolfram and zinc had receded substantially from their 1974 highs. By mid-September of 1975, tin had fallen slightly below $3.00 per fine pound, producer prices of zinc to $0.319 per unit, and prices of antimony, lead and copper to $1.41, $0.163, and $0.54 per pound respectively.

Institutions in the Mining Sector

Over the years, a number of specialized public institutions have been created to serve the mining and metallurgical sectors, in addition to the basic policy-making and regulatory agencies found in virtually all countries where mining is an important sector. There are

[d]"Official" prices are those established each 15 days by the Ministry of Mines for royalty and export tax purposes. "Official" prices are based on world prices in particular markets, including the London metal exchange.

Table 2-7. Average Annual "Official" Prices for Major Mineral Exports, 1968-1974 (US $ per commercial unit)

	Tin[a]	*Zinc*[b]	*Antimony*[c]	*Wolfram*[d]	*Copper*[e]	*Bismuth*[f]	*Lead*[g]	*Silver*[h]
1968	1.461	0.110	5.49	43.33	0.501	2.10	0.105	2.14
1969	1.633	0.130	8.28	48.79	0.607	2.50	0.120	1.79
1970	1.677	0.140	28.70	75.34	0.636	5.83	0.138	1.76
1971	1.580	0.153	6.73	53.95	0.480	3.77	0.116	1.53
1972	1.700	0.176	7.26	38.11	0.471	2.54	0.135	1.63
1973	2.110	0.236	11.77	41.37	0.744	3.46	0.190	2.36
1974	3.600	0.350	22.90	81.12	0.910	8.24	0.270	4.59
1975 (first half)	3.260	0.386	17.83	95.12	0.550	4.93	0.158	4.31

Source: Ministry of Mining and Metallurgy, *Bolivia: Minería en Números* (La Paz, 1975)

[a]Per fine pound of tin (L.M.E.). Average annual Metals Week price for *straits* quality tin in the New York market was as follows, for 1970 through 1974: 1970–$1.74; 1971–$1.67; 1972–$1.77; 1973–$2.27; 1974–$3.96. New York prices are always higher than L.M.E. prices due to freight charges.

[b]Per fine pound of zinc. Average annual *Metals Week* prices for delivered zinc (East St. Louis) were, for 1970-74: 1971–$0.163; 1972–$0.178; 1973–$0.207; 1974–$0.360.

[c]Per "long unit" of antimony (22.4 lbs.).

[d]Per "long unit" of wolfram.

[e]Per fine pound of copper. Average annual *Metals Week* (L.M.E. cash) prices for foreign refinery copper were: 1970–$0.628; 1971–$0.479; 1972–$0.465; 1973–$0.788; 1974–$0.904.

[f]Per fine pound of bismuth.

[g]Per fine pound of lead.

[h]Per troy ounce of silver.

also several well organized and vocal private-sector organizations representing the interests of distinct segments of the mining sector. These institutions are merely identified in this chapter. For a more detailed discussion of their structures and functions, the interested reader may consult a variety of official publications of the Ministry of Mining and Metallurgy.[13]

The Public Sector

Policy-Making, Regulatory and Related Agencies. The Ministry of Coordination and Planning (CONEPLAN) is responsible for coordination of planning in all sectors, including mining and metallurgy, and assumes the central role as well in the formulation of near- to medium-term macroeconomic policy. The Finance Ministry, in cooperation with CONEPLAN and with the Ministry of Mining and

Metallurgy, is responsible for the formulation and execution of tax policy for the mining sector (as well as all other sectors). Two agencies of the Ministry of Finance have particular relevance for the mining sector: Internal Revenue (Renta Interna) and Customs (Aduana). Renta Interna is responsible for collection of a series of minor taxes affecting the mining sector (see Table 6-1). The Aduana at present collects over 97 percent of all taxes paid by the mining sector. The most important of these are the *regalías* (royalties) and export taxes, described in detail in Chapter Six.

Direct responsibility for the Bolivian nonfuel minerals sector is vested in the Ministry of Mining and Metallurgy. Prior to 1971, the Ministry had overall responsibility for metallic and nonmetallic minerals as well as hydrocarbons. Now the Ministry is responsible for general policy toward the mining and metallurgical sector only, mining contracts, concessions and claims, and data-gathering and publication of mining statistics and the Mining Cadastre. The Minister of Mining is also the Chairman of the National Mining Council (Consejo Nacional de Minería) which meets on a weekly basis to follow and evaluate developments in the sector, and which often airs key policy issues before full cabinet meetings. The Ministry exercises varying degrees of control over the following dependent entities: (a) COMIBOL, the state-owned National Mining Company; (b) ENAFBOL (Empresa Nacional de Fundiciones de Bolivia), the state-owned tin smelter; (c) GEOBOL, the Geological Service (Servicio Geológico de Bolivia), responsible for geological mapping, evaluation of mineral resources, prospecting, exploration, mineralogical analysis and certain technical services; (d) COBOEN, the Bolivian Nuclear Energy Commission (Comisión Boliviana de Energía Nuclear), responsible for all aspects of exploration and development of radioactive ore; and (e) SIDERSA, the Iron and Steel Enterprise, an entity formed with equity inputs from COMIBOL and ENAFBOL, to develop and exploit iron ore deposits in Bolivia and to lay the groundwork for a possible future iron and steel industry in Bolivia.

Financial Institutions. The prime financial institution concerned with the mining sector is BAMIN, the state Mining Bank (Banco Minero de Bolivia), also a dependent institution of the Ministry of Mining. BAMIN functions as (a) a development bank for the small mining sector, extending loans and technical assistance to eligible firms; (b) an official mineral marketing agency for the small mining sector; and (c) a supplier of material inputs for small miners and cooperatives. In addition, CBF, the Bolivian Development Corporation (Corporación Boliviana de Fomento), provides finance to larger mining enterprises in the form of equity investment.

Research, Service and Training Agencies. The most important research agency for the mining sector is IIM, the Mining and Metallurgical Institute (Instituto de Investigaciones Minero-Metalúrgicas). IIM is responsible for various aspects of research on mining, beneficiation, smelting and refining. IIM also provides technical services such as assay and laboratory analysis. INSO, the National Occupational Health Institute (Instituto Nacional de Salud Ocupacional), is concerned with occupational diseases in all sectors, particularly silicosis and industrial accidents among miners. Finally, various universities such as the Technical University in Oruro, as well as universities in Potosi and La Paz, offer curricula in geology, mining and metallurgical engineering.

Private Sector Institutions

The twenty-five to thirty-odd private medium mining firms are grouped into the National Association of Medium Miners (Asociación Nacional de Mineros Medianos). The Association functions not only as a forum for discussing common problems faced by the subsector (tax policy, marketing arrangements, technical problems, public relations), but also formally channels the views of the member firms to government agencies. The National Chamber of Mining (Camara Nacional de Mineria) represents over two thousand small-mine owners, and functions in much the same way as the Medium Miners' Association. The Association of Cooperatives (Asociación Nacional de Cooperativas Mineras) represents the interests of both large and small cooperatives. Finally, the various Associations of Mining Workers (Federaciones de Sindicatos de Trabajadores Mineros de Bolivia) concern themselves with policies and issues affecting mine workers, including wage policy, health and safety and housing.

PRINCIPAL PROBLEMS FACING THE MINING SECTOR

The Bolivian mining sector in 1975 faced a number of short-term and long-term problems. Some of these problems are common to the sector as a whole; others relate more directly to specific components of the sector—COMIBOL, the medium mining enterprises, small miners and cooperatives, and the metallurgical industry. The more general problems are discussed first, followed by those that are more specific to each subsector.

Near- and Medium-Term Market Outlook

As of mid-September, 1975, the L.M.E. price of tin stood at $2.94 per fine pound. This was, in current U.S. dollars, nearly 80 percent higher than the average tin price that prevailed in 1970, but in constant dollars of 1968-69, only about 45 percent higher than 1970 tin prices. Mid-September 1975 zinc (producer) prices were some 129 percent higher than the average price in 1970, while those of tungsten, silver and lead were respectively 19 percent, 154 percent and 18 percent above 1970 levels. Antimony prices have slipped over recent years from the unusually high level reached in 1970, but in late 1975 were still substantially above the more indicative 1969 price. Finally, copper prices in mid-September, 1975, had fallen 15 percent below levels recorded in 1970. Nonetheless, except for copper, prices for Bolivian exports as a whole were in late 1975 significantly higher in nominal U.S. dollar terms than in 1969-70. However, a dollar's worth of tin, zinc, or other minerals in mid-1975 provided significantly less command over world consumer and capital goods than was the case in mid-1970. One measure of the decline in the international purchasing power of the dollar is the dollar/SDR rate.[e] From 1970 to the end of 1974 the dollar/SDR rate increased from $1:SDR 1 to almost U.S. $1.20:SDR 1, signifying a decline of nearly 20 percent in the value of the dollar relative to the SDR. The dollar/SDR rate increased further, to almost 1.25:1 and 1.20:1 in the first quarter of 1975, and from April through June vacillated between 1.25:1 and 1.20:1.[14] Thus, from the vantage point of Bolivian policy-makers at mid-1975, the real value of the dollar had declined by between 20 and 25 percent since 1970, offsetting a substantial share of the potential benefits of higher prices for Bolivian mineral exports since the beginning of the decade. The medium- to longer-term outlook, dependent as it is on prospects of world economic recovery, is somewhat more favorable, but hardly provides a basis for euphoria in view of the marked increase in mining costs in Bolivia (particularly costs of imported inputs) in recent years.

Imported Materials Cost

Costs of imported materials (which in 1974 accounted for about 85 percent of total materials costs and about 9 percent of total costs) increased dramatically from 1971 onwards. For example, the costs of three significant imported mining inputs (explosive ammonium nitrate, drill steel and bits and metal sacks) rose dramatically from 1972 to 1974. Together, these items constituted about 20 percent of total materials costs in both 1973 and 1974. For ammonium nitrate

[e]SDR's are "special drawing rights" of the International Monetary Fund, created in the late sixties to serve as an alternative unit of reserve money. The dollar/SDR rate at end September, 1975, was 1.162.

alone, c.i.f. costs per kilo rose, from November 1972 to April 1975, by almost 200 percent. (Costs per kilo at the mine site rose by almost the same amount.) The rise in ammonium nitrate unit costs from 1973 to 1974 alone was nearly 44 percent, while that for metal sacks and drill steel and bits[f] was 59 and 17 percent respectively (see Appendix B). A very detailed examination of import clearance documents for 1973 and 1974 for one large mining firm reveals the following recorded increases for materials costs and certain equipment items from 1973 through 1974[g]:

Input	*Percent Increase in per Unit Costs (1974 over 1973)*
1. Calcium carbide	79.9
2. Lead nitrate	108.6
3. Fuses	45.9
4. Metal sacks (polypropylene)	98.5
5. Water pumps	161.4
6. Compressors	160.4
7. Power shovels	11.6
8. Ore cars	0.0
9. Metal air ducts	142.1

The observed increases recorded in imported materials costs were not confined to COMIBOL and the medium mines. BAMIN acts as a distributor of such inputs to the small mining sector. While BAMIN's pricing policy toward such sales is not noted for great responsiveness to market conditions, we may observe (see Appendix B) that even on BAMIN warehouse sales, 1974 prices of major inputs such as dynamite, steel bars and metal sacks increased by 59 percent, 49 percent and 5 percent respectively over 1973 levels.

Recent developments in costs of imported materials inputs and in world prices have been particularly damaging to a number of the more marginal COMIBOL mines. Following the marked increases in ammonium nitrate prices and the 1975 decline in world tin prices, the costs of explosives *alone* accounted for about 30 percent of the

[f]By August of 1975, per unit f.o.b. costs for drill bits (nearly 6 percent of materials costs in 1974) had increased by about 55 percent over 1973 levels.

[g]The possibility of overstatement of c.i.f. costs on the import clearance documents (so as to transfer profits abroad surreptitiously) cannot be ignored, especially given the low import duty rates on these items. However, it seems unlikely that such overstatement would be to the extent of more than 10 to 20 percent, especially since almost none of the private mining firms purchase inputs from affiliated businesses abroad.

per unit value of tin leaving the Catavi concentration plant in August 1975 (see Appendix B). This leaves only 70 percent of value to cover costs of other material inputs, labor and energy costs, taxes, depreciation, transport, smelting and other costs. Since Bolivia was already the world's highest-cost tin producer before 1973, these developments were cause for some considerable concern within the sector. Developments in costs of materials, equipment and labor used in the mining sector are taken up more fully in Chapter Three.

General Stagnation in Exploration and Mine Development Activities

The general stagnation that has characterized the mining sector for the past two decades has proven to be resistant even to the tonic implicit in the surge in minerals prices in 1973 and 1974. This is readily apparent from production statistics and is particularly evident when overall investment in the sector is considered. Because data on total realized investment in exploration, mine development, exploitation and concentration are scanty at best, it is necessary to resort to partial, often anecdotal evidence on this point. For example, GEOBOL officials estimate that as far as exploration is concerned, probably no more than U.S. $10 million has been spent in the last decade. This compares unfavorably with exploration outlays on *single* claims spent by a number of individual companies in Indonesia over the period 1968-1975. Nevertheless, several points are fairly clear and are rarely seriously debated within Bolivia. First, it seems evident that the most dynamic element of Bolivian mining, the medium mining subsector, has not invested significant new capital in the sector since 1968-69, when optimistic expectations kindled by what was perceived as a favorable attitude by the Barrientos Government led several larger firms to undertake commitments for sizable investments in all phases of mining activity, particularly exploration and development.[h] Following the two years of political unrest after the death of Barrientos in 1969, the major private firms practically ceased new investment in capacity, except for a number of smaller mines which were quickly geared up in 1973 and 1974 to

[h]Exploration and development are the two phases of the economic process of mineral production up to the point of actual extraction. This report follows the conventions of the study of the Carter Commission in Canada in the definition of exploration and development. Exploration is divided into two stages: *primary* exploration (prospecting and reconnaissance) and *exploration proper*. In mining for solid minerals, the development phase involves outlays for capital expenditure and organization to prepare the mineral body for extraction of ore in commercial quantities. M.W. Bucovetsky, "The Taxation of Mineral Extraction," Study #8, *Studies of the Royal Commission on Taxation* (Ottawa, July 1974), pp. 3-4.

take advantage of high metal prices in those years. Since 1971, the medium miners have maintained that perverse incentives for exploration and development investment, particularly those alleged to be implicit in the tax system and in the labor law have served to substantially curb further investment in the sector. While data that might substantiate this fact are difficult to obtain, it is clear that nearly all medium mining firms ended virtually all exploration activities in 1974, and in some cases have even dismantled exploration teams that had been painstakingly assembled in 1967-71. Further, many public and private officials argue that much of what passes for exploration in this subsector is merely the purchase by medium miners of mining claims formerly owned by small miners.

Most of the important firms have recently begun to reinvest high 1973-74 earnings heavily, but in commercial agriculture, not in mining. In 1973 and 1974, private foreign investors began to show some tentative interest in exploration in the Bolivian mining sector, but with the exception of one major investment in zinc in 1973-1974, little concrete results had emerged by end-1975. In any case, only medium-sized foreign firms are likely to be attracted to Bolivia, given the narrow-vein nature of Bolivian deposits (see Chapter Nine).

Investment in exploration and mine development by COMIBOL has largely followed an erratic pattern, characterized by long delays and unforeseen difficulties in implementation. The most significant COMIBOL investment in exploration and development in recent years was a cross-cut at Caracoles, one of its largest mines. This successful project, costing over U.S. $1 million, resulted in the confirmation of over a dozen new veins with average widths of one meter (wide by Bolivian standards) and of tin content of between 1.5 and 3.0 percent (versus present grades in Catavi of less than 0.50 percent). However, this project had been initially approved back in 1963, and was originally scheduled to require two years (instead of twelve) for completion. Another major COMIBOL exploration project at Corocoro involved several million dollars, but went awry when large and unforeseen water courses were encountered, flooding the shafts. Exploration and development in and around COMIBOL's Catavi mines, potentially a rich source of new deposits, has been stagnant for years owing mainly, it is said, to the additional risks of investing in a mining zone beset with continuing labor difficulties.

In spite of the fact that virtually all the major deposits now being worked were discovered by small mining firms, almost no exploration and insignificant amounts of development outlays are being made in this subsector. The Small Miners Association cites high, "blind" taxes, lack of investment credit, political uncertainty and their inability to explore the nation's "fiscal reserves" (see Chapter Nine) in explaining the absence of exploration activity in recent years.

After several years of relative inactivity, GEOBOL, with United Nations and World Bank financing and technical assistance, has been engaged in a number of exploration efforts. These include exploration for gold in northeast Bolivia and a search for alluvial deposits of tin near the Brazilian border. A substantial body of opinion, within and outside Bolivia, holds that the nation still possesses a sizable untapped potential for many metallic and nonmetallic minerals. But by 1972, it has been reliably reported by GEOBOL that only 25 percent of the country's land area had been mapped geologically, and that less than five percent had been explored on any systematic basis. However, since the completion of a GEOBOL pilot survey project for the Southeastern Cordillera and the initiation of a new survey project (to cover about 8.5 percent of the land surface of Bolivia) for the western part of the nation's "mining belt," the former ratio has probably increased somewhat. The extent of untapped mineral potential of Bolivia is, in any case, still very much an unsettled question. Certain areas that were reputedly rich, such as large sectors of the Department of Beni in the northwest, have proved disappointing.[15] Other knowledgeable observers of the Bolivian mining sector point out that most of the easily found outcrops have already been claimed and worked, and that even though less than five percent of national land area has been explored intensively, this represents more than 15 percent of the mineralized area of the nation. Nonetheless, there is a widespread view that the mineral wealth of Bolivia has been rarely scratched, and many point to the fact that almost all important deposits now being worked were known before 1911 as indicative of the lack of serious exploration efforts in the past half-century.

Without doubt the immobilization of a sizable portion of the mining belt in the nation's fiscal reserves has contributed to the lack of exploration activity in Bolivia, especially over the past 25 years, particularly the last 10 years. Nearly 33 percent of the total area of the country has, in recent years, been assigned to the fiscal reserves. These reserves have not been generally open for private sector exploration, and COMIBOL exploration in these areas has been minimal.

Taxation Issues

The type of complaints over taxation characteristic of Bolivian mining enterprises (particularly private sector firms) are not materially different from those voiced by textile firms in Chile, gold producers in Ghana, toy makers in Hong Kong, beverage producers in Indonesia, rubber producers in Malaysia, or coffee growers in Colombia. As seen from the private sector, enterprise taxes are always too high and/or inequitable, or else they are a principal cause of resource misallocation, and constitute the underlying reason for sectoral stagnation, or some combination of all of the above.[i] Although COMIBOL has appeared indifferent toward tax issues until the very recent past (when the enterprise first began to meet its principal legal tax obligations), the small and medium mining firms have for years focused their dissatisfaction upon perceived defects in the structure as well as the level of tax rates. Although there may be very good reasons for wanting higher taxes on mining profits than on, say, profits from garment manufacturers or service activities (see Chapter Four), it is indeed clear that the design and implementation of existing taxes on the mining sector have suffered from a number of real as well as (largely) imagined shortcomings. These are discussed in considerable detail in Chapter Seven and need not be covered in great detail here. Briefly, however, it may be noted that:

1. The present system does not allow the government to share in much of the benefits of high profits when market conditions are good, but is overly burdensome to mining enterprises when market conditions turn sour;
2. The present system provides perverse incentives for rational exploitation and concentration decisions, leading to socially wasteful "high-grading" of ore deposits and excessive losses in the concentration process.
3. The existing system, relative to a tax on profits, tends to severely dampen investment in risky activities such as exploration and mine development; and
4. The present system treats equally enterprises that are vastly unequal in cost structure, so that (among other defects) large, mechanized operations tend to be significantly undertaxed while smaller, more primitive operations tend to be overtaxed, relative to the situation that would prevail under a profits tax.

There are other shortcomings as well, all discussed in Chapter Seven. But the existing system is not, it should be noted, completely without redeeming features. It is doubtless simpler to administer

[i]A common view among Bolivian mining firms, and also often heard in Canadian and Australian private mining circles, is that the tax system serves to convert "ore into rock," whereas the sensible aim should be to convert "rock into ore."

than almost any alternative system that might be realistically proposed and does generate a sizable volume of fiscal revenues. Further, some of the basic underpinnings of the present structure provide a fairly good basis for sensible reform that will not require radical new departures in the short to medium term, and can serve as a good foundation for more ambitious long-term reform.

Technological Issues

As indicated in an earlier section, Bolivia is the world's highest-cost tin producer. Bolivia's cost disadvantage relative to Far East producers arises from several sources. First, those producers (Malaysia, Indonesia, Thailand) primarily exploit alluvial and eluvial deposits using dredges, hydraulic processes and other techniques suitable for large, low-grade tin ore deposits. Because virtually all of Bolivia's important deposits are narrow-vein underground deposits, Bolivian tin ore must be very high-grade to make exploitation feasible, and ore grade has declined continuously since 1930. Second, the tin minerals of Southeast Asia possess certain marked advantages in the concentration process. There, concentration processes based on gravimetric methods allow tin ore to be easily processed to almost pure cassiterite. Bolivian ores typically contain not only cassiterite but sulfides and other compounds that cannot be completely separated in the concentration process.[16] A significant share of Bolivian ore contains colloidal tin that can be separated only by resort to expensive pyrometallurgical processes (such as roasting in volatilization plants). In addition, some tin is contained in stannite (Ca_2FeSnS_4) and is not acceptable to many smelters. As a result, the average grade of Bolivian tin concentrates is only 39 percent for COMIBOL exports, 32 percent for the medium mines, 35 percent for the small mining sector and 20 percent for "other" exporters,[17] compared to an average grade of concentrates of over 70 percent for both Indonesia and Thailand.[18] Another consequence of the presence of sulfides, and particularly of stannite, is lower recoveries during concentration. Furthermore, every additional upgrading step, as in volatilization plants, introduces one more source of metallurgical loss, thus reducing overall recoveries. The problems of Bolivia's cost disadvantages in tin are heightened by the need to produce high-grade concentrates that can be sold to ENAFBOL, as this requirement generally leads to higher metal losses during the concentration process.

For other metals, it is pertinent to note that Bolivia's competitive position in copper is not good because most of its production of this metal comes from relatively thin beds ("*mantos*") at Corocoro and similar "red-bed" ore deposits, as opposed to the porphyry copper deposits mined by open pit and block caving elsewhere in the world. Corocoro-type deposits do have an advantage in that they allow the production of high-grade copper concentrates, because the ores are either native copper or chalcocite. Bolivia also has a competitive disadvantage relative to other countries like Peru in regard to its potential for finding new lead-zinc ore deposits, since the geological makeup of the former has conspicuously few limestone units, with which lead-zinc ores are commonly associated. Closely related to the foregoing is the problem of the high content of iron in many Bolivian sphalerites, which considerably complicates marketing of zinc concentrates. The solution to this problem may be the construction of a zinc smelter/refinery designed for treatment of this type of concentrate.

Another technological problem faced by Bolivia derives from the complexity and diversity of its ores.[19] In order to be able to recover many of the companion and byproduct metals it would be necessary to erect a smelter complex like that of La Oroya in Peru, which can take advantage of interlocking and complementary metallurgical technologies and processes. This is particularly important since Bolivia possesses such a wide variety of metals.

Perhaps one of the greatest problems faced in realizing Bolivia's mineral potential in the future is the extensive cover of geologically relatively recent volcanic rocks and alluvial plains (including the *salares*). Most of the bedrock exposures have been prospected long ago and any indications of ore in them have been tested. In the future, it will become increasingly necessary to locate ore deposits under the volcanic and alluvial cover, a task that will require many more trained geologists, geophysicists and geochemists than Bolivia has at present.

Infrastructure Problems

Bolivia is a high-cost producer of minerals for reasons other than the technical problems cited in the preceding section. This is particularly evident in the case of tin. Owing to the isolation and great distances between Bolivian tin deposits and export ports, relative to those of Malaysia, Indonesia and Thailand, and due to the less developed and more costly transport infrastructure in Bolivia (because of the more mountainous terrain), internal transport costs per ton of tin-in-concentrate are significantly higher in Bolivia.[20] Bolivia also encounters some difficulties with regard to costs of ocean freight to major markets, particularly with regard to high conference charges for small-lot shipments on such metals as copper.[21]

Although internal transport costs do not appear to be nearly so high as the figure of U.S. $1.00 per ton/kilometer often cited by mining firms, such costs are nevertheless a substantial portion of value, particularly for low-grade concentrates.[j] Although approximate figures for the relative proportion of total concentrates shipped by rail and by truck are apparently not available, it is known that firms make extensive use of both modes. Shipment by truck involves high costs due to the generally very poor condition of main feeder and trunk roads (the major exception being the Oruro-La Paz highway), which in turn is a function not only of altitude and climatic conditions, but of low levels of maintenance. Under such conditions, vehicles deteriorate rapidly and losses due to delays and accidents are considerable.

The railway and electric power network is in poor condition, generally owing to a much-depleted capital stock. On the railway network, concentrate losses-in-transit due to poor repair and upkeep of rolling stock have been as high as 12 percent in some cases involving bulk mineral shipments. Physical conditions for the rail network promise to improve over the next five years, as a recent IBRD loan for upgrading the railway system is implemented; but the short-term costs of the upgrading process have come as a rude shock to the mining sector. The overall structure of railway tariffs was maintained for several years at levels insufficient to prevent heavy operating deficits, let alone allow positive rates of return on railway investment. The mining sector has long maintained that while freight tariffs for imports have consistently resulted in heavy subsidies for such items as flour and other foodstuffs, tariffs for mineral exports have been so high, owing to the unavailability of volume discounts on rail shipments, as to heavily penalize shipments of relatively low value-to-bulk minerals such as zinc and lead. Thus, small and medium-scale mineral producers were particularly dismayed by a 68 percent across-the-board increase in railway tariffs in 1975.[k] Capacity of the rail network was, until late 1975, at least as serious a problem for the mining sector as the rundown condition of the roadbed and rolling stock. The problem appears especially serious in Potosi Department, where two large zinc mines are considering a six-fold expansion of production before 1979. In this area alone, an

[j]For example, in the Andean mining area of Peru, transport costs for concentrates from Julcani to Huancavelica are only U.S. $0.034 per ton/kilometer. In Bolivia, costs for zinc are about $0.06 per ton/kilometer.

[k]Interviews with managers of several small and medium mining firms were replete with references to planned investment projects that were halted as a result of this tariff adjustment.

investment of over U.S. $5 million would be required just to meet the rail transport requirements of *one* of the two firms.

Special Problems of the State Enterprise Sector

Large to medium-sized public and quasi-public mineral enterprises in other countries (such as PERTAMINA and P.N. Timah in Indonesia, CVRD in Brazil and the Ghana state gold mines corporation) have at one time or another shared some common problems. Some of these problems have to do with a lack of clearly formulated long-term and even short-term goals for the enterprises; others are related to a history of political interference in investment, personnel and operating decisions. At one time or another, most large state enterprises, in mining and outside mining, have been treated as "milch cows" available for exploitation for various and sundry political purposes, some worthwhile by national standards, others perhaps inadvisable on all counts.

COMIBOL has experienced some or all of the above types of difficulties since its creation in 1952, and in addition shares with the rest of the mining sector the problems noted in the previous section, as well as some difficulties peculiar to itself. This section attempts to focus on the special problems that appear to be more critical for the state enterprise than for the rest of the mining sector. Many of these problems (organization and management issues, inventory control) are further discussed in Chapter Nine.

The special problems of COMIBOL appear to fall into two distinct categories: (a) organization and management problems, and (b) labor and related problems.

Organizational and Management Problems

COMIBOL is a multimineral enterprise, unlike many of its counterparts in other countries, where state mining enterprises are more specialized by mineral.[1] Thus, COMIBOL management faces highly diverse, often extremely complex technical issues, solutions for which are not easily devised by large multinational firms with several more decades of experience in mining. To deal with technical issues, as well as intricate marketing and investment issues, on a rational basis would require a highly streamlined, well integrated management and accounting structure. By 1975, COMIBOL did not possess either, although very substantial progress has been made on both fronts since 1965.

[1]For example, P.N. Timah in Indonesia is involved in the exploitation of tin only. Another enterprise, P.N. Aneka Tambang, handles bauxite, nickel and a few other metals. In Ghana, the state gold corporation works only in gold, and in Brazil the state iron enterprise is involved only in iron.

The internal problems arise from a number of causes. First, the organization structure appears to be centralized to an extent inconsistent with running such a large, diversified firm in an efficient manner. The various levels of the hierarchy do not have clearly defined roles in various types of decisions that would enable the organization to act quickly and effectively, and the budgeting and control processes are not well suited for providing motivation and means for evaluating management performance. Thus, adequate control of subordinate levels of management is rendered difficult, and managers cannot be adequately evaluated during or at the end of budget periods. Other, related, internal problems are discussed further in Chapter Nine.

Labor and Related Problems

Any analysis of labor and related problems in COMIBOL (and, for that matter, management-accounting problems) should be made in the light of the enterprise's own self-perception. There has been a tradition within COMIBOL of viewing the organization as a major social institution, as well as the major economic institution in the country. Much of this tradition survives today, and top management of the enterprise tends to stress COMIBOL's substantial social responsibilities in the fields of employment, education, nutrition and hospital care. Indeed, if the fulfillment of these responsibilities is viewed by a consensus of Bolivian society as matching public preferences for essentially public (or quasipublic) goods, and if COMIBOL can perform these functions at lower social costs than the responsible government entities, a strong case could be made for continued adherence to the tradition noted above.

On the other hand, if the "social responsibility" of COMIBOL is defined as that of making optimum use of scarce mineral endowments for the benefit of Bolivian society as a whole, and if this requires COMIBOL to make the maximum possible contribution to public savings consistent with its fulfillment of this function, then the government (as the only shareholder in COMIBOL) might wish the enterprise to place a higher premium upon more effective cost control and higher before-tax profits, leaving to the government proper the tasks of satisfying public wants for education, health and social development, with the greater fiscal resources that might be generated by a differently oriented COMIBOL.

However defined, COMIBOL's total outlays for labor costs of all types (wages, bonuses, education, health and commissary losses) account for a far larger share of total revenues than in the private mining sector.[23] Commissary losses alone reached U.S. $11 million in 1974 (25.5 percent of total labor costs in COMIBOL as a whole and more than 33 percent of labor costs in mining activity alone) and were expected to rise to U.S. $14 million in the following year.[24] Despite several attempts (particularly in 1966-68) to pare COMIBOL's payrolls by eliminating surplus labor, several thousand redundant *surface* workers (but not underground miners) continue to draw full wages. As a result, only 35 percent of COMIBOL laborers work in the interior of COMIBOL mines, as opposed to about 60 percent in the medium mining sector. Various studies, ranging from that of a special Presidential Commission on COMIBOL (1974) to the labor syndicate's (1975) study, estimate that at least 3,500 surface workers could be pensioned off (for old age or intractable infirmities) with little or no reduction in output, but with significant results for productivity.

A final major related problem in COMIBOL is the widespread theft of ores, particularly from high-grade tin veins (such tin requires no concentration). This problem is particularly serious at the large Catavi mine, where reported losses from mineral theft often reach 30 percent of the value of minerals bought from outside sources.[25] For COMIBOL operations in general, theft in 1974 may have run as high as 8.7 percent of COMIBOL's own production of ore.[26]

Special Problems of the Medium Mining Sector

As noted earlier, a significant degree of heterogeneity characterizes the 24-30 firms in the medium mining sector at any given time. The five or six large firms at one end of the spectrum are significantly different in several respects from the eight to ten firms at the other end of the spectrum. (See Appendix A.)

Almost all the enterprises have encountered some managerial and technical problems in recent years, and virtually all possess some significant, but potentially remediable, gaps in internal record-keeping practices. Both such problems, however, are more serious for the smaller medium miners, particularly given the greater *internal need* for better accounting practices in the larger firms. Further, most medium mining firms have experienced some degree of difficulty in obtaining short- and medium-term credit in recent years, although again this has been more serious for the smaller firms not having access to finance abroad. The apparent greater availability of agricultural credit (and far lower taxation, both *de jure* and *de facto*, in commercial agriculture) has reportedly been a significant factor contributing to substantial investment by medium mining firms in cattle-raising in Beni and large-scale agriculture in Santa Cruz.

Among the other major concerns of the larger medium mining

firms have been fear of nationalization on the one hand, and the extreme difficulty of going out of business in Bolivia on the other. From all accounts, both factors were responsible for much of the private firms' reluctance to reinvest in the mining sector during the sixties and early seventies. With regard to the first factor, public officials and private individuals on numerous occasions identified fear of nationalization (especially in the period 1969-72) as a principal factor that has inhibited growth in firm size in mining, although the size threshold at which nationalization became a distinct possibility is rarely stated. As for the second factor, it is apparently true that closure of firms has been extremely rare (not only in the mining sector), as past governments have been extremely reluctant to deal with unemployment problems with other policy instruments. Rather than allow closure of bankrupt firms, successive governments have tended to prop them up with continuing credit subsidies, forgiveness of past tax liabilities and other measures. Thus, many private sector mining firms have developed an aversion to hiring additional labor when market conditions warrant, as they are reluctant to assume the added responsibilities for workers in the event of business downturn or failure.

Other problems of the medium mining sector include scarcity of skilled workers and a self-admitted lack of sophistication in international markets. The scarcity of highly skilled labor has not only resulted in production bottlenecks in a number of mines, but to expensive equipment damage (particularly in dredges) due to worker inexperience. Lack of sophistication in international markets has reportedly led in some cases to subpar exploitation of the bargaining positions of Bolivian firms with consequent losses of potential profits.

Special Problems of Small Firms and Cooperatives

As noted previously, the number of "small" mining firms active in the sector fluctuates considerably from year to year, from as low as 2,500 to as high as 4,000, depending on market conditions. This group of firms shares a number of the serious problems implicit in small-scale mining activity everywhere: primitive technology, minimal or nonexistent health and safety standards, absence of exploration, low productivity, and lack of access to credit. But unlike small miners in Malaysia, Thailand, Indonesia, the Philippines and even Chile, small miners in Bolivia work at extremely high altitudes (often in excess of 4,800 meters), at long distances from mineral collection points and with a grossly deficient transport infrastructure. Further, the labor supply to the great majority of small mines is highly seasonal, as many miners leave during harvest periods and return thereafter.

A much-discussed problem specific to small miners is the difficulty they have encountered in marketing their minerals, particularly low-grade tin concentrates, copper and antimony. Although this problem has a number of technical and economic dimensions, much of the blame in the past, particularly before 1975, was typically attributed to BAMIN. The small miners, as well as numerous individuals knowledgeable in the mining sector, allege that the benefits of lower taxes granted to small mining firms were more than counterbalanced by the administrative difficulties and higher charges involved in the mandatory marketing of their minerals through BAMIN. In response to these complaints, the institution was recently reorganized, with the intention of making it more responsive to the needs of the small-scale mining sector.

NOTES TO CHAPTER TWO

1. For a cogent discussion of the notion of "retained value," see R.F. Mikesell, *Foreign Investment in Copper Mining* (Washington, D.C.: Resources for the Future, 1975), p. 66 ff.

2. Total taxes as a percent of gross export values averaged 4 percent from 1900 to 1920, 1.5 percent in the period 1921-1930 and 10.1 percent in the 1931-36 period. From 1936 to 1947, taxes, including implicit taxes through differential rates of exchange, averaged 16.6 percent of export values (exclusive of oil) and were 18.2 percent in 1973 and 19.6 percent in 1974. The effective tax rate (taxes/gross export values) for 1973 and 1974 do not appear, at first glance, to be very much higher than those prevailing in, say, 1936. It must be recalled, however, that in the twenties and thirties enterprises were exploiting tin deposits of much higher quality than in 1973 and 1974, with consequent lower costs per ton and higher profits per ton, relative to value of output. For example, in what is now COMIBOL's (the state mining enterprise) Catavi group of mines (then the jewel of the Patiño interests), which still accounts for nearly one-third of COMIBOL output, the tenor (metal content) of ore extracted was 6.65 percent in 1925 (source: Walter Gomez D'Angelo, *Mining in the Economic Development of Bolivia* [PhD dissertation, Vanderbilt Univ., 1973], p. 141). By 1970, the tenor of the ore had fallen to 0.89 percent, and by 1975 averaged only 0.45 percent.

3. According to one source, the labor share in total export value in the twenties was about 13 percent in the large mines (now part of COMIBOL) and between 13 and 30 percent in the medium-sized mines (Gomez D'Angelo, *op. cit.*, pp. 51-52). By 1974 the labor share of total mine revenue was around 30 percent in COMIBOL and slightly above 21 percent in the medium mines. (See Chapter Three.)

4. Chapter Three shows that even in 1974, after some degree of import

substitution in mining materials inputs, about 85 percent of all materials used in the sector were imported.

5. This is obvious in any visit to almost all COMIBOL (formerly Patiño, Aramayo and Hochschild mines) and many medium mines. Vintage of COMIBOL equipment clusters in the 1915-1930 era, as do almost all buildings.

6. The national accounts figures above are based on National Planning Office (CONEPLAN) data for 1970-73 and CONEPLAN preliminary estimates for 1974. The usual caveats regarding the reliability of national accounts apply with particular force in Bolivia. CONEPLAN began (in 1976) a thoroughgoing effort to revise the National Income Accounts.

7. We were unable to obtain yearly fatality figures for COMIBOL mines as a whole. It would be interesting to compare fatality rates in these mines with those for other countries. On the whole, the annual fatality rate for underground accidents per 100,000 tons of ore extracted could be substantially lower than for the U.S., as Bolivian mines have not had the catastrophic occurrences that occur in, say, American coal mines every few years. In 1974, for underground (non-coal) mines in the U.S., there were 48 fatal accidents, 3,257 non-fatal disabling accidents, and 1,169 nondisabling accidents. Injury frequency rates per mission work-hours were 44.7 for disabling (including fatal) accidents and 15.40 for nondisabling accidents (U.S. Department of the Interior, Mining Enforcement and Safety Administration, *MESA Safety Reviews*, November, 1975, Table 4-A.

8. Figures from a special, presidentially commissioned report on COMIBOL in November 1974.

9. U.S. Department of the Interior, Bureau of Mines, *U.S. Mineral Resources 1974* (Washington, D.C.: U.S. Government Printing Office, 1974), p. 48.

10. International Bank for Reconstruction and Development, *Appraisal of a Mining Credit Project in Bolivia* (Washington, D.C., 1973), Annex 7.

11. United States Department of the Interior, *Minerals Yearbook*, 1973, vol. I, p. 215.

12. United States Department of the Interior, *U.S. Mineral Resources*, p. 313.

13. The most recent of these is: Ministerio de Minería y Metalurgia, *Diagnóstico del Sector Minero-Metalúrgico*, 1965-1974 (La Paz, November 1975), Section III.

14. See Chase Manhattan Bank, *International Finance*, June 2, 1975, p. 1.

15. Interviews with medium mining firms.

16. See Gomez d'Angelo, *op. cit.*, p. 34; and William C. Kelly and F.S. Turneaure, "Mineralogy, Paragenesis and Geothermometry of the Tin and Tungsten Deposits of the Eastern Andes: Bolivia," *Economic Geology*, vol. 65, no. 6, September/October 1970, pp. 609-679.

17. Ministry of Mining, *Boletín Estadístico*, July 1975, p. 1.

18. Figures for Thailand are from International Bank for Reconstruction and Development, *The Mining Sector in Thailand* (Washington, D.C., 1975), Appendix 1. Figures for Indonesia are from personal interviews in the Ministry of Mines, Jakarta (June 1975).

19. See, for example, Kelly and Turneaure, *op. cit.*, pp. 609-612.

20. According to CONEPLAN, in many cases more time is required for

Bolivian mineral exports to arrive at maritime ports in Chile and Peru and be dispatched from them than to go from the maritime ports to their final destinations in Europe, the U.S.A. and Japan (CONEPLAN, *Plan Operativo*, p. 159).

21. See Arthur D. Little and Co., *Commercialization of Bolivian Non-Ferrous Metals and Minerals, Addendum to Vols. I and II* (Cambridge, Mass., April 1973), p. 48 ff.

22. Interviews at mining enterprises, May, July and August of 1975. An unknown part of these losses occur in ocean shipping.

23. See Chapter Three.

24. See CONEPLAN, *Plan Operativo 1975*, p. 119. "Pulpería" losses arise from the pricing of foodstuffs in COMIBOL commissaries at prices far below domestic market prices (such items include several kilos of meat per worker per month, bread, and flour).

25. Interview at Catavi, August 1975.

26. Derived from information on p. 8 of U.S. Department of State, *Industrial Outlook Report: Minerals 1973-1974* (La Paz, Bolivia, July 7, 1975).

Chapter Three

The Macroeconomic Impact of the Mineral Exporting Sector on a Developing Economy[a]

Glenn Jenkins and
Malcolm Gillis

OVERVIEW

In this chapter we identify and attempt to quantify the direct and indirect links that the mineral exporting sector has with other economic activity in a developing country. We find that the traditional macroeconomic models used for analysis in developed countries may not be appropriate for some developing countries because the effective constraints and linkages between sectors are significantly different. Keynesian macroeconomic models would lead us to expect that the principal channel for the mining sector's impact on aggregate economic activity would be through its direct expenditures on labor, materials and transport services. To the extent that unemployed resources characterize these markets, an expansion of mining activities will generate additional real income. Real income may be further increased when the income created initially is spent on domestic goods that are also produced with inputs that would otherwise be unemployed. In the case of Bolivia, which may be typical of many other less developed countries, we find that this linkage between the mining sector and the rest of the economy is relatively less important than three other ways in which the mining sector interacts with the rest of the economy.

First, in a developing economy the mining sector often is a major earner of foreign exchange. The nature of the (static) comparative

[a]Without implicating them in any way, the authors wish to thank several persons for extensive and valuable comments on earlier drafts of this chapter. They include Ralph Beals, Michael Roemer, Joseph Stern and Brian Wright.

advantage of many primary-sector-oriented developing economies is such that they could only produce at significantly higher cost many of the commodities they now import. Generally, a large proportion of these imported goods are intermediate inputs required for production or capital goods needed for future growth.

Second, in many developing economies with primary-oriented production and trade, the mining sector often is a major source of government revenue. These tax revenues provide a potential source of savings for the economy if the government chooses to channel these revenues towards the financing of investment projects, or they provide the means of affecting the distribution of income through current expenditures on wages, salaries, or transfer payments.

The third way in which mining sector activity influences macroeconomic activity is through the way that cyclical fluctuations in the mining sector's revenues and tax payments alter government revenues and the financing of the government budget. Because it is not feasible or desirable for government revenues to be reduced to correspond to short-run decreases in either international mineral prices or mineral production, the resulting government deficit must be financed in other ways, most often through central bank borrowing. Thus, a decrease in the mining sector's activity may ultimately result in an aggravation of inflation rather than creating the deflationary pressures that traditional Keynesian macroeconomic models predict would occur.

In the following sections we attempt to evaluate the direct impact of the mining sector on aggregate economic activity and conduct an analysis of the other channels, outlined above, through which the mining sector can influence economic activities in Bolivia.

THE DIRECT IMPACT OF THE MINING SECTOR ON DOMESTIC AGGREGATE DEMAND

Because of the capital-intensive production technology used in the mining sector, this industry generally accounts for a much smaller proportion of total employment in a less developed country than it does of the country's total production. In Bolivia the mining sector has normally been responsible for the production of approximately 11 percent of the country's gross national product,[1] but it employs less than three percent of the country's labor force.[2] The impact on the domestic aggregate demand of less developed countries from the mining sector's demand for capital and material inputs is usually also minimal because most of these items must be imported.

In the analysis of the direct effect that the mining sector has on the level of aggregate demand in an economy, it is necessary to distinguish between changes in mining sector revenues that arise because of changes in the quantity of output produced and those due to changes in mineral prices. When there is a change in the quantity of production, there will normally be a corresponding change in the quantity of labor employed, materials and energy purchased, leading to further changes in income flows to the mine owners and tax revenues to governments. Alternatively, for a change in the price of a mineral commodity, the initial impact will be realized on the income accruing to the mine owners and the government's tax revenues. If this change in price exists for a significant period of time, then it may result in changes of the same direction in the real wages received by miners and in investment expenditures in new mining capacity.

To begin our analysis of the direct impact of the mining sector on aggregate demand, we will consider the case of an increase in mineral production that is sold abroad. From the examination of the expenditure flows that are generated by this expansion of mining sector output, we wish to evaluate the magnitude of its "first round" impact on domestic aggregate expenditure.

The sale of additional mineral output (dQ) will result in an increase in gross sales revenues of (dS) when the price of the mineral remains equal to (P_0), i.e., $dS = dQ \cdot P_0$. Gross sales revenues in turn can be broken into the national mineral revenues (B) that will accrue to the country exporting the mineral product and the smelting, transportation and insurance costs (F_c) that are purchased outside the country. If the proportion of these foreign costs to gross sales revenue is denoted by (A_f) then the change in national mineral revenue can be expressed as follows:

$$dB = (1 - A_f)dS \qquad (3\text{-}1)$$

This change in national mineral revenue (dB) can be broken down further into its expenditure components, which include payments for labor (L), materials (M), purchases of other services (O), energy (E), domestic smelting and transportation costs (D), taxes (T), profits, interest and rentals (I), and capital recovery (K).

To derive the initial impact on an economy's aggregate demand, an estimate must be made of the foreign trade leakages that will occur when the expenditure components of national mineral revenue (B) are increased. In this analysis we will first outline the factors that will determine the size of the foreign trade leakages for the mining sector. This framework will then be used to estimate the direct impact of the mining sector on aggregate economic activity in Bolivia.

Labor Expenditures

To evaluate the impact that the labor income component of national mineral revenue has on aggregate expenditures it is necessary to estimate (a) the change in labor income that occurs because of the change in the level of mineral production and national mineral revenue, (b) the extent to which an increase in labor used in the mining sector decreases the amount of labor employed in other sectors, and (c) the proportion of the incremental income generated by additional employment in the mining sector devoted to expenditure on traded goods. If we denote the proportion of national mineral income that is spent to purchase labor as A_L, and also assume fixed input-output coefficients in production, then the change in labor expenditure can be expressed as

$$dL = A_L \, dB. \tag{3-2}$$

We also assume that the supply of labor to the mining sector is elastic and, in addition, the sectors from which this labor is transferred suffer negligible losses in output because of the decrease in their labor forces. In many less developed countries unskilled labor is plentiful; however, these assumptions will serve to overstate the net impact of changes in mining sector labor income on the economy.

The income accruing to the mining sector's labor force will be either spent on domestically produced consumption goods (C^d), imported goods (C^m), or saved for future consumption. If we denote the marginal propensity to purchase imports out of mining labor income as (m_L), and the marginal propensity of mining labor to save as (b), then the change in consumption of domestically produced goods for a change in mining labor income can be expressed as

$$dC^d = (1 - m_L - b) \, dL. \tag{3-3}$$

By combining Equations (3-1), (3-2), and (3-3) we can derive an expression that relates the "first round" change in consumption from labor income to the change in gross value of sales as

$$dC^d = (1 - m_L - b) \, A_L \, (1 - A_f) \, dS. \tag{3-4}$$

Expenditures on Materials and Other Services

Typically, mining activities in less developed countries import a large proportion of their material inputs. In addition, many of the services such as foreign travel, legal fees and information services have also large foreign expenditure components. If we let A_M and A_O represent the proportions of total national mineral revenues that are used to purchase materials and other services, respectively, and let m_M and m_o denote the marginal propensity to import materials and other services, then the change in the domestic expenditures on materials and services (dM^d and dO^d) resulting from a change in mineral output can be expressed as,

$$d(M^d + O^d) = [(1 - m_M)A_M + (1 - m_o)A_o]\ (1 - A_f)\, dS. \quad (3\text{-}5)$$

Expenditures on Energy

Several less developed countries that are net exporters of petroleum (Indonesia, Venezuela, Colombia before 1975) have had a policy of keeping their domestic price considerably below the f.o.b. price for export sales. Bolivia is such a country which produces its own petroleum products and sells some of the production domestically at below world prices.[b] Because of the low sulphur content of Bolivian crude oil, any reasonable increase in Bolivian production of crude oil could be exported. Therefore, if the mining sector expands and uses additional petroleum products, the Bolivian economy loses the difference between the price the mining company is charged for the petroleum and what the country could get by exporting it.

We will denote A_E as the proportion that energy costs are of national mineral revenue, e_p as the share of petroleum costs in total energy costs of the mining sector, P^d as the domestic price of petroleum, and P^w as the f.o.b. price of Bolivian petroleum. The expression for the change in domestic expenditures resulting from the domestic consumption of petroleum products in the mining sector (dE_p^d) can be written as follows:

$$dE_p^d = A_E\, e_p\, \left(\frac{P^d - P^w}{P^w}\right) (1 - A_f)\, dS \quad (3\text{-}6)$$

Electricity is the second major source of energy used by the mining sector in most countries. If the mining sector increases its demand for electricity, then additional resources will have to be invested in this sector. Part of these resources will be from the domestic economy, but in developing countries most of the generation equipment, transmission lines, and distribution equipment will be imported. If the proportion of total energy costs that is made up of electricity costs is written as e_e, and the share of imported inputs in electricity generation is m_E, then the impact on domestic expenditures from a change in electricity consumption by the mining sector can be expressed as,

[b]Details of this arrangement are discussed in a subsequent section of this chapter.

$$dE_e^d = (1 - m_E)\, e_e\, A_E\, (1 - A_f)\, dS \tag{3-7}$$

Domestic Smelting and Transportation Costs

The purchases of smelting and transportation services that are made from domestic enterprises are similar to the expenditures on energy, as they often involve the use of artificially low-priced petroleum products and are also capital intensive. To estimate the impact on domestic expenditures of a change in transportation and smelting activity, it is necessary to break the expenditures on inputs used in this sector into the proportions spent on local labor plus domestically produced goods (D_L), imported inputs (m_D), and energy (D_E). Energy çonsumption in the smelting and transportation sectors must be further broken down into the proportion supplied by petroleum (D_{EP}) and the proportion from other domestic sources (D_{EO}). If the proportion of national mineral revenues spent on smelting and transportation purchased from inside the country is A_D, then the net change in domestic expenditures on these items from a change in mining sector sales is,

$$dD^d = \left[(A_D)(D_L) + (A_D)(D_E) \left((D_{EO} + D_{EP} \left(\frac{p^d - p^w}{p^w}\right)\right) \right] (1 - A_f)\, dS. \tag{3-8}$$

In Equation (3-8) the first term measures the changes in domestic expenditures brought about by the change in labor income and domestic purchases of goods by this sector. The second term measures the net impact of change in energy consumption by the smelting and transportation sectors on domestic expenditures.

Profits, Interest and Rentals

The impact of this income flow on domestic aggregate demand is determined by the proportion spent on domestically produced goods. As this income stream will accrue to upper-income domestic or foreign owners and lenders, we would expect that a large proportion of this income is spent on imports or paid out abroad. If the proportion of national mineral revenue that accrues in the form of profits, interest and rentals is expressed as A_I, and the marginal propensity for these income streams to be spent abroad is m_I, then the change in domestic expenditures on these items from a change in mineral sales can be written as,

$$dI^d = A_I (1 - m_I)(1 - A_f)\, dS \tag{3-9}$$

Capital Recovery (Depreciation Expenses)

The fixed capital that is depreciated through time by the mining sector consists primarily of machinery, equipment and buildings. The labor-intensive sinking of shafts is largely currently expensed and therefore not depreciated. While depreciation expense is itself not a cash flow, these funds can be either used to repay loans, reimbursed to the owners of the mine through dividend payments, spent on consumption goods, or invested. If we express the share of national mineral revenues that is allocated to depreciation expense as A_K, and the marginal propensity for these funds to be spent outside of the country as m_K, the change in domestic expenditures on these items from a change in mineral sales can be written as,

$$dK^d = A_K (1 - m_K)(1 - A_f)\, dS. \tag{3-10}$$

Tax Revenues

Tax payments are often an important item in the cash flow of the mining sector. In Bolivia they equaled approximately 32 percent of total mineral revenue in 1974. The manner in which these tax revenues are spent is a function of government policy and thus the proportion spent on domestic goods and foreign goods may be altered greatly through time. However, from historical data the propensity of the central government to spend its tax revenues on imported goods can be estimated. This propensity for the government purchase of imported goods will be written as (m_T), and the proportion of mineral revenue that is paid out as taxes denoted as (A_T).

In this analysis the government is treated as an extension of the general public in that government expenditures made on behalf of the public are included in the total level of well-being of the domestic residents. Thus, under such circumstances changes in government tax revenues that are matched by an equal amount of expenditures will not change the total level of savings in the economy. A change in government taxes and expenditures will be treated as a rechanneling of normal income by domestic residents; they will adjust their savings rate on their remaining net-of-tax income to reflect the fact that government either spends more or less on consumption goods than their desired level. This assumption is chosen because it is a more neutral assumption concerning the value of government expenditures than is the usual treatment found in most macroeconomic models that assume government expenditures have a total utility value of zero when summed over all the residents of the economy.

Combining the marginal propensity for the public to save from government expenditures of (b) with the marginal propensity for the government to purchase imported goods of (m_T), the change in direct domestic aggregate expenditures due to an increase in government tax revenues from the mining sector (dT^d) is estimated as follows:

$$dT^d = A_T (1 - m_T - b)(1 - A_f)\, dS \tag{3-11}$$

EMPIRICAL ESTIMATION OF IMPACT ON DOMESTIC EXPENDITURES

By combining Equations (3-4) to (3-11), the total "first-round" impact on domestic aggregate expenditures can be written as the following expression:

$$\begin{aligned}\text{Change in Domestic Expenditures} = \Big[& A_L(1 - m_L - b) + A_M(1 - m_M) + A_0(1 - m_0) \\ & + A_E e_p \left(\frac{p^d - p^w}{p^w}\right) + A_E e_e(1 - m_E) + A_D D_L \\ & + (A_D D_E)\left(D_{EO} + D_{EP}\left(\frac{p^d - p^w}{p^w}\right)\right) + A_I(1 - m_I) \\ & + A_K(1 - m_K) + A_T(1 - m_T - b)\Big](1 - A_f)\, dS.\end{aligned} \tag{3-12}$$

For the estimation of the parameters contained in Equation (3-12) and for the analysis of the cost structure of Bolivian mining, a special survey was made of both the medium mining firms and COMIBOL for the years 1973 and 1974. Differences in questionnaire format for the medium mines between 1973 and 1974 made the 1973 data less useful than that available for 1974. Therefore, the analysis contained in this section will draw more heavily on the 1974 information.[c]

[c]The survey was designed for purposes of this study by the Departamento de Estudios Económicos of the Ministry of Mining and the authors, and distributed under the auspices of the Instituto Nacional de Estadística. The survey was implemented by Ministry of Mining officials through a series of interviews in mining enterprises all over the country, and was analyzed and interpreted by Ministry of Mining staff and the authors.

The objective of this survey was to obtain detailed information on the use of inputs by type and their source, production by type of mineral, revenues from both domestic and foreign sales, taxes by type, and capital structure of the firm. Table 3-1 summarizes the different information items requested through this special mining sector survey.

Twenty medium mining firms completed questionnaires for 1973 and twenty-six completed them for the year 1974. These data were then extensively cross-checked with other government sources of information and for internal consistency. In the 1974 survey the data from seventeen of the firms passed the accuracy level that was

Table 3-1. Summary of Information Requested by Special Bolivian Mining Sector Survey

Item	*Description*
1. *Inputs:*	
a. *Labor*	Number of proprietors and active partners (for private firms only), workers and others.
	Number and total salaries and wages for: professionals administrative technicians, production technicians, employees, skilled workers, and occasional or commissioned workers.
	Value of family subsidies, social expenses, indemnifications and other fringe benefits.
b. *Materials* (used during period and held in inventories)	
	Quantity, value and source (imported or domestically produced) of ammonium nitrate, calcium carbonate, sacks for concentrates, drill bits, carbon lamps, clothes and rubber boots, chemicals, explosives, wood products, cement and other inputs.
c. *Energy*	Amount of electricity generated for own use, amount and value of electricity sold, and expenditures on electricity purchased from utilities, fuels and lubricants.
d. *Other Costs*	Interest payments on debt, mine rental payments; insurance, realization costs (smelting and transportation), hired local transportation, and miscellaneous expenses.
2. *Taxes*	Amount paid during year in royalties (*regalías*), export taxes, land rents and other taxes.
3. *Assets*	Value of existing stock, current investment and current depreciation expense for following: machinery and equipment, transportation vehicles and mining properties.
4. *Production and Sales*	Quantity of production during year of each type of mineral. Volume and value of sales of each type of mineral during year broken down by domestic and foreign sales components.

required in order for their information to be used in this analysis. For the 1973 survey sixteen of the firms presented data that were useful when aggregated, but they are not as reliable as the 1974 information.

One of the most important methods used in checking the accuracy of information provided in a survey of firms is to calculate certain key ratios that should have considerable consistency across the industry. These key ratios may then be compared firm by firm to check for inconsistent reporting of information. This was done extensively in the case of the special survey of Bolivian medium mines and for the COMIBOL data. Examples of the key ratios calculated by each firm are the relationship of total labor cost to total costs, total labor cost to total mine revenues, wage cost to total labor cost, total material cost to total cost, total material cost to total mine revenue, and foreign material purchases to total material purchases. When a ratio for a particular mining enterprise appeared to differ significantly from those for similar firms, then a second inquiry was made to the firm to determine the cause of the discrepancy. If no satisfactory explanation or correction could be made for the questionable information, then the firm was eliminated from the survey.

The values for the variables obtained through the special survey were also checked for their accuracy by making comparisons with other information for the same variables available from the Ministry of Mines. Export sales data by mineral provided to the special medium mining survey by enterprise were checked against data collected by customs on special "*poliza*" forms which are completed at the time minerals are exported. Domestic sales to the state smelter (ENAF) recorded in the mining survey were compared with data from ENAF concerning mineral purchases from the medium mines. This information was obtained from the Ministry of Mines.

Because taxation is the central issue of this book, particular care has been taken to verify the amount of taxes that were reported to have been paid by each firm. The payments in *regalías* reported in the mining survey were checked against the data obtained by customs on the *poliza* forms. Export tax payments were verified in the same manner. In the case of COMIBOL, a special effort was made to reconcile the data for tax payments made available by COMIBOL with the data available from other sources, particularly the Department of Finance. It was especially important to determine the amount of expenditures made on behalf of the government by COMIBOL in lieu of paying *regalías* that are actually true taxes and which should be included in the normal costs of operation for a mine.[d]

"Realization" costs[e] for internal sales were compared with the appropriate costs as recorded by ENAF. The foreign realization costs recorded in the special mining survey were compared with other independent sources of information on the costs of foreign transportation and smelting to determine accuracy of the information obtained through the survey.[f] The results of this survey for the medium mining firms and comparable data for COMIBOL are presented in Appendix C. In addition, a number of the important ratios that are presented will be used here and in other chapters of the book.

Table 3-2 contains a summary of the distribution of national mineral revenue for the medium mining sector and COMIBOL. These distributional shares are also presented in Table 3-2 for COMIBOL, along with a weighted average of these shares for these two major subsectors of Bolivian mining.

The distributional shares from Table 3-2 provide some of the key parameters needed for the estimation of the impact of the mining sector on domestic aggregate demand as given by Equation (3-12). For the remaining variables, supplementary information was obtained from separate studies on the Bolivian economy and, in some cases, parameter values were chosen by the authors on the basis of estimations that have been made for these variables in other similar developing countries.

Marginal Propensity of Labor to Import and Save

At the present time no reliable survey information on consumption patterns of Bolivian workers exists. However, by examining the purchases of commissary goods by COMIBOL, we obtain a fairly accurate picture of the source of a large proportion of the goods

---

[d]This subject (of determining actual mining taxes paid by COMIBOL) is discussed in Chapter Six.

[e]"Realization costs" include those incurred for transportation of concentrates from central collection points in Bolivia, smelting charges, insurance, etc. See Chapter Eight for a detailed discussion of this cost concept.

[f]An interesting discrepancy appeared in the information provided to the special mining survey regarding the value of the existing plant and equipment. The private firms apparently feared that this information might be used to impose a wealth tax on the value of the capital invested in the mines. As a result, most of the information gathered on the current values of the capital stocks for the mines are systematically biased downward. This information was not necessary for the purpose of this study and thus this systematic error did not impede our analysis of mining taxation in Bolivia.

Table 3-2. Distribution of Bolivian Mineral Revenue (1974)[a]

Component	Medium Mines	Weight	COMIBOL	Weight	Total Sector
Total domestic smelting and hired transportation cost (A_D)	.060	.3	.060	.7	.060
Total labor costs (A_L)	.188	.3	.300	.7	.268
Materials used (A_M)	.084	.3	.122	.7	.111
Other expenses (A_O)	.150	.3	.026	.7	.063
Energy (A_E)	.020	.3	.020	.7	.020
Capital recovery (depreciation) (A_K)	.038	.3	.020	.7	.025
Total taxes (A_T)	.310	.3	.320	.7	.317
Profits, interest and rentals (A_I)	.150	.3	.132	.7	.137
Total distributions	1.000		1.000		1.000

[a]Bolivian mineral revenue is defined as the gross value of sales less foreign realization costs, foreign transportation costs and insurance costs. These deductions, expressed as a proportion (A_f) of gross sales, amounted to approximately 0.20 in 1974.

consumed by Bolivian mining labor. COMIBOL's accounts show that at least 25 percent of the commissary goods it sells to workers are imported.[3] Using this information on the import content of consumption by mining labor and assuming a marginal propensity to save (b) of 0.10, the marginal propensity to import out of labor income (m_L) will be approximately 0.225.

Propensity to Import Materials and Services

Material purchases in 1974 were equal to 11.1 percent (A_M), and "other expenses" costs were 6.3 percent (A_O), respectively, of national mineral revenues. Information from the survey also indicated that at least 85 percent of the material purchases are imported. Therefore, the value of m_M is set equal to 0.85. Because "other" goods and services includes foreign travel and various other foreign expenses, it is assumed that the imported content of these expenditures (m_O) is approximately 40 percent.[g] Thus, "other" services have slightly less than one-half the import content of materials.

Import Content and Subsidization of Energy

The total outside purchases of energy by mining firms in Bolivia has been equal to approximately two percent of total Bolivian

[g]Assumption based on interviews with enterprise management.

mineral income. Of this two percent, approximately one third represents purchases of electricity (e_e) and two thirds represent purchases of petroleum products, inclusive of lubricants (e_p).

When examining the effect of an increase in purchases of petroleum products on aggregate demand, we must consider the pricing policies in Bolivia for these goods. Bolivia produces its own petroleum products and exports its surplus. In 1974 the domestic price of crude oil (p^d) was approximately U.S. \$3.00 per barrel while the export price (p^w, f.o.b. Bolivia) was approximately U.S. \$12.60. Because of the low sulphur content of Bolivian crude oil, any reasonable increase in Bolivian production of crude oil could be exported at the price of U.S. \$12.60 net of export costs.[h] Therefore, if the mining sector expands and uses additional petroleum products, the Bolivian economy loses the difference between the price the mining company is charged for the petroleum and what the country could get by exporting this product.

If the mining sector increases its demand for electricity, then additional resources will have to be invested in this sector. Part of the additional resources invested in the electricity sector will be from the domestic economy, but the majority of the generation equipment, transmission lines, and distribution equipment will be imported.

In Bolivia hydroelectric power is the primary source of electricity generation. This means that the fuel costs are a relatively small part of the total capital cost. The thermal component is largely oil-fired, which means that if additional oil is consumed domestically, through an increase in thermal electricity generation, there will be a decrease in aggregate expenditures because of the decrease in petroleum exports. For the purposes of this analysis we will assume that any such increases in oil-fired thermal generation are small. This, of course, will impart an upward bias to our estimate of the total increase in aggregate domestic expenditures arising from increases in mining sector output.

Elsewhere, approximately 90 percent of total annual costs in a hydropowered electrical system are capital-related and 10 percent are labor costs.[i] Of the 90 percent which are capital costs, approximately

---

[h]In November 1975 the domestic price of Bolivian crude oil was raised to \$10.00/bbl., while the average f.o.b. price (inclusive of \$0.60 in costs of preparing crude for export) had moved up to \$13.45/bbl. Given that the costs of transporting, refining and distributing a barrel of processed crude was U.S. \$4.60, the opportunity cost of a barrel of crude used domestically was U.S. \$16.20 in 1974 and U.S. \$17.45 in 1976. See Charles E. McLure, Jr., "Taxation of Domestic Consumption of Petroleum" (paper prepared for Musgrave Commission on Fiscal Reform for Bolivia, August 1976), p. 5. This paper also discusses the welfare costs involved in the subsidy.

[i]These percentages are based on the Canadian experience. In Bolivia the capital cost will tend to be an even greater proportion of total costs because of the lower wages paid to labor.

80 percent of the capital items such as generators, turbines, transmission and distribution lines are imported, and 20 percent of the capital costs are either labor costs or goods which are produced domestically. Thus of the total annual costs, approximately $0.90 \times 0.80 = 0.72$ are for imported goods (m_e), and 0.28 are domestically purchased inputs ($1 - m_e$).

Import Content of Domestic Smelting and Transportation Costs

In the production of these services approximately 25 percent of the value comes from the value-added of local labor and the use of domestic goods (D_L), 50 percent comes from the services of imported capital and 25 percent of the value is derived from the use of fuel (D_E). At the present time both charcoal and fuel oil are the energy sources used in the tin and antimony smelters. For purposes of this exercise, it is assumed that 50 percent of the energy valued at current Bolivian prices comes from each of the two fuels (D_{EO}, D_{EP}).

Profits, Interest, Rentals

Approximately 13.7 percent of the value of mineral revenue in Bolivia during 1974 was appropriated in the form of profits, interest and rental payments. As this income stream will accrue to either relatively high-income owners of mines or domestic and foreign lenders, it is reasonable to assume that a large proportion (m_I) of incremental income will be spent on imports. For this estimation it is assumed that 70 percent of the profits, interest and rental payments are either saved or spent on foreign goods.

Propensity to Import with Capital Recovery Funds

In the case of COMIBOL, these funds have generally been reinvested in mining activities. In the case of private owners of the medium mines, it is assumed that they will spend approximately 70 percent of their incremental income on imports of either consumption or investment goods. Therefore, in this evaluation m_K is set equal to 0.7.

Propensity for Government to Import from Tax Revenues

Table 3-3 contains a comparison of the foreign exchange purchases of the central government with the total central government expenditures, inclusive of foreign aid receipts, for each year from 1970 to 1974.

Table 3-3. Expenditures and Foreign Exchange Purchases of the Central Government (in millions of B. $ at current prices)

	1970	*1971*	*1972*	*1973*	*1974*	*Average*
Central government expenditures[a]	1,261.3	1,405.5	1,856.4	2,954.9	5,338.3	2,563.3
Foreign exchange purchases of the central government[b]	102.0	81.7	179.8	298.7	542.6	240.9
Proportion of imports and foreign payments	0.08	0.06	0.10	0.10	0.10	0.09

[a]Source: Ministry of Finance. Data include foreign aid-financed expenditures.
[b]Banco Central de Bolivia, *Boletín Estadístico*, June 1975, Table B, p. 42.

Approximately 10 percent of the total government revenues inclusive of foreign aid receipts over the past five years have been spent on foreign purchases or payments of interest and principal on foreign loans. Because this proportion has been almost constant as the quantity of government expenditures has changed, the marginal propensity for government to import from tax revenues (m_T) has been set at 0.10 for this analysis.

Table 3-4 summarizes the values of the parameters to be used in the estimation of the "first-round" impact of a change in mineral production and sales on the domestic demand for goods and services. Solving Equation (3-12) with the parameter values shown in Table 3-4, we find that approximately 45 percent of gross sales revenues from increased mineral production are spent on domestic goods and services. Therefore, the initial ("first-round") expenditure leakages in terms of foreign purchases and savings are approximately 55 percent of the change in the mineral sales that arises from a change in mineral production.

IMPACT OF A CHANGE IN MINERAL PRICES ON DOMESTIC AGGREGATE DEMAND

In this section an attempt is made to evaluate the impact on Bolivian aggregate demand of a change in mining sector income arising from a change in the world prices of minerals sold. When the selling prices of these minerals increase, there will be an increase in the income flowing to the owners of the mines and to the government in the form of additional tax revenues. In the past, labor has also shared in any real increase in income arising from higher prices. Likewise, during periods of falling prices the income to labor has not increased

Table 3-4. Parameter Values for Estimation of Impact of Change in Mineral Production and Sales on Domestic Aggregate Expenditures

Parameter	*Value*	*Parameter*	*Value*
A_L	0.269	m_L	0.225
A_M	0.111	m_M	0.850
A_o	0.063	m_o	0.400
A_E	0.020	m_e	0.720
A_D	0.063	D_{bL}	0.250
A_I	0.137	m_I	0.700
A_K	0.025	m_K	0.700
A_T	0.317	m_T	0.100
A_f	0.200	p^d	\$3.000
b	0.100	p^w	\$12.600
e_p	0.667	D_E	0.250
e_e	0.333	D_{Eo}	0.500
		D_{Ep}	0.500

and has perhaps fallen in real terms, through the effect of inflation. Other costs of mining operations should not be significantly altered by changes in mineral prices.

For most minerals (lead and zinc are the exceptions) the combined rate of royalty (*regalía*) and export tax is (except at very low minerals prices) approximately 50 percent of presumed income per unit of minerals exported.[4] As presumed income is the putative base of the most important mineral tax,[5] approximately 50 percent of any increase in income that results from a price increase will accrue as government revenue.

Considering the past history of wage increases, we assume that labor in the Bolivian mining industry will obtain the same share of any increase in income stemming from price increases as they do presently for increases in income that arise from increased output. From Table 3-2 we know that labor received 26.7 percent of total Bolivian mineral income in 1974. The residual 23.3 percent of the additional income generated by a price increase will accrue to the owners of the mine in the form of increased profits.

Therefore, when evaluating the impact on domestic expenditures from an increase in mineral prices we will set $A_T = .500$, $A_L = .267$, $A_I = .233$, $A_f = .200$ and all the other shares will be set equal to zero.

Substituting these shares and values for the other parameters from Table 3-4 into Equation (3-12) we find that the change in domestic expenditures is equal to 52 percent of the change in mining incomes caused by a change in mineral prices. These first-round domestic expenditures of 52 percent for increases in mineral prices, as compared to the previous case of 45 percent for increases in mineral output, lead us to conclude that when the income of the mining sector increases because mineral prices have increased there will be a larger impact on domestic economic activity than if income increases because production has increased. The principal reason for this result is that when prices rise, government revenues increase by a greater amount than the proportional increase in price, because of the slight progressivity with respect to price of the major mineral tax when prices are in their 1975-76 range.[j] As a larger proportion of government revenues tend to be spent domestically than is the case for other parts of mining income (e.g., materials purchased), a larger impact will be made on Bolivia's internal aggregate demand when prices rise than when mineral output increases.

The high level of foreign expenditure leakages for either an increase in mineral prices or production indicates that the multiplier effect of the mining sector on the Bolivian economy must be quite low. This analysis has been concerned only with the initial effects of an increase in mineral revenue. Evaluation of the full general equilibrium impact on aggregate demand and output from the expansion of the mining sector would require the estimation of an econometric model with appropriate supply constraints for each of the sectors of the economy. Data necessary for the development of such a model are not available in either the quantity or quality to warrant effort in this direction at the present time.

There is some further evidence which indicates that the rate of foreign leakages may be even greater than estimated here. When year-to-year changes in foreign exchange earnings are compared to changes in foreign exchange reserves from 1969 to 1974, we find that the level of reserves do not seem to have been very sensitive (except in 1974) to changes in foreign exchange earnings.[6] If the foreign exchange leakages are less than 100 percent of the mining sector income, then if foreign exchange earnings were to increase, foreign exchange reserves should also initially rise. A reason for the failure of foreign exchange reserves to respond to foreign exchange

[j]Although the combined "bite" of the *regalía* and the export tax is indeed progressive with respect to price, the pattern of their joint impact is highly regressive with respect to income at low minerals prices and moderately regressive at high minerals prices (see Figures 6-1 through 6-3).

earnings may be found in the fact that the level of amortization of the government's foreign loans appears to have been closely correlated with the changes in foreign exchange earnings.[7] This lends some support to the hypothesis that the additional foreign exchange earnings are not only used up by the increased demand for imports brought about by an increase in aggregate demand, but that in years when there is a significant increase in foreign exchange earnings, there is an exogenous increase in the demand for foreign exchange brought about by the government's desire to repay a larger quantity of its foreign loans.

The conclusion from the above analysis is not that the Bolivian mining sector has only a modest impact on the Bolivian economy. Rather, it indicates that the importance of the Bolivian mining sector to the economy must arise in ways other than through its role as a purchaser of domestic inputs.

THE MINING SECTOR AND THE SUPPLY OF FOREIGN EXCHANGE

In several developing countries the mining sector is the primary source of foreign exchange.[8] Foreign exchange from the mining sector has allowed these countries to enjoy a standard of living significantly higher than what otherwise would have been the situation. From 1968 to 1974 the Bolivian mining sector has contributed 55 percent of the total foreign exchange available to the economy. The petroleum sector has supplied an average of nine percent of the total over this period, but accounted for 22 percent of the total foreign exchange available in 1974. Imports made up 86 percent of the uses of foreign exchange with consumption imports consisting of 20 percent, primary and intermediate goods 27 percent and capital goods 39 percent of the total uses of foreign exchange. The sources and uses of foreign exchange in Bolivia are shown in Table 3-5.

While the mining sector has been a major foreign exchange earner, it is also a user of foreign exchange, as it requires substantial volumes of imported materials and capital goods. However, after allowing for the use of imported inputs in mining production, we find that from 1968 to 1974 the mining sector alone has still generated 52 percent of the total net supply of foreign exchange available to the Bolivian economy.

One way to measure the benefits of the mining sector is to see what adjustments would be required if the mining sector were smaller. With a smaller mining sector than exists at present, the

Table 3-5. Distribution of Sources and Uses of Foreign Exchange in Bolivia, 1968-1974

Sector	*1968*	*1969*	*1970*	*1971*	*1972*	*1973*	*1974*	*Average*
	Sources of Foreign Exchange[a] (percent)							
Mining	58	56.0	63	57	52	51	50	55
Petroleum	2	.3	6	10	9	9	22	9
Other Bolivian sources	19	27.7	28	25	17	27	24	24
Foreign sources	21	16.0	3	8	22	13	4	12
Total percentage	100	100.0	100	100	100	100	100	100
	Uses of Foreign Exchange[b] (percent)							
Imports			82	84	81	94	89	86
Consumption goods			(17)	(18)	(16)	(24)	(23)	(20)
Primary and intermediate goods			(25)	(25)	(28)	(26)	(31)	(27)
Capital goods			(40)	(41)	(37)	(44)	(35)	(39)
Other uses			18	16	19	6	11	14
Total percentage			100	100	100	100	100	100

[a]Banco Central de Bolivia, *Boletín Estadístico*, June 1975, p. 42, Table A.

[b]Banco Central de Bolivia, *Boletín Estadístico*, June 1975, p. 42, Table B; p. 52, Table 10; and documents of the Dirección de Planificación Global y Política Económica, Ministerio de Planeamiento y Coordinación.

equilibrium exchange rate (number of pesos per U.S. dollar) would have to be greater, therefore causing imports to be more expensive in terms of the value of the resources Bolivia would have to give up to purchase these goods. Thus exports from the mining sector have caused substantial gains from trade to accrue to the Bolivian economy as a whole: they have generated producer and consumer surplus by lowering the resource cost of imports and generating economic rents in the mining sector.[9] In the following pages an attempt has been made to calculate an "order of magnitude" estimate of this gain in economic welfare. While the data used in this estimation are specific to Bolivia, the model and qualitative conclusions are directly applicable to any country which relies heavily on the mining sector as a supplier of foreign exchange.

Bolivia, like most other developing countries, does not have any influence on the world prices of the imports it purchases and has a negligible independent influence on the world price of the products it exports. These characteristics allow us to express the supply of mining output for Bolivia (measured as its value in U.S. dollars) net of direct mining sector imports as a positive function of the Bolivian exchange rate. This supply of foreign exchange from mining is shown in Figure 3-1. The area under this curve for any quantity of output may be viewed as a measure of the total variable cost (in pesos) of producing that quantity of mining output. Similarly, in Figure 3-2 the supply of *other exports* generating foreign exchange is shown as a positive function of the exchange rate, with their quantity measured in terms of their U.S. dollar value. In Figure 3-3, the total supply of foreign exchange net of direct intermediate-good imports is derived as the horizontal sum of the two supply functions shown in Figures 3-1 and 3-2. Also in Figure 3-3, the demand for imports and other uses of foreign exchange (also expressed in terms of U.S. dollars) can be shown as a negative function of the exchange rate. On the average over a period of time, the demand for foreign exchange must equal the supply of foreign exchange, thus allowing us to derive an equilibrium exchange rate, written as E.

If the mining sector did not exist, then the total supply of foreign exchange would be illustrated by the supply curve $E_0^x S_{NM}$. This is the supply curve of nonmining exports. In order for the demand and supply of foreign exchange to be equated, the exchange rate would have to rise. This will decrease the demand for imports by increasing

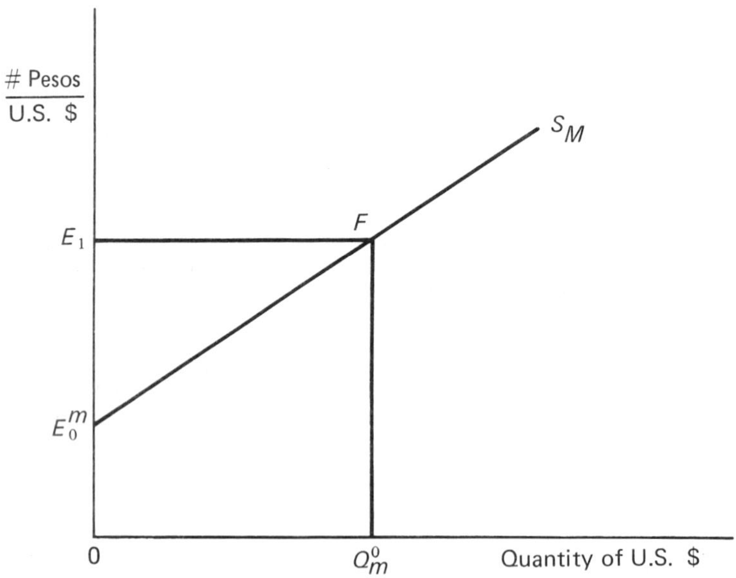

Figure 3-1. Supply of Foreign Exchange from Mining

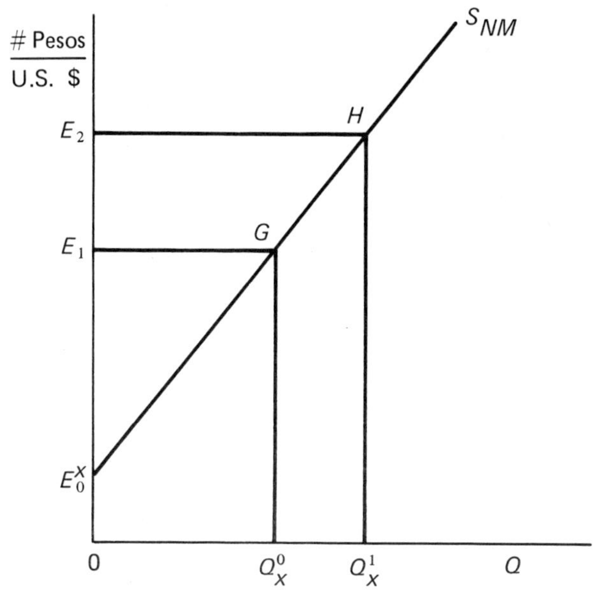

Figure 3-2. Supply of Foreign Exchange from Non-Mining Exports

their domestic price. Also, the supply of nonmining exports would increase because of the higher price (measured in pesos) they would receive for their nonmineral sales on the world market.

If the initial mining sales were equal to Q_M^0, then their elimination would release resources per year measured by the area $0\ E_0^m\ F\ Q_m^0$ in Figure 3-1. However, the higher exchange rate would induce additional resources to be spent to increase production of other exports equal to the area $Q_x^0\ G\ H\ Q_x^1$ in figure 3-2. This area represents the incremental domestic resource cost of producing an additional $(Q_x^1 - Q_x^0)$ U.S. dollar's worth of nonmining exports. At the same time as the higher exchange rate (number of pesos per dollar) induces further exports of nonmining goods, it would also make imports more expensive and thus cause a smaller quantity of them to be consumed. The value lost by the economy through the reduction in the consumption of the imports is measured by the area under the demand curve $Q_x^1 B\ A\ Q_t^0$, in figure 3-3.

To summarize, if the mining sector were to shut down, the net loss in economic welfare as measured by this simple model would be equal to a benefit of released resources from the mining sector equal

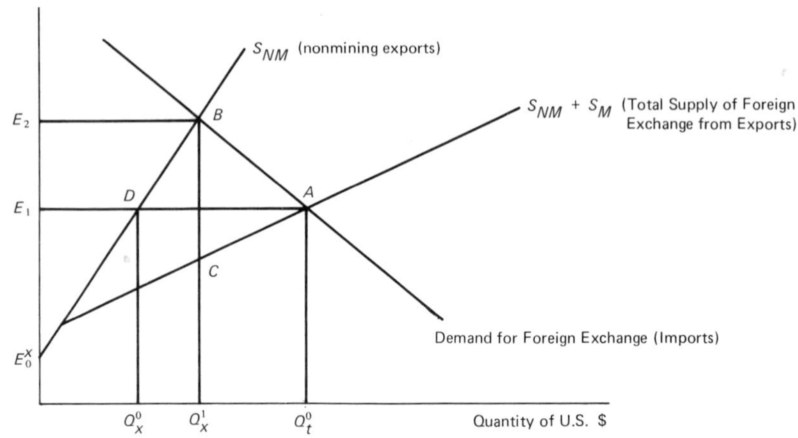

Figure 3-3. Supply and Demand for Foreign Exchange

to $0\ E_0^m\ F\ Q_m^0$ in Figure 3-1, a loss from increased resources used in production of nonmine exports equal to $Q_x^0\ G\ H\ Q_x^1$ in Figure 3-2, and a loss from the reduction in the quantity of imports demanded of $Q_x^1\ B\ A\ Q_t^0$ (figure 3-3).

To estimate values for the areas outlined in the theoretical model, the following information is used: The average total amount of imports during 1973 and 1974 was U.S. $363 million. Approximately U.S. $50 million of this amount was used to purchase inputs for production in the mining sector (20 percent of mineral sales). Also, approximately U.S. $20 million of this foreign exchange was used to purchase direct inputs for production in the nonmining export sector (15 percent of sales). Deducting these amounts from the gross value of imports, the value of Q_t^0 (which represents the total amount of imports initially purchased by Bolivia) in Figure 3-3 is equal to U.S. $293 million. Since the mining sector has produced 52 percent of total foreign exchange used by the nonexport sectors, the value of Q_0^M in Figure 3-1 is equal to U.S. $152 million and the value of Q_x^0 in Figure 3-2 is equal to U.S. $141 million. The current exchange rate (E_1) of 20 pesos per U.S. dollar is assumed to be the equilibrium exchange rate that prevails before the impact of the elimination of the mining sector is measured. It is assumed that at the current time no mineral exports would be made if the market exchange rate went as low as 10 pesos per U.S. dollar. Therefore, the value of E_0^M in Figure 3-1 is set equal to 10 pesos per U.S. dollar. Using linear supply and demand curves as an approximation, the only further information required before carrying out the estimations is the elasticity of supply of nonmining exports in Bolivia at the current level of production and the elasticity of demand of imports by Bolivia. Considering the types of goods exported and imported by Bolivia, illustrative values for the elasticity of supply of nonmining exports (S_0) and the elasticity of demand for Bolivian imports (N_d) would be 1.5 and −1, respectively.

Using the above information, the first step in this estimation is to calculate the new equilibrium exchange rate if the mining sector were eliminated. From the definition of the elasticity of supply and demand, we have the following:

$$S_0 = \frac{\partial Q_x^s}{\partial E} \frac{E}{Q_x^0} = 1.5$$

$$\therefore \quad \frac{\partial Q_x^s}{\partial E} = (1.5)\left(\frac{141}{20}\right) = 10.6$$

$$N_d = \frac{\partial Q_t^d}{\partial E} \frac{E}{Q_t^0} = -1$$

$$\therefore \quad \frac{\partial Q_t^d}{\partial E} = (-1)\left(\frac{293}{20}\right) = -14.6$$

We also know that:

$$(Q_t^0 - Q_x^0) = 293 - 141 = \text{U.S. \$152 million.}$$

Solving for the change in the exchange rate dE:

$$\left[\frac{\partial Q_x^0}{\partial E} - \frac{\partial Q_t^d}{\partial E}\right] dE = (Q_t^0 - Q_x^0)$$

Substituting values for the variables and solving for dE we have:

$$[10.6] + 14.6\, dE = 152$$

$$dE = \frac{152}{25.2} = 6.0$$

For these parmeters the Bolivian exchange rate would move from its current value of 20 pesos per dollar to 26 pesos per dollar if the mining sector were eliminated, i.e., E_2 would equal 26 (figure 3-3). For a change in the exchange rate of 6 pesos per dollar, the change in the supply of nonmineral exports is:

$$dQ_x^s = \frac{\partial Q_x^s}{\partial E}\, dE = (10)\,(6) = \text{U.S. } \$63.3 \text{ million.}$$

Solving for the change in economic welfare:

$$\text{Area } 0\, E_0^m\, F\, Q_m^0 - Q_x^0\, G\, H\, Q_x^1 - Q_x\, B\, A\, Q_t^0$$

$$= \frac{10 + 20}{2}\,(152) - \frac{(20 + 26)}{2}\,(63.6) - \frac{26 + 20}{2}\,(152 - 63.6)$$

$$= 2280 - 1462.8 - 2033$$

$$= 1216 \text{ million pesos.}$$

Thus, if the mining sector were not in the national economic picture, the change in Bolivian economic welfare is estimated as a loss of 1,216 million pesos per year, or approximately U.S. $60 million per year (evaluated at an exchange rate of B. $20 = U.S. $1).

These calculations are meant to be only illustrative of the magnitude of the additional contribution to Bolivian economic welfare that is realized annually by virtue of the existence of the Bolivian mining sector. This additional producer and consumer surplus of U.S. $60 million annually is approximately equal to one third of the net addition to the supply of foreign exchange generated by the mining sector.

It is very important to note that while this analysis points out the major impact that the Bolivian mining sector has on the economic welfare of the economy, it in no way justifies the subsidization of a marginal mine which has resource costs in operation greater than the value of sales gross of taxes that it is producing. For the mining sector taken as a whole, the producer and consumer surplus generated may be large, but the additional economic welfare generated by the marginal (high-cost) mines may be negative.

THE MINING SECTOR AND CENTRAL GOVERNMENT FISCAL AND MONETARY POLICY

From 1965 to 1974 identifiable tax revenues received by the central government from the mining sector have averaged 25.4 percent of total central government tax revenues. During this period the relative importance of mineral taxes in the overall revenue picture increased from an average of 23 percent of total central government tax revenues for the years 1968-1972 to an average of 35 percent during 1973 and 1974 (see Table 3-6). The opposite trend is observed in the revenues collected through import duties and foreign exchange taxes. During the period 1968 to 1972 the import duties averaged 38 percent of total government revenues, but fell to an average of 26 percent for the years 1973 and 1974.

Although the mining taxes made up 25.4 percent of total central government tax revenues over the period under review, this underestimates the total impact that the mining sector has had on government revenue. As was shown in Table 3-5, the mining sector's foreign exchange revenues have financed approximately 55 percent of total imports purchased by the Bolivian economy during the period 1968 to 1974. Therefore, a proportion of the import taxes should be attributed to the mining sector, because if the mining sector's earnings of foreign exchange had not existed, then these imports could not have been purchased. In Table 3-6, Row C, the import taxes generated because of the mining sector earnings of foreign exchange are calculated using the assumption that the other export sectors would not have expanded if the mining sector had declined. However, from the example presented in the previous section, we found that under the assumptions posited, when the mining sector declined by an amount of U.S. $152.0 million the other export sectors expanded by U.S. $63.6 million. Approximately 40 percent of the reduction in mining sector exports might be replaced by the expansion of other export sectors.

If it is assumed that no response occurs in the supply of other export sectors when the mining sector declines, then the imputed import taxes generated by mining sector exports are equal to approximately 22 percent of total central government revenues during the period 1968 to 1972, falling to a 13 percent average for 1973 and 1974.

Table 3-6. Total Mining Sector Taxes and Import Taxes Paid to Central Government as % of Total Central Government Tax Revenues (Adjusted) 1968-74 (millions of Bolivian pesos)

	1968	*1969*	*1970*	*1971*	*1972*	*1973*	*1974*	*Average 1968-72*	*Average 1973-74*
A. Total identifiable mining sector taxes paid to the central govt. (from Table 6-1 of this book)[a]	100.0	168.8	339.3	155.4	240.8	821.0	1,515.4		
B. Total import taxes including Foreign Exchange Tax[b]	367.5	385.7	457.1	416.5	454.7	617.8	1,125.6		
C. Imputed import taxes generated by mining exports (Row B X Row 1, Table 3-5)	213.2	215.9	287.9	237.4	236.4	315.1	562.8		
D. Total central govt. tax revenue[a]	877.0	946.5	1,241.8	1,107.6	1,289.2	2,038.0	4,655.2		
E. Mining sector taxes paid to central govt. as % of total tax revenue (A ÷ D)	11.4%	17.3%	27.3%	14.0%	18.7%	40.3%	32.6%	23%	35%
F. Import taxes paid to central govt. as % of total central govt. tax revenue (B ÷ D)	41.9%	40.8%	36.8%	37.6%	35.3%	30.3%	24.4%	38%	26%
G. Import taxes imputed to have been financed by mineral export sales as % of total (C ÷ D)	24%	23%	23%	21%	18%	15%	12%	22%	13%

[a]Derivation of Row A (total central government tax revenues): Total central government tax revenues including *regalias*, as reported in the Plan Operativo, 1975, Table 13, were as follows (in $B): 1968 – 873.9 1969 – 941.6, 1970 – 1,138.1, 1971 – 1,042.7, 1972 – 1,202.9, 1973 – 1,822.1, 1974 – 4,625.7. However, these figures do not include certain payments in lieu of central government taxes by COMIBOL. The COMIBAL payments in lieu of taxes have been accounted for in the notes to Table 6-1. In addition, ENAFBOL received the revenues from earmarked royalties on the first 100 tons of tin produced per month. This should be viewed as an equity input by the central government into ENAFBOL, that would have had to be financed by some other means. Thus, the amounts of the earmarked tax should be counted as central government tax revenues.

[b]Import tax figures obtained from Department of Finance.

In Table 3-7 the imputed import taxes generated by the Bolivian mining sector's foreign exchange earnings are recalculated using the assumption that 40 percent of any reduction in the mining sector's exports will be offset by the expansion of other export sectors. The total taxes attributable (directly and indirectly) to the Bolivian mining sector are shown in Row C of Table 3-7. They averaged 31 percent of total central government tax revenues over the period from 1968 to 1972, and 45 percent during 1973 and 1974.

The fact that the mining sector bears such a large proportion of the total burden of central government taxes and yet only accounts for 11 percent of Gross National Product provides one indication that a general tax reform may be required in Bolivia to give the government a wider tax base. Because the central government has to rely so heavily on the mining sector for revenues, governmental activities are directly influenced by the fluctuations in the international market for its principal mineral exports.[10] We next analyze how this reliance by the central government on tax revenues generated by the mining sector has also interfered with the activities of the Central Bank in its management of the money supply and bank credit.

Table 3-7. Central Government Tax Revenues Generated Directly and Indirectly by the Mining Sector (millions of Bolivian pesos)

	1968	*1969*	*1970*	*1971*	*1972*	*1973*	*1974*
A. Total Mining Taxes Paid Directly by Sector[a]	100	168.8	339.3	155.4	240.8	821.0	1,515.4
B. Import Taxes Imputed to Mining Sector	127.9	129.5	172.7	142.4	141.8	189.1	337.6
C. Total Actual and Imputed Taxes Generated by Mining Sector (A+B)	227.9	298.3	512.0	297.8	382.6	1010.1	1,853.0
D. Total Actual and Imputed Taxes Generated by Mining Sector as a % of Total Govt. Tax Revenue[b]	25.9%	31.5%	41.2%	26.9%	29.6%	49.6%	39.8%

[a]See footnotes in Table 3-6.

[b]Row C, Table 3-7 ÷ Row D, Table 3-6.

The Bolivian Central Government, like those in most developing countries, operates within a number of serious fiscal and monetary constraints. It relies very heavily on the mineral sector for tax revenue and foreign exchange, and with a very underdeveloped domestic capital market it is unable to sell bonds to the general public, so budget deficits are covered by borrowing directly from the Central Bank. As in most countries, the level of government expenditures is not flexible downward in real terms. However, this problem is greatly magnified in the case of a country such as Bolivia, because its tax revenues are very heavily dependent on the current international market for its mineral exports. Tax revenues from this source can decline quickly in both real and nominal value.

When the central government has a budgetary deficit and borrows from the Central Bank, this institution has two options. It can expand the amount of bank credit available to the public sector and reduce the amount of bank credit available to the private sector, or it can increase the amount of bank credit available to the public sector without decreasing the amount available to the private sector. If the first option is chosen, then when the government deficit expands, its financing will have a depressing effect on the private sector of the economy due to the decrease in the amount of bank credit available to them. If instead the Central Bank expands the total amount of credit available to the economy, the nominal money supply must also increase. If the government deficit is large, then the increase in the money supply will be large, which can potentially lead to an inflationary situation.

Because the mining sector has been responsible for generating such a large proportion of the government tax revenues, the fluctuations in the revenues of this sector have generally been reflected in corresponding fluctuations in the amount of the government deficit. In years when mineral revenues are not growing as rapidly as expected, the tax revenues from this sector will not keep up with central government expenditures and a larger than expected fiscal deficit will arise. If this deficit is financed by the Central Bank and it chooses to expand the total amount of credit available, then inflationary pressures are created in the economy. This observation, that a decline in a major exporting sector may generate inflationary pressures, is the opposite to what is predicted by most macroeconomic models.[11] Traditional Keynesian models emphasize the role of exports in determining aggregate expenditures but do not include the government fiscal and borrowing constraints that exist in the case of a developing country with an export sector serving as the major source of tax revenues.

Alternatively, the Central Bank could finance a central government deficit by restraining the expansion of credit available to the private sector. If the Central Bank chose this option, a decline in the mining sector could cause a significant restriction in the total amount of bank credit available to the rest of the economy and generate a depression in private sector activity in Bolivia. As bank credit made available to the private sector is largely used to finance working capital, a reduction in this type of credit will force private firms to restrict their activities or risk bankruptcy.

During a period when mining sector revenues are rising rapidly, tax revenues will increase and the government deficit will be relatively low. With low demands from the central government for credit, the Central Bank will not be forced to expand the money supply. When the government deficit is small, the Bolivian money supply has usually risen at a very slow pace with the objective of offsetting the inflationary effect of previous government deficits financed by expansionary monetary policies. Bolivia has generally not allowed increased foreign exchange earnings to accumulate as reserves in the Central Bank and create a monetary expansion. An important exception to this statement was 1974, when significant reserves were allowed to accumulate, but their effect on the money supply was partially sterilized through increases in the basic and marginal reserve requirements. In most years when the mining sector has had a strong growth in revenues, the net result has been that the Central Bank has had to finance a smaller proportion of the central government's expenditures. In this situation it is less likely to create unduly large increases in the money supply which would lead to inflation.

This overview of the effect of the mining sector on the central government's fiscal and monetary policy suggests that the taxation and expenditure constraints of the government are the most important factors in this relationship. Because the government has had to rely, directly and indirectly, so heavily on the mining sector for revenue, any fluctuation in the sector will be reflected in the central government's deficit. The final constraint that central government deficits must be financed by the Central Bank is the link that causes mining sector cutbacks to be potentially inflationary. In such a situation mining sector expansion will allow the Central Bank more freedom to manage the money supply in the way it feels is most appropriate.

In contrast to traditional macroeconomic models, the principal impact on the economy of the leading export sector in Bolivia is not through its demand for domestic goods and services, as the level of foreign exchange leakages from the mining sector is very high. At the same time, the mining sector is a critical variable in the determination of the level of economic activity in Bolivia through its function as the major supplier of foreign exchange and tax revenues. Because of the taxation and capital market constraints, the level of activity of the mining sector has been the primary determinant of both monetary policy and expenditure policy in Bolivia during the past decade.

NOTES TO CHAPTER THREE

1. See Chapter Two.
2. See Chapter Two.
3. COMIBOL, *Memoria Anual 1974* (La Paz, April 1975), p. 75. Table: "Compras Sociales."
4. See Chapter Six.
5. This is discussed in detail in Chapter Six.
6. Banco Central de Bolivia, *Boletín Estadístico*, June 1975, pp. 42-43.
7. *Ibid.*, p. 4.
8. See Chapter One, Table 1-1.
9. For a discussion of the meaning of the concept of "economic rents" in the mining sector, see Chapter Four.
10. This relationship between mining revenues and government expenditures has existed for at least the past 25 years. Walter Gomez-D'Angelo, *Mining in the Economic Development of Bolivia, 1900-1970* (PhD dissertation, 1973, Vanderbilt University).
11. This type of macroeconomic model is also suggested, but not specified, in Arnold B. Baker and Frank Falero Jr., "Money, Exports, Government Spending, and Income in Peru, 1951-1966," *Journal of Development Studies*, July 1971.

Chapter Four

The Design of Mineral Tax Policy[a]

Meyer W. Bucovetsky
and Malcolm Gillis

INTRODUCTION

This chapter is concerned with concepts peculiar to mineral taxation and with relating those concepts to the goals of mining tax design and to the different means of attaining those goals. In short, the focus is upon criteria for an "ideal" mineral tax policy.

However, we emphasize at the outset that the ideal is incapable of precise implementation even if one were designing a tax policy in a situation where none had existed before. It is that much more difficult when one is dealing with a tax system "in being," and one in which, as in Bolivia, the mineral sector must for the foreseeable future provide a substantial share of central and subnational government revenues. This chapter deals primarily with theoretical issues, and with some illustrations of how they may be implemented. Chapter Seven relates these concepts to the realities of present-day Bolivia.

It should also be observed that mineral tax policy is a subset of an overall tax framework and is influenced by the quality of the entire tax system. That quality may be judged against a number of criteria, some of which may be in conflict with each other. Hence the appraisal of any tax system, or a subset of that system, depends on

[a]This chapter has benefited significantly from comments made by several people on earlier drafts. The authors are especially grateful to Ralph Beals, Richard Bird, Michael Roemer, Carl S. Shoup and Joseph Stern. But all these commentators are absolved from responsibility for any errors that may remain.

the weights that the appraiser attaches to these sometimes conflicting objectives. Among the many criteria that have been propounded, the following set of five requirements has enjoyed wide acceptance as a basis upon which any mixed economy tax system may be judged.[1]

1. The distribution of the tax burden should be equitable.
2. Taxes should be chosen so as to minimize interference with economic decisions in otherwise efficient markets.
3. The tax structure should facilitate use of fiscal policy for stabilization and growth objectives.
4. The tax system should permit efficient and nonarbitrary administration and it should be understandable to the taxpayer.
5. Administration and compliance costs should be as low as is compatible with other objectives.

Such rules of thumb as to what constitutes a "good" tax structure have been challenged in recent years, first by writers working with the theory of the "second best,"[2] and then by criticisms in the growing literature of optimal taxation.[3] In particular, many of the recent writings on optimal taxation of goods and income have called into question some of the basic premises of traditional tax design, but have also reinforced a number of others. But in spite of these developments, general acceptance of the five requirements listed above does not appear to have appreciably diminished in recent years, particularly since the literature on optimal taxation has not yet produced alternative rules and criteria that are readily understood by framers of tax policy.

It is also important to note that whereas an overall tax structure may come fairly close to satisfying all of the five objectives listed above, it is not necessary—if indeed it were possible—that each and every separate component of a tax system do so. This is particularly true for taxes applicable only to a specific economic sector, such as minerals taxes. For example, while such taxes, suitably designed, can contribute materially to the goals of stabilization and growth, and may be structured in such a way as to permit efficient administration at low costs of taxpayer compliance, it is difficult to come to any definitive judgments as to the implications of mineral taxes for equity unless one is also prepared to examine the purposes for which minerals tax revenues are designated on the expenditure side of the budget.

SOME CONCEPTS AND DEFINITIONS

Minerals as a Distinct Tax Base: The Concept of Resource Rent

Shoup likens a mineral tax to "an installment sale of the mineral by the government."[4] Due likens a country's mineral endowment to a "great pyramid of gold bars . . . provided by nature."[5] Inherent in both these characterizations is the view of minerals as a gift of nature whose exploitation yields a return beyond the necessary factor payments to the labor and capital required for its discovery and extraction. Because this additional return is due to the limited quantity of mineral resources, a country's mineral endowment is said to yield a *scarcity rent.* The rent element stemming from mineral exploitation is a prime target of mineral taxation, particularly in societies in which property rights in mining are not vested in individuals, but in society as a whole (i.e., in virtually all less developed nations).

The analysis of scarcity rents derives from the classical economists' concept of the pure surplus arising from agricultural land. In the classical view, agricultural land is limited in quantity, has no alternative uses, and is virtually indestructible. But its quality in terms of fertility and accessibility may vary. Consequently, cultivation will extend only so far as those parcels of land that are just capable of generating returns that cover the competitive earnings of the required labor and capital (i.e., at the margin of cultivation the value of output is just sufficient to pay labor and capital enough to retain them in agriculture). On marginal land there is no surplus left over for the landowner. On superior land, a surplus over the cost of labor and capital is generated, and the landowner can be expected to preempt the excess as scarcity rent. The higher the quality of the land, the higher the *differential rent* that parcel commands.

To the classical economists, a tax on land rent was attractive because it was a tax without "excess burden" (i.e., without additional burden, over and above the amount of revenue collected from the tax). Other taxes which impinge on the returns earned by human and capital resources in particular activities create incentives to change behavior patterns from what they would be in the absence of the taxes. Hence, they involve some dead-weight loss. If efficient resource allocation is obtained prior to the imposition of the taxes, such tax-induced deviations from optimal (efficient) behavior reduce society's real income.[6] Because the landowners will not alter the quantity of land supplied, a tax on land rent induces no distortion of factor supplies (i.e., supplies of labor, capital and land) and therefore no welfare loss.[7]

Modern economic analysis has invalidated the stark classical distinction between payments to landowners and payments to owners of the other productive inputs. On the one hand it is recognized that land has many uses and that it is both augmentable and destructible. On the other hand, particular kinds of labor and capital may be more productive in one use than in others. What remains of the rent concept is that scarcity rents (and differential rents) exist whenever a payment to any factor owner in a particular occupation exceeds the minimum payment required to retain a factor in that use.

A *resource rent* may be viewed as a particular kind of scarcity rent that attaches to a depletable natural resource. On the analogy to agricultural land, differential resource rents arise because some deposits are richer, are closer to markets, or have lower extraction costs than do others. But even "at the margin of cultivation" a positive surplus, beyond costs of exploitation, is likely to accrue to an active mine. This is because a potential developer has the option of postponement. His decision as to the timing, rate, and intensity of his mining activity depends on his expectation of cost and revenue trends and on the rate at which he discounts the future. A gap between price and his cost may normally be expected to open up and widen over time as richer (or lower-cost) ores are depleted. The developer is best off to defer activity on a particular deposit when the expected future price/cost gap, discounted to present value, is wider than is the current gap. He should start exploiting the deposit only when the current gap has grown enough to offset any advantage from further delay. It follows that, in a world where the future is known with certainty, the gap between revenue and cost of the least lucrative deposit being mined today will be positive and growing at the market rate of interest. A progression of successively leaner (or higher-cost) deposits would come on stream over time. The certainty assumption gives the concept precision. But what distinguishes resource rents from other scarcity rents is that their amount is dependent on the choice of a time profile for exploration and exploitation.[8]

For present purposes we may conclude that, imprecise though the amount may be, minerals are capable of providing a yield beyond that necessary to attract human and capital resources to exploit them. That yield may be privately appropriated by a land-owner or mining company, or, under certain conditions, by the consumer. But if pure rent is requisitioned by the government, it has the attraction of being "costless" in the sense of imposing no penalty on potential output. For the reason, then, of economic efficiency as well as what most would probably regard as "natural justice," the right of a community to the fruits of its national patrimony, minerals are a legitimate source of government revenue, apart from "normal" taxes on the income of the capital and labor used to exploit them.

There is another kind of "rent" which is not confined to natural resources, but which is often associated with them. Given that mineral markets are subject to a high degree of seller concentration (relatively few firms marketing mineral commodities), sellers may earn a "monopoly" profit. Such a surplus return is above that generally available, and stems from the power of sellers to set noncompetitive prices. Skillful tax policy is capable of appropriating to the producing country some of the *monopoly returns* otherwise enjoyed by the mining companies.

Beyond that, recent experience has shown that a nation with a monopoly on a given mineral, or, more likely, a group of nations that collectively dominate supply and can act in concert, can itself levy a tribute on the buyers of the mineral through its tax policy.[9] Its success depends on how costly it is for users to substitute other materials for the commodity in question. To the extent that such a policy was unanticipated, it creates windfall profits for producers of the mineral in other countries not a party to the initial tax increase. Or, to the extent that such a policy causes a reflected increase in the price of substitute materials, unanticipated windfalls are again created for the producers of the substitutes. There is then another kind of potential rent: an opportunity to tax the windfall profit. The United States, Canada and Bolivia, although not parties to the OPEC cartel, have an opportunity to tax the windfall rent received by their domestic oil producers, if they allow domestic prices to rise to world levels. Alternatively (and this has been done to some extent in all three countries mentioned), they may cause part of the implicit windfall rent to be distributed to domestic consumers and processors, by holding down the price of petroleum to them. Similarly, as world prices of oil rise, unanticipated windfall gains may be realized by producers of natural gas, taxed away by host governments, or distributed to domestic consumers and processors.

It should be noted that taxes that impinge with greater than ordinary severity on the particular yield to mineral capital and labor (i.e., taxes whose burden is not confined to capturing rents or to taxing income at normal rates) may be sustained for long periods without apparent ill effect. Because of the longevity and geographical intractability of mining investment, the return to mineral capital (and labor if that too is immobile) has the quality of a "quasi rent" once it is committed. In such a situation, these factors may be heavily taxed in the short run without causing them to shift out of mineral extraction. However, in the longer run a tax which falls with greater than normal severity on resources employed in mining will inhibit further investment in mineral exploration and exploitation. A tax which alters rational long-run decision making may have a high immediate payoff but will reduce the present value of ultimate mineral returns to both the mining company and the country.

Less obviously, if the tax takes a form that increases costs of production at the output margin, for a given plant, output rates are likely to be reduced. There is good evidence that unit costs of production rise with increases in the rate of output from any given mine. Profit maximizing behavior implies that the output rate selected is the one where incremental costs attributable to the "last" unit equal incremental revenues attributable to it.[b] A tax or royalty based on output imposes an increase in cost on all units including the "last," impelling a decrease in the output rate selected. Here, assuming that the operator's rate of time preference is the same as the government's (a subject to be discussed in a subsequent section), the rate of exploitation will be slower than that desirable from the standpoint of maximizing the present value of the social return. In particular, a tax which increases unit costs on all units mined (including the least profitable unit) may force the operator to abandon or to defer mining some ore which it would otherwise have paid him to mine; i.e., an output royalty will force the operator to high-grade—to mine selectively. Depending on the nature of the ore body and the type and quality of exploitation methods, it may be impossible ever to retrieve the abandoned ore, so that some potential mineral wealth—and natural resource rent—may be permanently lost.

---

[b]The statement above, while true of profit maximizing behavior generally, is not literally true in mining. The fact that a mine is finite means that output rates selected will stop short of equalizing costs and revenues at the margin in any one period. The correct rule for profit maximization in mining is that a positive spread between costs and revenues at the margin should be equalized, in discounted present-value terms, for each period of the mine's anticipated operation. This well known proposition is developed in Herfindahl, "Some Fundamentals of Mineral Economics," *Land Economics*, vol. 31 (May 1955), pp. 131-138. Indeed, it is precisely this "non-equalization" of costs and revenues at the margin that gives rise to a present-valued surplus on marginal ore, a natural resource rent. This complication does not destroy the validity of the statement above, since an output royalty reduces the spread at each annual margin by an equal undiscounted amount. In present-value terms the spread at more distant margins is reduced by less than the spread in nearer periods. This implies a shift in the preferred output from present to future.

This is especially likely when underground deposits are worked by small-scale enterprises using primitive methods, and with little attention to mine safety.

Risk, Time Preference, and Taxes[10]

It will be apparent from the foregoing that the rents attainable by a government from its mineral resources—assuming they are privately developed—will depend crucially on the minimum rate of return that is required by the developer. In turn, that rate of return depends partly on the risk perspective of the developer. The risks associated with the enterprise are mainly those related to exploration, future price, future cost, political outlook and management.

Let us assume that we are dealing with a country that is virgin territory so far as mining is concerned: there has been no detailed geological mapping and no previous exploration. On the other hand, let the geology be such that there is a reasonable chance of mineral occurrence. The government decides to partition the country into blocks and to auction off permanent mining rights for each block, with no further taxes to be imposed. Let us also assume that there are a large number of noncolluding bidders for the mining rights. Each bidder will presumably make an appraisal of the odds of his finding and successfully operating a mine, based on his own experience and his knowledge of the terrain. He would then estimate what his costs (capital and operating) and their time profile might be, as well as the likely gross revenue stream. Obviously there is a wide range over which the outcome might fall. But there is some expected stream of revenues and costs that is the most likely, and to which the bidder would apply his required rate of return. If the expected net revenue stream discounted by the rate of return is positive, that expected net present value represents his best estimate of the lump-sum worth of future rent on the property. That expected amount is the maximum he would bid for the property right. The fact that the amount is subject to a probability distribution—a range of dispersion—is evidence of risk. The assumption that most would make is that the wider the dispersion of possible outcomes (the greater the risk), the more the bidder would hedge himself by raising the discount rate he applies, demanding a larger "risk premium." (Thus, we assume that most people are risk averse; that is, are repelled by risk.) However, note that the bidder might choose a development strategy that promises less than the maximum possible return, if the lower-yield alternative also promises a smaller risk exposure.

The most obvious risk (or uncertainty) in this case has to do with whether or not an ore body will be found—and of what size and quality. A second facet of exploration risk has to do with the uncertain time lags between initial exploration effort and ultimate production. Exploration risk is peculiar to mineral extraction (although analogous to investment in industrial research and development) and is what is usually meant by saying that mining is a "risky business." But it should be observed that risk of this sort is reduced when chances are pooled, either by increasing the number of bets accepted by one taker or reducing his interest in any one bet.

Risk of wrong estimates of future prices and costs is not peculiar to mining, but is particularly hazardous in that industry because of the long-term fixity of the capital committed. These kinds of risk are not likely to be much reduced by pooling because price or cost changes are likely to be universal when they occur. But it should be noted that price risk embraces the possibility that future prices will be higher than anticipated as well as lower. Risks regarding the future efficiency of management are again not peculiar to mining. They also have relevance to the following discussion on the alternative of government operation of mines. The greatest risk of all, to the private developer in the present-day atmosphere, is the political unknown: the possibility of future expropriation or adverse change in the terms of the original contract.

Let us now return to the auction of virgin territory. That form of competitive bidding puts the entire risk on the private developer. The alternative of granting title to potential mining land on a "first come, first served" basis, with a stipulated future royalty on mine output, puts more of a risk on the government; here the government shares in the exploration risk but not in the others. (It also eliminates the risk of collusion among the bidders.) Again, if rights are granted "first come, first served" with the government taking a stipulated share of future net income, part of the management and marketing risks are shifted from the company to the government.

In principle, and speaking in expectational terms, if the government has the same risk perspective as do potential developers, and if it shares in the risk, say through an income tax, then its average expected return should be the same as if it accepted none of the risk (as in the competitive auction). The choice of best taxing device then depends in part on the risk premiums required by developers and the government. If the government has a larger number of prospects, if it is pooling a larger number of "bets," it might require a lower risk premium than would a private developer. In that case it will achieve a larger expected return if the taxing device chosen throws a larger proportion of the risk on its own shoulders (i.e., the residual rent will be greater, since the developer's rate of discount will be lower than otherwise). Conversely, in developing countries, and particularly small ones, the government presumably has fewer schemes to pool than do mining companies, and therefore it might do better by opting for taxing devices that place less of the risk on itself: specific output royalties, for example, rather than taxes on net income.

No country, to our knowledge, has ever used the "totally blind advance bid" used here for illustration. The risk premium applied would be too great to yield a positive net present value. In oil and gas, where expectations can be confined within a narrower dispersion than for metals once the geology is known, variants of the open competitive bid are used (as in Alberta and Alaska), but involving a "package" consisting of lump-sum initial payment and a commitment to future royalties. In mining, concessions are common enough—more often on a one-to-one negotiation basis than by competitive bidding—usually carrying a definite commitment to spend some amount on exploration. It remains true that the more definite is a particular financial undertaking on the part of the developer the more risk he bears, and hence the higher must his expected return be. In any event, where the government can remove some of the exploration risk at lower cost than can private developers—as, for example, in detailed geological mapping of the country, where economies of scale may apply—it would be well advised to do so, in the interest of increasing its potential revenue by more than its actual outlay.

These conjectures are overshadowed by the risk of expropriation, or, more generally, the political risk that hangs over a foreign mining company. Herfindahl observes that anything that increases the likelihood of expropriation is likely to decrease the government's long-run take on its mineral lands, whether or not it does expropriate.[11] Further, he argues that the auction of concessions with a low level of information is likely to lead to a demand for expropriation. This is because (assuming that the return from all bids is the same as if the level of information were higher) the dispersion of outcomes may be greater; e.g., one "lucky" company may obtain a bonanza for a low price although all winning bidders together, lucky and unlucky, may have bid an adequate total sum. A low-cost bonanza makes the company a likely target for a public outcry for expropriation. Companies are aware of this and will tailor their bids accordingly. Under a bidding system, then, the yield to the government will be higher if it auctions the mining rights *after* an ore body has been found. As a recent illustration, the Government of Panama used this strategy for the Cerro Colorado copper deposit, one that appears to be the biggest undeveloped copper deposit in the world. The property was delimited by Canadian Javelin. But once its value was established the Panamanian Government put the winning rights up for auction. In May 1975, three major mining companies were reported as anxious to bid; in February 1976 an agreement to proceed with development was signed with Texasgulf Inc. under a 15-year management contract and an 80 percent government equity participation.[c]

This case also illustrates the desirability of open competitive bidding when concessions are granted. A company that obtains a concession, instead, through one-to-one negotiation is more susceptible to the later accusation that the concession was unjustly granted. A case in point is the move by the Government of Thailand, in 1975, to revoke a group of major offshore tin mining concessions held, but not yet exploited, by the foreign-owned Thailand Exploration and Mining Company.[12] This was the result of mounting public clamor that stemmed from the clandestine circumstances under which the concessions had originally been granted. As a result, three other major offshore tin projects—which Thailand badly wants—were (in 1976) being reappraised by the developers. In its own interests a government is well advised to extract its mining revenue in a manner that minimizes the mining company's vulnerability to political risk.

Since political risk, whether or not a government wills it that way, is bound to decrease the rent it can obtain from private developers, the question arises as to why it would not pay the government to develop its mining industry on its own.[d] That is, if the government chooses to develop its mining industry under public ownership it bears the exploration, marketing and management risks itself. But to the extent that it feels no political risk, other things being equal, it could obtain not only the residual rent it might have received from the private developer but also the differential return for political risk that the private investor would have required, after allowing for the

[c]It should be emphasized that the Panamanian case was not an expropriation in the sense of a seizure that broke a previous commitment. The exploration and technical studies carried out by Canadian Javelin were made under a concession that from the beginning did not include the right to bring any discovery to production, although apparently it did include a right to compensation. In late August 1975 Canadian Javelin was reported to have signed an agreement with the Government of Panama whereby it would be paid $23.6 million as compensation for its expenditure. The company claimed its total expenditure to have been $25 million.

[d]The argument here is couched in terms of a government launching its own mining venture *ab initio*. However, it also would apply to a government buying out an existing mine. If fear of expropriation hangs over the company, other things being equal, its offer price would be less than the present value of the ownership rent stream to the government.

cost of capital. Countering that argument for public ownership is the management efficiency consideration—reasonably well documented—that in many cases absence of the profit imperative has severely dampened the incentive for efficient operation, that potential rents have often been dissipated by government enterprises in the form of swollen payrolls, and bad project selection.[13] Other factors that argue against government operation of mining relate to inability to attract qualified managers due to defects in management incentive systems often found in government enterprises, inadequate access to development capital, and lack of access to markets. The latter is particularly important for metals like aluminum and iron, where mining is the initial stage of a vertically integrated chain of supply with formidable barriers to entry downstream.

For completeness, it should be added that a different risk perspective is not the only factor giving rise to different rates of discount applied by governments and mine operators. Analysts such as Shoup contend that government—as custodian for the future, and because it is "more certain about the future tax structure"—will have a lower rate of time preference; i.e., it will place relatively more weight on future revenues than would a private operator.[14] If that is true, a scheduled increase in tax rates will be of more importance to the government than to business. Both may gain by altering the time flow of a revenue stream.[e] This is also an important rationale for tax holidays, such as the five-year exemption from all non-income taxes (i.e., from output taxes and royalties and export duties) to new mines granted by the Philippines. But as the recent Irish experience as well as that of Bougainville Copper in Papua New Guinea illustrates,[15] a tax holiday is an invitation to a public clamor for renegotiation. In the future mining companies are likely to be highly skeptical of such tax gifts. A device with similar effect, but one with less obvious political exposure, is that of granting current cost treatment to all initial capital costs where a tax on mining profits is levied. That method is used in Zambia and in a modified way under the present Canadian income tax.[f] Finally, if Shoup's conjecture is correct, a government will do better if it negotiates a time stream of

[e]Mexico reduces income taxes payable by new Mexican-controlled mines for the first 12 years of their operation. Zambia reduces its mineral profits tax over a mine's initial three years if after-tax profit fails to reach a 12 percent return on equity.

[f]In Canada, exploration expenditure may be expensed against any income source. Development—including land costs—is depreciated at 30 percent of the declining balance. Other capital costs may be expensed against income from the mine itself, or depreciated at 30 percent (thereby creating a loss carry-forward). See Chapter Five.

payments rather than a single lump-sum concession sale. The conclusions would still hold, even if the "pure" rate of time preference of both parties were the same, given that risk of expropriation loomed larger in the eyes of the developer.

TYPES OF PAST PREFERENCES TO MINING

Implicit in the foregoing analysis is the assumption that mining returns are a legitimate separate source of government revenue, apart from the "ordinary" taxation of net income originating in the mining sector (after payment of special taxes or royalties). For example, if a well designed device is used to extract the economic rent of mining operations, the residual return to invested capital should then be taxed on the same basis as are the returns to other investment vehicles.[g]

There is, however, an opposite view—now on the wane—that mining should, on balance, be taxed less onerously than other forms of endeavor. The argument seems to be that the faster minerals are extracted the better, that because minerals are "there" it behooves society to dig them up. Hence minerals should be taxed lightly, if at all; to the extent that a tax inhibits or delays mineral extraction it is to be deplored. The implicit premise of this line of advocacy is a view of minerals as a "gift of providence"; a nation's mineral endowment is imparted with a quasireligious mystique.[16]

On a less rhetorical plane, there are two devices in particular that have been advocated and used to decrease the burden of income taxation on mining: the rapid writeoff of exploration (and development) expenditures, and the depletion allowance or other reductions in the income tax rate applied to mineral profits. In terms of efficient resource allocation, the first of these has rather more to commend it.

The rationale for rapid writeoff of exploration is three-fold. First, as noted earlier in this chapter, if a government takes a "longer view" than does the explorer (or places a lower premium on risk) such a device may actually increase the present worth of potential mineral levies. Second, it must be recognized that, since exploration is a sampling process, the real cost of successful mineral discoveries must take account of the wasted cost of unsuccessful exploratory ventures. What follows from this is that *all* costs of exploration, successful or not, should be deductible against the income tax base, if taxation is to be neutral across industries and is to avoid allocative

[g]The trick, of course, is to confine the additional tax to the potential rents, which is much easier said than done.

distortions. Provided that condition were met, there would be no reason from the viewpoint of tax neutrality for accelerating the rate of writeoff. But it is also true that the tax offset of abortive exploration is less than perfect, even when permitted. This is particularly true in small countries, where the operator of a "failed" exploration project may have no local income otherwise taxable. For that reason, the acceleration of exploration writeoffs to successful explorers may restore the mean expected return to all explorers.

The third reason for accelerated writeoff of exploration relates to the concept of externalities. Exploration is an exercise in information gathering. The argument goes that what one explorer learns (even in a negative sense) becomes a "free good" to the next explorer. In other words, the potential benefit from an exploration project exceeds the private return. In keeping with the precepts of optimal resource allocation, an activity, such as exploration, whose societal yield is greater than its perceived private yield, should be pushed beyond the limit determined by individual private profit expectations. Accelerating the income tax writeoff of exploration expenditure is one way of accomplishing that purpose. This reasoning would be unexceptionable provided that information (on failures as well as successes) were in fact public knowledge. It is less than totally convincing when explorers are able to keep their secrets. The argument really favors an imposed obligation to release exploration information more than it does the desirability of rapid writeoff.

The other main type of tax preference to mining, the depletion allowance, also arose originally from a desire to grant the extractive industries a tax allowance for their total real cost. An example will serve to illustrate. Suppose ten drillers have each drilled an exploratory oil well for which each has spent $100,000, for land rights and drilling costs. Nine of the wells are dry holes, but the tenth effort yields a well whose discounted present worth is $1 million net of all further expected costs. In the absence of taxation, the entire effort will have yielded a normal return. But in the presence of taxation, if the nine dry holes cannot be written off against income, the net return from the entire operation will have been reduced to below the normal rate of return (the latter now taken as the post-tax rate prevailing elsewhere). One way of compensating for this effect, of course, is to allow "cost depletion" on all the wells on the same basis as other assets are depreciated. The method chosen in the United States in the early part of this century was to allow the successful well "discovery depletion"—the right to depreciate the net value of the discovery—as rough compensation for the cost of the entire program. Discovery depletion proved unworkable, and was quickly converted to percentage depletion, the right to deduct a stated percent of taxable income from the tax base. (In the United States from 1970 to 1974 the deduction was 23 percent of gross income from oil wells subject to a limit of 50 percent of net income; there was a downward-ranging scale for other minerals.) As such, percentage depletion is at one remove further from the original rationale. Even taking it as a valid proxy for discovery depletion, the defect in logic is that a large part of the abortive expenditures (particularly those related to intangibles, which could be expensed) were already tax deductible, so that duplicate compensation was being granted. In Canada, a similar percentage depletion device had even less logic because virtually *all* exploratory spending—including even land costs in oil and gas—were in fact deductible, and on an accelerated basis. Thus the percentage depletion allowance became nothing more than a reduced income tax rate applied to mineral extraction.

Similar tax reductions have been available to mining in other countries including South Africa, Australia, New Zealand, and the Philippines, whether or not they were termed a "depletion allowance." The 20 percent Australian allowance was recently removed, as was the depletion allowance in the United States for the major oil companies. The allowance in the Philippines is being phased out, that in Canada was converted into an incentive that must be "earned" by further exploratory and development spending.

SOME METHODS OF IMPLEMENTATION

In this section we appraise the major types of mineral taxes and some of their variations. Primary emphasis is placed on the output royalty, not because it is the "ideal" mining tax, but because until quite recently it was the preeminent type, and because in appraising it the merits and limitations of its rivals also emerge.

Gross Output Royalties

The simplest form of mineral levy, the gross royalty on output, may be either specific (a stated amount per unit of output)[h] or ad valorem (a proportional or progressive rate applied to the value of output). The fixed (specific) royalty is generally held in low esteem for two related reasons: it interferes in wasteful ways with the miner's production decisions, and it is an insensitive gauge of resource and monopoly rents.

[h]Specific rates most often are applied to "weight of contained mineral." Sometimes, however—tin and bauxite royalties in Indonesia are an example—the rates are applied to "weight of saleable material," regardless of purity. The purpose is to provide incentive for domestic beneficiation of ore.

As was noted above, a tax which impinges with equal severity on all units of output is economically inefficient, since it distorts the operator's calculations at various decision margins. It tends to discourage investment and to stretch out the time pattern of production.[i] Furthermore, the grade of ore that is just profitable to extract is higher with the royalty than without it. Thus, if an undeveloped ore body is not susceptible to selective mining the royalty may inhibit, or at least delay, development of that mine. Alternatively, if the ore body does lend itself to selective mining, the royalty encourages bypassing of low-grade material which would otherwise have yielded a return in excess of direct costs of extraction. There is yet another way that a fixed royalty may reduce metal recoveries—and that is at the concentration, rather than the extraction stage. The royalty reduces the profitability of reprocessing material that was discarded at the initial milling stages, because it adds an extra tax cost to the direct cost of additional (and less lucrative) reprocessing of waste.[j] The higher the royalty the fewer concentration steps that will be justified and the more contained metal that will end up on the scrap heap.

Since the specific royalty encourages waste or delay in mining it reduces the present value of expropriable rent. Moreover, since it does not distinguish between "ordinary" deposits and bonanzas, it is incapable of retrieving the maximum natural resource rent even given the reduced discounted surplus available. Again, unless rates are sensitively adjusted both among different minerals and in different time periods, the fixed output royalty is incapable of appropriating monopoly and windfall rents.

The ad valorem royalty suffers from the same criticism insofar as it influences decision margins related to any given ore body; it adds to the cost of all units of output including the least lucrative. However, it does have the virtue of discriminating among rich and lean ore bodies. Further, particularly if rates are progressive with respect to mineral prices, it is capable of capturing some of the monopoly rents and windfall profits. On the other hand, in inflationary times, a progressive rate structure that is geared to gross output

[i]To the extent that a tax on "net income" impacts on the opportunity cost of capital as well as on mineral rents, it will alter decisions made on capital investment, and hence over the long run it, too, will reduce exploration, extend the production period, and/or raise the cutoff grade. However, the inhibiting effect is much less severe than under an output royalty, which affects output as well as capacity decisions.

[j]It should be noted that while any tax would reduce the profitability of reprocessing, the royalty has particularly severe effects relative to an income tax, not only because it (unlike the income tax) increases the direct costs of processing, but it could (unlike the income tax) turn profit negative.

values will capture a "windfall profit" that may be spurious. To the extent that costs—as well as revenues—are rising, a progressive output tax will impinge more and more on quasirents instead of on genuine surplus returns.

Given the foregoing strictures, why is the output royalty so popular? Maximum time-discounted revenue yield and noninterference with decision margins are not, of course, the only criteria for judging a tax. The output royalty has the supreme virtue of administrative simplicity. It is readily understood by the most unsophisticated of taxpayers. It has the simple but powerful attraction of being hard to "cheat" on both by miners and by government administrators. Moreover, since richer ore would ordinarily be mined before leaner ore, the waste due to high-grading and a slower pace of extraction, measured in terms of a diminished present value, is much lower than their undiscounted amounts.[k]

In less developed countries, where most of the mineral output is exported, the output tax is even simpler to collect if it takes the form of an export levy, thus reducing the number of collection points. But if the royalty is imposed only on exports it will miss the opportunity to tax minerals consumed locally—usually the heavier-weight nonmetallics.

Comparing across countries, rather than across activities, to the extent that mining costs in one country are lower than those of its competitors, an output tax may be used to capture differential rents without discouraging investment. (The short-run distortions regarding rates of output and grade would, however, remain.)

Another desirable characteristic of a tax, particularly for poor countries, is revenue stability. Borrowing may be difficult. Moreover, as noted in Chapter Three, where mining dominates an economy, fluctuations in mineral revenue may initiate a cycle in the economy as a whole. The most stable of mineral revenue sources is the specific output tax. Ranking somewhat lower on this criterion—but still ahead of taxes based on net income—is a proportional tax on output values. A dramatic illustration of the greater variability of income taxes compared to those on output is afforded by the experience of Zambia in 1971. In that year Zambia converted its previous ad valorem output taxes (of which there were two: a royalty and an export tax) into the theoretically more efficient mineral profits tax (based on net income). The rates were selected to approximate the

[k]However, a mining company in practice might choose to delay exploitation of richer ore, in order to show its stockholders a more steady rate of return year by year, or to mine more high-grade ore when prices are low. This is irrational from the economist's point of view, but it has happened.

same revenue yield. Unfortunately, in the first year of the new tax, 1971, Zambia suffered from both a decline in the price of copper (its major mineral output) and a fall in the volume of production.[l] As a result, the proportion of copper taxes in total government revenue fell from an average of 59 percent in 1966-70 to 39 percent in 1971. Precisely because income taxes are more sensitive to ability to pay, which makes them more acceptable as rent collection devices, they also amplify fluctuations in the revenue collected.[m]

Another objection to output royalties—in contrast to income taxes—is a purely institutional one which, nonetheless, has important implications for countries whose mining industry is largely foreign-owned. This is the fact that income taxes paid to the host government are generally creditable against income tax otherwise payable by the mining company to its home government. In contrast, royalties are merely deductible against the home country tax base. Thus the company can "afford" to pay a higher amount of income tax in the mining country than it can of output taxes.

A difficulty with ad valorem output royalties—which they share with income taxes—is that revenues can be falsified, particularly where sales are not made by the mine at arm's length. This is a problem that is avoided by the specific output tax. The difficulty may be obviated in the case of both the income tax and the ad valorem output tax, by basing the levy on quoted open-market prices. It is argued that using a presumptive revenue may work a hardship on some mines, but it is equally true that use of a presumptive value gives an additional incentive to the mine to seek a higher price.[17]

An alternative approach to the proportional royalty (which is exemplified to some extent by the Bolivian *regalía*) is to base it not on gross output value but on value net of an allowance. The allowance can be set to account for a notional level of "cost," in which case the output royalty shades into a "true" income tax, the main difference being that costs are presumptive instead of actual. In contrast with the income tax, this form of taxation has the virtue of

[l]The fall in production in 1971 was mainly caused by a disruption in operations at the Mufulira copper mine in September 1970. Value of copper output in Zambia fell from $908 million in 1970 to $620 million in 1971. (Copper accounts for over 90 percent of Zambia's mineral export values.)

[m]On the information available we cannot tell what would have been the effect on tax collections, after the fall in price and production, if the original ad valorem output tax had been retained. Undoubtedly the revenue decline would have been of smaller proportions than that resulting from the profit tax. Under a profit tax, collections vary more than proportionately with value of output because fixed costs are deducted from the tax base.

greater administrative simplicity and lower compliance costs. It eliminates the problem of cheating or disputes over permitted deductions and may be a spur to management efficiency. In contrast with the income tax, however, such a royalty cannot allow for differences in cost, either between different mines or within a single ore body. Hence the adverse effect on decision margins and economic efficiency remains, although in attenuated form.

If the floor price is indeed to approximate average costs of an efficient producer, it should take account of movements of those costs over time. If costs are rising and the floor price remains fixed, a proportional tax on net revenue (and *a fortiori* a progressive one) will fall increasingly on quasi-rents with enervating effect on the industry.[n]

An adaptation of the principle of the net output royalty, with the effect of a steeply progressive tax on gross values, is a surcharge superimposed on normal royalties. The intent is to capture windfall profits from unexpected price increases. As recently applied by both the province of British Columbia (Canada) and the Philippines, a proportional surcharge is applied when mineral prices exceed an arbitrary norm. In the Philippines a premium duty of 30 percent is applied to the difference between a price established monthly for selected export products (by the Bureau of Customs) and 80 percent of the relevant price in February 1974. The surcharge is not confined to minerals. Among the latter it applies to copper, gold and silver, for which the "basic" prices (i.e., 80 percent of the February 1974 value) are 90 cents a pound, $120 an ounce, and $2.30 an ounce, respectively. In the case of British Columbia, when the "gross value" of the mineral (international metal price less smelting and refining costs, but not excluding transportation) exceeded 120 percent of "basic value," that excess was taxed at 50 percent. ("Basic value" for 1975 was $0.65 per lb. for copper, $92.10 per ounce for gold, $3.35 per ounce for silver, and was also established for ten other minerals.) The Philippine tax makes no allowance for increases in cost, and hence is subject to the criticism made above. British Columbia did attempt to allow for escalating costs by increasing the "basic value" annually to reflect half the year's increase in the wholesale price index.

Jamaica was recently able to overcome the shortcomings of the specific output tax and turn it into a potentially effective device for

[n]If prices and costs are both rising, and the floor price remains fixed, a proportional tax on revenue net of the floor becomes in effect a progressive tax on gross revenue.

extracting natural resource rents, and presumably also to improve its terms of trade with the rest of the world (i.e., to shift some of the tax to foreign users, at least over the short run). It did this by raising its bauxite production taxes in one bold stroke by 700 percent.[18] Since bauxite and alumina "markets" are internal transactions within vertically integrated companies, the tax is based on the price of aluminum ingot. Jamaica was able to impose this tax without immediate loss of output, first because it enjoys a transportation advantage to its markets (a differential resource rent) and, second, because its actions were rapidly emulated by its Caribbean neighbors, Guyana, Surinam, the Dominican Republic, Guinea and Haiti.[o] To forestall transfer of operations to other countries in the short run, the tax is based on a presumed minimum level of production set at about 90 percent of capacity. In other words, the production royalty becomes a lump-sum tax if production falls below the stipulated value; thus changes in output levels below that value carry a zero incremental cost in virtue of the royalty.

What has been implied in our discussion of output royalties is the assumption that royalty rates are fixed by law or regulation and apply to all producers of a given mineral. In principle, however, it is quite possible for the royalty to be negotiated in advance in much the same manner as our hypothetical lump-sum auction. A royalty of this sort may be regarded as simply the time-distributed value of the initially calculated net rent on the deposit. It differs from the lump sum mainly in that some of the risk is placed on the government. The comments in this chapter on the desirability of competitive bidding and of delaying the auction until an ore body has been delimited continue to apply.[p] A negotiated royalty stream should be a more sensitive barometer of potential resource rents than is a fixed royalty schedule.[q] The negotiated royalty also avoids the adverse effect on

[o]Despite the higher production taxes, and the subsequent coercive purchase by the Jamaican Government of 51 percent interests in three of its concessionaires, the government had hopes of a near-term increase in alumina production on the island. However, the hope may not be realized. In August 1975 Revere Copper and Brass closed down its Jamaican bauxite mining and alumina refining operations. See *World Mining*, February 1976, p. 66.

[p]An important problem has to be resolved if a government moves to ad hoc open negotiation on all new discoveries. This, of course, is the question of motivating exploration. A government that takes this approach would presumably have to take over the exploration function itself, either directly or under contract. The alternatives of giving "first preference" to the private discoverer or his nominee creates a number of new uncertainties and, if the commitment is kept, cannot be reconciled with open bidding.

[q]The Newfoundland Royal Commission on Mineral Taxation recently advocated that the province negotiate on all future mining concessions for a specific output royalty designed to capture all projected returns in excess of a normal return on the necessary investment.

output and grade decisions attributed to the universal, arbitrary royalty. However, since separate negotiation is required for each concession, this procedure presupposes a considerable measure of technical and market information and of negotiating skill on the part of the government. A negotiating government may deem itself less disadvantaged if the object of negotiation is not to determine its share of future output but rather its share of some more sensitive gauge of realized profits.

A quite similar effect—one that guarantees a government a share in windfall profits (concomitantly increasing the share of risk it bears)—is achieved by a tax device that, superficially, does not look like a royalty, but is directly derived from the royalty concept. That is the *in situ*, or annual property tax, levied on the "net realizable" present value of ore in the ground.[19] In its simplest form, an annual appraisal is made of the value of a deposit, net of capital cost allowances and operating costs, and discounted to the time of appraisal. A percentage rate is then applied to the appraised value. This tax then, again, is a way of prorating the rental value of the mine over its life, this time allowing for "price" and "cost" surprises. It suffers from the element of arbitrariness, since, unlike the contracted royalty, it is not subject to negotiation. Unfortunately it also requires a sophisticated administrative apparatus. A tax of this type was recently applied by the Canadian province of Saskatchewan to its potash mines. The tax is roundly detested by the industry, and some firms have refused to pay it pending a court decision as to its validity. Saskatchewan probably has a sufficient differential rent advantage that if the tax is upheld it will not much inhibit development of the industry. The province, in any event, has now embarked on a policy of public ownership.[20]

Income and Excess Profit Taxes

The strengths and weaknesses of income (or profits) taxes are implicit in the extensive foregoing discussion of royalties. Thus we will pass over the ordinary income tax quickly. As compared with output royalties, income taxes (while not wholly neutral with respect to investment decisions) impose fewer constraints on the miner's operating decisions, including cutoff grade. As compared with a net royalty (on revenue over a well selected presumptive cost), a "true" income tax can make allowance for exploration and development expenditures, if incentives are desired in that direction. On the other hand, the income tax is more costly to administer and comply with, and presents more opportunity for evasion.

Although the income tax is more sensitive to resource, selling-power and windfall rents than is the output royalty, it fails as an instrument for precisely delineating those rents. If the rate is set at 100 percent, there is, of course, no provision for a return on the investor's capital (including his required risk premium). Moreover, the higher the tax rate the greater the disincentive toward managerial efficiency. To the extent that the tax rate exceeds that applicable to other enterprise income, it must inhibit some exploration and discovery, since there is some mineral investment, at the margin, that will yield no net rent. But if the rate is the same as that applied to capital income from other sources the tax will have failed to capture surplus mineral returns.

In an attempt to delineate and appropriate mineral rents, proposals have been made for a special tax on mineral profits that exceed normal returns to capital.[21] The essence of these schemes is that they apply a high rate of tax to profit in excess of an arbitrary rate of return on invested capital.

However, it would be a mistake to regard excess profits taxes as having resolved the conundrum of defining and collecting economic rent. Primarily, this is because rent is a surplus over cost, and cost must include the investor's allowance for risk, a variable and unknowable amount. A fixed general formula that allowed a developer a satisfactory rate of return after an ore body had been proven would be inadequate to motivate a risk-averse explorer.[22,23]

The excess profit tax is one alternative—among many—for exacting mineral rents in ad hoc negotiation over a specific known ore body. However, given any particular anticipated profit stream, present worth of expected government revenue under the excess profit tax could be duplicated by some rate of ordinary income tax (presumably at a rate higher than that on other income, but less than 100 percent—and less than the applicable excess profit tax rate). In those circumstances, the advantage of the excess profit tax is that it delays tax payment until the investor has retrieved his investment compounded at the prescribed rate of return.[24] The operator's perceived political risk would be lower under the excess profit tax alternative. For reasons noted above, this may make both parties better off. On the other hand, the rate of tax on a marginal dollar of profit would necessarily be higher under the excess profit form than it would be under the equivalent-yield income tax. Hence the incentive toward cost cutting and efficient management would be that much weaker under the more complex tax. Similarly, the temptation to falsify revenues and costs would be the greater, as would administrative complexity. No clear-cut verdict can be made between the two approaches in regard to particular negotiated contracts. As a legislated general method of mining taxation, the excess returns approach is cumbersome and has no obvious advantage over an equal-expected-yield income tax. It is of some interest to note, in this regard , that the Province of Manitoba (Canada) recently adopted a supplementary tax that applies to profit exceeding a stated rate of return on invested assets. The apparent reason for preferring this to a straight increase in the ordinary profit tax rate is that the province's two major taxpayers own plants of vastly different vintage. The ordinary provincial mine profit tax provides for historic cost depreciation; the excess profit tax base allows a return on the current value of assets. Hence its application is deemed more equitable than an extension of the normal profit tax. However, the required result might have been obtained under the ordinary tax alone with appropriate revaluation of assets.

OPTIMAL MINERAL EXACTIONS

It will be apparent that there is no single "correct" mineral tax design. Each nation must determine its objectives and choose that tax (or package of taxes) that ranks highest in terms of the goals it fulfills and lowest in terms of those it violates. Even on the narrower criterion of maximizing present-valued rent, no precise answer can be given. Whatever the method of taxation, there will be some effect on investment, discovery, and output. Resource rents are easier to describe than to identify and collect. A tradeoff dilemma exists between the total amount of surplus achievable through mining operations and the proportion of that surplus that may be appropriated by government. Moreover, what other governments are doing helps to determine what one government can collect.

Considering the various taxes discussed, on revenue and on efficiency grounds preference must be given to income taxes. However, there are other methods of mineral revenue collection that preserve the spirit of the income tax, retain the advantages of private operation and reduce the inhibiting political risk that increasingly faces mine developers. These newer forms of revenue collection—profit sharing, production sharing, and contracting—are discussed more fully in Chapter Five. Here we may note that because they provide the appearance if not the actuality of national control, they are coming to be preferred by the mining companies themselves.[25] In the sense of reducing risk premiums, these devices increase the potential rent that is appropriable by government.

It is also likely that an informed government can achieve more by ad hoc negotiation or renegotiation than by general rules. However, for reasons of administrative simplicity and to motivate the small explorer, it would be preferable to establish a viable general rule to apply to the bulk of mining ventures with negotiation to apply only in the case of major projects.

NOTES TO CHAPTER FOUR

1. R.A. Musgrave and P.B. Musgrave, *Public Finance in Theory and Practice* (New York: McGraw-Hill, 1973), p. 193.

2. See R.G. Lipsey and K. Lancaster, "The General Theory of the Second Best," *Review of Economic Studies*, Vol. 24, Dec. 1956.

3. Recent developments in the literature of optimal taxation are capably summarized in D.E. Bradford and Harvey Rosen, "The Optimal Taxation of Commodities and Income," *American Economic Review*, May 1976, p. 94-102.

4. C.A. Shoup and others, "An Analysis of the Taxation of Mineral Resources in Developing Countries," mimeographed (Geneva: United Nations Secretariat, 1972).

5. J.F. Due, "The Developing Economies, Tax and Royalty Payments by the Petroleum Industry, and the United States Income Tax," *Natural Resources Journal*, vol. 10, Jan. 1970, p. 10.

6. The term "efficiency" as used here may be unfamiliar to noneconomist readers. Economists view resource allocation as efficient if resource use is such that no readjustment is possible that would make someone better off without making someone else worse off. The conditions under which efficiency can be achieved are stated in any basic textbook in public finance. For example, see Musgrave and Musgrave, *op. cit.*, pp. 462-463.

7. In a recent paper Feldstein has shown that the classical conclusion that a tax on rent could not be shifted is wrong; indeed, a tax on land rent would, in his framework, induce more saving, more investment and, hence, a reduction in the return to capital. Nonetheless, he concludes that a tax on pure rent remains attractive, because it does not impose a tax wedge between the gross and net returns of variable factors of production. See Martin Feldstein, "The Surprising Incidence of a Tax on Pure Rent," Harvard Institute for Economic Research, Discussion paper #424, July 1975.

8. As Scott first pointed out, the issue is not that minerals are a nonrenewable stock, but rather that rational exploitation of even renewable natural resources—forests, fisheries—requires explicit choice with respect to investment timing and a consequent rate of depletion. What may be harvested in any one year depends on what was harvested in other years. See Anthony Scott, *Natural Resources: The Economics of Conservation* (Toronto: McClelland and Stewart, 1973), Chapter One, pp. 8-10.

9. The best-known instance of exercise of monopoly power over natural resources by producing countries is, of course, the OPEC-imposed quintupling of levies on petroleum since 1973. Among nonfuel minerals, which are the concern of this study, the outstanding example is the increase in bauxite levies by six Caribbean nations in 1974-75. See M. Gillis and C.E. McLure, Jr., "The

Distributional Implications of the Taxation of Natural Resources," *Rice Studies*, vol. 61, no. 4, fall 1975.

10. This section draws inspiration from O. Herfindahl, "The Value of Mineral Surveys to Economic Development," pp. 236-251 in D.B. Brooks (ed.), *Resource Economics: Selected Works of Orris C. Herfindahl* (Baltimore: Johns Hopkins Press, 1974), and more particularly from Shoup, *op. cit.*

11. O. Herfindahl, "The Value of Mineral Surveys," p. 244 in David Brooks (ed.), *op. cit.*

12. *Far Eastern Economic Review*, April 4, 1975, p. 248. A somewhat similar case—involving Canadian Javelin—occurred in the Canadian province of Newfoundland at about the same time. In July 1975 the Newfoundland Government proposed to strip the company of its und eveloped Julienne Lake iron-ore deposit in Labrador (granted by an earlier government).

13. See Chapter Two and Chapter Nine for some examples in COMIBOL. These problems are not confined to COMIBOL, as the recent experience of PERTAMINA (the National Oil Company of Indonesia) so clearly testifies. See *The Economist*, April 5, 1975, pp. 82-83.

14. Shoup, p. 14.

15. See Chapter Five.

16. The "gift of providence" view of minerals is in the ideological strain of 18th century physiocratic economics which regarded land as the only source of "real" economic wealth. Such a view of minerals can lead to two diametrically opposed policy prescriptions. On the one hand it may lead to a narrowly "conservationist" view: keep the minerals in the ground whatever their value; on the other, it may lead to the viewpoint discussed in the text above: that it is "sinful" to leave minerals in the ground .

17. Use of presumptive revenue—and also of presumptive cost—shifts risk from the government to the miner with the attendant virtues and vices noted earlier in this chapter.

18. See Gillis and McLure, *op. cit.* An interesting sidelight (*ibid.*, p. 26) is that the Jamaican tax was increased in one stroke rather than being phased in, because it was feared that technological progress will make it possible to extract alumina from lower-grade materials, a possibility that the action by Jamaica and its neighbors will no doubt hasten. This illustrates a point made earlier in this chapter, that the precise calculation of economic rent requires the gift of prophecy.

19. This type of tax receives a favorable appraisal by W. Vickery, "Economic Criteria for Optimum Rates of Depletion," and by Gaffney in his "Editor's Conclusions" in M. Gaffney (ed.), *Extractive Resources and Taxation* (Madison: University of Wisconsin Press, 1967).

20. The *in situ* property tax on net ore values should not be confused with area rentals—such as the Bolivian *patentes*—which are discussed elsewhere in this work. The purpose of the latter is not mainly that of revenue or rent exaction but rather as incentive to explore and mine without delay. For reasons why landholders would ordinarily be motivated to develop at less than the socially optimal pace, see Gaffney, *op. cit.*, pp. 391-399.

21. A recent example of such a proposal is presented in R. Garnaut and A.

Clunies Ross, "Uncertainty, Risk Aversion and the Taxing of Natural Resource Projects," *Economic Journal*, vol. 85 (June 1975), pp. 272-287.

22. For further elaboration of this point, see R. Mikesell, in Mikesell (ed.), *Foreign Investment in the Petroleum and Mineral Industries* (Baltimore: Johns Hopkins Press, 1971), pp. 36 and 433. For further critical discussion of a tax on profit over normal returns on mineral investment, see Herfindahl, *op. cit.*, p. 249.

23. The inhibiting effects of an excess profits tax on the motivation to explore might, in the right national circumstances, be countered by coupling the tax with a marketable loss carry-forward for those ventures that turn out to earn less than a normal profit. (We would like to thank Carl Shoup for pointing this out to us.)

24. This conclusion applies to the particular scheme advocated by Garnaut and Clunies Ross. A modified version of that scheme was adopted in the 1974 renegotiation of the Bougainville Copper agreement in Papua New Guinea. The company is to be taxed at 33 1/3 percent on profit up to the point where it is left with a 15 percent after-tax return (on total assets). Profits beyond that level are to be taxed at 70 percent.

25. The principals of a knowledgeable firm of international geological consultants recently wrote: "It is fairly clear now that most multinational and large mining companies have come to the conclusion that partnership with government is the only answer to operating mineral deposits in developing countries." D.R. Derry, C.E. Michener, and J.K.B. Booth, "World Mining Exploration Trends," *World Mining*, March 1974, p. 50.

Chapter Five

Comparative Mining Taxes

Meyer Bucovetsky,
Malcolm Gillis and
Louis T. Wells

OVERVIEW OF TRENDS IN THE TAXATION OF MINERAL PROJECTS

Up until the early nineteen-sixties, fiscal revenues from mining and hydrocarbon taxes in most developing countries were derived from output taxes (typically called royalties in most of the countries examined in this chapter), income taxes, or some combination of the two. In the 1960s, new forms of revenue collection appeared. Some of these amounted to little more than royalties or income taxes, although the new trappings made them more politically acceptable than the old methods. Other forms of taxation did represent more fundamental change.

Royalties and income taxes have in no way disappeared as a source of revenue from mineral projects in developing countries, but there have been major changes in the relative importance of the two types of taxation. The shift has been a strong one, away from royalties as the principal source of income, toward heavier reliance on income taxation. The shifts have been made in countries of varied political persuasion. The evolution of fiscal development is similar not only for countries with a strong private enterprise approach (Liberia, for example) but for those with a major state enterprise sector (Indonesia), or with strongly socialist economies (Guinea, for example).

Royalties were the principal source of governmental mining tax revenues for a time. In early oil agreements, Middle East governments typically collected four gold shillings per ton of oil exported. In these early agreements, governments depended on ad valorem (a percentage of value) or specific (a certain payment for unit of mineral) royalties for most government revenues. Some of the royalties varied progressively with the price of the material; others were proportional. Even several decades ago the Malay States imposed a royalty on tin which varied with the price of tin. Liberia, even in the late 1940s, collected a royalty of five cents per ton for iron ore, plus an additional royalty if price was over a certain base figure, in its LMC (Liberia Mining Company) Agreement with Republic Steel.

Whatever the formula, royalties had their advantages. Most important among these was the ease of administration. In fact, early attempts to impose an income tax in Persia (the d'Archy oil agreement) had failed largely because of the administrative problems.

In spite of their advantages, royalties had clearly given way to income taxes as the principal source of revenue for hydrocarbons and for metallic minerals by the early 1960s. In agreement after agreement, the fiscal rules that had prevailed were changed, and almost all new agreements placed much heavier reliance upon the income tax for revenue. The shift was almost universal.[a] For example, about the time of independence, Malaysia introduced an income tax on the mining sector and the LMC agreement of Liberia, mentioned earlier, was amended to provide for the payment of taxes based on income (the actual arrangement was somewhat more complex than this simple description suggests).

Of course, income taxes have differed from one agreement to another. In some cases, they were based on a presumed revenue, for example. This has been typical of many copper agreements which calculated revenue on the London Metal Exchange price, but used actual costs of production. Sometimes the tax rates were proportional, while at other times they were progressive. Agreements differed in the deductions allowed in calculating taxable income. However, all were based on an effort to collect a percentage of the difference between gross income and costs.

The general shift from heavy reliance on royalties to income taxes was based on sound reasons. The average revenue that could be obtained from an income tax was higher than that which could be expected from royalties. The shift of risk away from the entrepreneur, implicit in the change to income taxation, enabled him to

[a]Although the general trend over the past two decades has been toward heavier reliance upon income taxes on metallic minerals (the case of hydrocarbons is much less clear), several nations turned to royalties in 1973-74 as a means of appropriating a high share of profits from the minerals boom of those years. This was true in the case of Malaysia and the Philippines (see later sections of this chapter).

accept a higher expected rate of taxation. Moreover, the income tax was usually creditable against a foreign investor's tax in his home country. And the income tax gave the desired incentive for the rational exploitation of marginal ores.[1]

In virtually no arrangements did royalties disappear completely. As noted in later sections of this chapter, royalties continue to account for an appreciable share of total mineral taxes in a number of countries that adopted income taxes on mining in the 1950s and early 1960s (e.g., Malaysia, Thailand, Indonesia). Royalties were retained for several reasons: they serve (a) to reduce the risk that the government would have practically no income when prices of minerals are low; (b) to provide a tax that can be adjusted to the potential of the individual mine; (c) to give a source of revenue to state and provincial governments, or (d) to appropriate "windfall" profits when prices are very high (cf. the royalties enacted by British Columbia in 1974).

Although the shift from royalties to income taxes as the principal source of revenue was made in virtually every country and for nearly every mineral, there were exceptions, to be sure. The exceptions, such as that for bauxite, were based on rational reasons.

In the case of Guyana's bauxite arrangements with the Reynolds Company, the agreement called for an income tax. However, Guyanan officials maintain that Reynolds never paid an income tax above the small floor tax payment set in the agreement. Since bauxite (or alumina) is typically sold by the mine to a smelter owned by the same parent, the transfer price could be set such that no profit is shown for the mining operation. And for bauxite, there has been no world price upon which price adjustments could be made by the host government to calculate income. Not surprisingly, bauxite mining has, as a result, retained a heavy reliance on royalties. With gross revenue difficult or impossible to ascertain, the pattern has been to base taxes on a percentage of the price of aluminum ingot, the first product in the vertically integrated chain to have a market price. The procedure has characterized the recent measures (1974-75) taken by the International Bauxite Association producers.[2]

Newer Forms of Revenue Collection

In the mid to later 1960s, new forms of revenue collection in the mining sector appeared. Profit sharing, production sharing, and service contracts, to name a few, became the labels attached to the new arrangements. But the labels have been used inconsistently. The following sections describe "ideal" types under these categories of mineral agreements, even though in reality the labels are sometimes used for agreements that are rather different from the descriptions that follow.

Profit Sharing. Profit sharing (equity sharing or "participation") agreements generally call for the government to receive its revenues in some dividend-like form. In many cases, the government actually holds shares, as in the Liberian-American Mining Company (LAMCO) agreement in Liberia. In others, it simply receives payments much like those a shareholder would receive, but without actually owning shares. The final LMC agreement in Liberia was of this type. No shares were actually held by the government. The purpose of not holding shares was to make the arrangement suitable for Republic Steel to receive credits against U.S. income tax.

It has been tempting for governments to believe that profit sharing is equivalent to income tax: that 50 percent of dividends is equivalent to 50 percent of taxable income. Of course, this is not so. Dividends are paid after the repayment of principal on debt and provision for reinvested earnings. Income taxes take their share before these deductions, and are thus almost always larger. To be sure, special arrangements could be made such that profit sharing is actually exactly equivalent to income tax. However, profit sharing is now often included in an arrangement as a *supplement* to income tax. In such cases, the government usually takes a certain percentage of the taxable income as an income tax and then shares in the distributed earnings as well.

Production Sharing. Production sharing formulas also vary greatly in form. One might consider any agreement in which the government can take part of its payment in kind as a production sharing agreement. The copper agreement between Overseas Mineral Resource Development (OMRD) and the Government of Ecuador would then qualify, since the government has the right to take the (small) royalty portion in kind, if it desires. But that agreement retains the essential structure of an income tax arrangement with a small royalty. In fact, most production sharing agreements, it can be argued, are little more than royalty or income tax agreements in a politically more attractive form.

Consider typical second-generation oil agreements in Indonesia. Beginning in 1967, these production sharing agreements initially provided that all petroleum produced above that required for the firm's costs be split, 60 percent to the government and 40 percent to the firm.[b] Basically, the initial underlying agreement was little

[b]Renegotiations in January of 1974 increased the incremental take of the government to 85 percent when oil prices are more than $5 per barrel.

different from a 60 percent income tax, although there are some distinctions. First, title to equipment is in the hands of the national oil firm, and the government does not have to worry about calculating the sales price of oil transferred by the firm to affiliates. (That is, if the government actually takes its share in oil; in fact, until the oil crisis, the government usually turned the oil over to the firm to sell for the former's account). Second, the treatment of deductions under the Indonesian agreements differed from those found under a true net income tax. Up until 1976 allowable costs were limited to a certain percentage (40 percent) of the oil produced.[c] If the actual costs were higher than 40 percent, then the agreement took on characteristics more like those of a royalty. Third, the creditability of taxes paid under production sharing agreements of the Indonesian type has been called into question (at least so far as the U.S. Treasury is concerned). Prior to April of 1976, taxes paid under the agreements were thought to be fully creditable against U.S. income taxes. However, an April 1976 Treasury opinion (requested by a contracting firm) has given rise to serious doubts as to creditability of the taxes paid by American firms under the agreements as they stand in 1976. The issue is now scheduled for negotiation in 1977.

Service Contracts. Service contracts (or management contracts) generally call for a certain payment to be made to the firm, rather than for the firm to make a certain payment to the government. The contracts vary considerably as to the form that payment is to take. The payment may be a fixed annual fee, it may be based on output, it can be a percentage of profits, or it may be a combination of these forms.

Summary of Present Practices

Today, mining and hydrocarbon agreements present a rather bewildering maze of income tax, income tax with a small royalty (typically ranging between three and six percent of gross mineral values), profit sharing, production sharing, and service contract arrangements. As pointed out, there still remain a few pure royalty agreements, but only in rather special circumstances, as in Bolivia. In

[c]This feature of the Indonesian production-sharing agreements was under renegotiation in June-July of 1976. At that time the Indonesian Government sought to convert the 40 percent limitation (first enacted when oil prices were about $1.50 per barrel) to a system that would allow firms to amortize eligible costs over seven- and fourteen-year periods (depending on the reserves of the contracting companies). The government reasoned that with 1976 crude prices at a level about five to six times that prevailing when most of the agreements were signed, the 40 percent limit allowed writeoff of costs at an accelerated pace not foreseen by either the companies or the government, nor allowed by oil nations elsewhere in the world.

the new forms of agreements and in many of the modern income tax arrangements, structures have evolved that allow the elements of an agreement—financial flows, control, and ownership—to be packaged in ways that were not common before, say, 1960.

Consider financial flows, control, and ownership in the older arrangements. Typically, the government would take some revenue with royalties or income taxes. The rest of the financial flows would belong to the firm and would be split among its owners largely according to their fraction of ownership. Control would, at least theoretically, be similarly divided. Government, without ownership, would have no say in the internal management decisions of the firm.

New agreements have broken these links. In Indonesian production sharing agreements, for example (and in Eastern European coproduction arrangements), the private firm usually has no ownership rights. Yet, in many cases, the firm has a great deal of control and enjoys satisfactory financial flows. In Liberia's LAMCO arrangements, the government has 50 percent of the shares, but less than 50 percent of the places on the board of directors of the firm. On the other hand, for the Hanna nickel project in Colombia, the government has minority ownership, but a strong position on the board of directors, and the government board members have the right to veto decisions in a number of specified areas.

Today, the possibilities of working out an allocation of financial flows, control, and ownership that satisfies the needs of both the investor and the host government are much more varied than they appeared to be in the past. It is the repackaging of these elements that has made attractive new forms of agreements for mining and petroleum.

IN-DEPTH COMPARISONS OF MINERAL TAX STRUCTURES

In subsequent sections, the mineral tax legislation of eight countries other than Bolivia, plus seven Canadian provinces, is examined in detail. In addition, Table 5-3 presents basic features of minerals tax laws of all these jurisdictions plus nine other nations, studied in less detail. The first eight countries were chosen for more intensive scrutiny for a variety of reasons. Malaysia, Thailand and Indonesia were closely examined since they are major competitors for Bolivia in international metal markets, particularly in the case of tin. In addition, both Thailand and Malaysia, like Bolivia, have numerous small-scale mining operations. Chile, Peru and Colombia were included because they are nearby South American countries; further,

Chile and Peru are major copper exporters and have significant numbers of small-scale mining operations as well. Peru is also a competitor in other metals (zinc, tungsten, lead). Canada (and seven of its mining provinces) and Eire were chosen for in-depth study since both have recently enacted significant changes in mining tax legislation. The nine other nations studied in less intensive fashion (Gabon, Jamaica, Liberia, Mexico, New Caledonia, Papua/New Guinea, the Philippines, Zaire, Zambia) were included in Table 5-3 by virtue of having important mining sectors and to provide more geographical dispersion for the comparative study of mining taxes.

Major Tin Producers

Mineral Exactions in Malaysia.[d] Malaysia is the world's largest miner of tin, accounting for 37 percent of western world production in 1974. Furthermore, Malaysia is generally acknowledged to have a cost advantage over all the other tin producing countries. The amount of tin (in concentrates) produced in 1973 was 72,259 tons, or about two and a half times the output of Bolivia, the second largest producer. Exports of tin account for about 90 percent of the nation's mineral exports, but only about 20 percent of total export value. The total amount of Malaysian tax revenue raised from mining cannot be broken out from the official statistics. However, it is known that in 1972 tin export duties accounted for about eight percent of government revenue.

The Malaysian tin mining industry is divided into three sectors. First, there is the capital-intensive dredge mining sector, which accounts for about 43 percent of tin production. As of the end of 1973, there were 59 dredges in operation, but these were owned by a relatively few mining companies, under foreign control. In particular, three large foreign companies (with interlocking directorates) dominate the sector. The small-scale gravel pump mining sector accounts for upwards of 50 percent of production, but is declining in importance relative to dredge mining. There were 940 gravel-pump operations at the end of 1973, mostly controlled by Malaysian nationals. The balance of production is accounted for by a miscellaneous group of 46 mines using underground, open cast, or hydraulicizing methods. There is only one important underground lode mine.

[d]Malaysia is a confederation of the 11 states that formerly constituted the Federation of Malaya (now known as West Malaysia), and of Sabah and Sarawak (now known as East Malaysia). Laws regarding mineral land tenure—and to a lesser degree regarding taxation—vary among West Malaysia, Sabah and Sarawak. Since all of the country's tin is mined in West Malaysia—and the present discussion is mostly confined to tin—only the laws of West Malaysia will be described.

Exploration and Exploitation. Mineral rights are separate from surface rights, and the state owns all minerals. Three kinds of rights to explore or produce are applicable. A prospecting license is a right to explore a specific area for six to twelve months, or for a further period if progress to date is approved. A mining license is permission to mine a specified area for a single year, issued to individuals (only). A mining lease is the award of mining rights for a negotiated (renewable) period. It carries an obligation to begin operation within a year of the award. Leases lapse if work ceases for 12 months or more.

Taxation

Output Taxes on Tin

• Export Tax. This tax is mildly progressive in structure, levied at the export stage. It is deductible from the income tax base. The tax varies with price in such a way that as prices rise, the effective tax rate on the metallic content of concentrates (and on exported metal) approaches 16 percent ad valorem.

The export duty on tin concentrates is collected when the miner sells the ore, and is calculated according to a formula expressed in terms of the ex-smelter Penang price per *picul* of tin, in Malaysian dollars.[e] At a Penang price for tin metal equivalent to U.S. $1.28 per lb., the tax on a pound of concentrates is about 15¢. It rises thereafter by 12¢ per lb. of concentrates for each U.S. dollar increment in the price of tin. It is presumed that the average metallic content of exported concentrates is 75 percent tin. Hence, the burden of tax on exports of refined metal is equalized by taxing tin metal exports at a rate one-third higher than that on the same weight of concentrates.

• Export Duty Surcharge on Tin. In March 1974, as part of a broader economic stabilization package, Malaysia imposed a surcharge of tin exports. The surcharge takes the form of a progressive ad valorem tax. Again, for convenience, expressed in U.S. dollars per pound, when the price of tin is above $2.25 the surcharge is one percent of the export value. When the price rises to $3.85, the surcharge jumps to a maximum of five percent. Thus, at August 1975 L.M.E. prices for tin of $3.08 per lb., the sum of export taxes and the export duty surcharge is equivalent to about 16.85 percent of Penang tin export values (see Table 5-2).

[e]The *picul* is the Malaysian unit of weight, and is equal to 133.1/3 lbs. One U.S. dollar exchanged for 2.34 Malaysian dollars in June-July of 1975. The average Penang price in 1974 was about 96 percent of the average London Metal Exchange (L.M.E.) price for tin.

Output Taxes on Other Minerals

The export tax is also applied to other minerals. Rates, however, are proportional ad valorem, amounting to 15 percent for iron ore, 10 percent for other metalliferous ores, 5 percent for gold, and 2.5 percent for nonmetallics.

Income Taxes

• Corporation Income Tax. This tax is actually a tax on undistributed profits (of *all* companies) since, although a withholding tax is collected on dividend payments at the company level, the dividend recipient receives credit for the tax withheld. The tax is levied at a flat rate of 40 percent.

In calculating taxable income, a mining company receives an initial depreciation allowance of 60 percent (instead of the usual 20 percent) for investment in mining plant and machinery. The rest of expenditure on mining investment (including preproduction) is amortized over the expected life of the mine. Abortive prospecting expenditure, incurred in eligible areas of Malaysia, qualifies for writeoff, provided prospecting is abandoned without proceeding to the development stage.

Finally, an investment tax credit of from 25 to 40 percent (the amount depends on the area, and on the degree of Malaysian value added) has been in effect since 1968, on approved new corporate capital expenditures.

• Personal Income Tax. This is a graduated tax, ranging from 6 percent to 50 percent applying to personal income (including dividend income) and income from unincorporated enterprises. Rules for computing mining income subject to tax are the same as those summarized above for incorporated mines.

• Development Tax. This levy was introduced in 1967 and takes the form of a flat-rate, five percent additional charge on income from a business or profession. The charge applies to both companies and individuals (dividends are exempt), but companies pay a minimum tax of about U.S. $214. In contrast with the corporate income tax, the development tax applies to total profit (whether or not distributed).

• (Excess) Profit Tax on Tin Mining. This tax was introduced in January 1965, at a time when tin prices were rising. It was intended to capture for the government a portion of windfall gains realized at the time, but it has remained in effect since. The tax is computed with reference to the excess over a "standard profit" which amounts to approximately (U.S.) $0.32 per lb. of contained tin metal. To be more precise, the standard profit is defined as the number of *piculs* of tin metal contained in the concentrates marketed during a year, multiplied by one hundred Malaysian dollars (M$100). When the actual profit from tin mining exceeds this "standard" amount by more than M$10,000, a profit tax of 10 percent is levied on the whole of the excess.[f]

This form of tax was chosen in preference to an increase in the export tax rate because it falls only on mining operations that make large profits when the price of tin is high, and does not impinge on marginal deposits. But the "standard" is quite arbitrary and is unrelated to long-term changes in mining costs.

Final Comment on Malaysia. Tin mining in Malaysia is subject to the same income taxes and development tax as are other enterprises. Provision for capital cost recovery is somewhat more favorable than in other industries. However, tin production is also subject to three other distinctive taxes: a progressive, specific export tax, a tax on net income over an arbitrary base, and, most recently, an export surcharge based on tin prices.[3] Hitherto Malaysian mining has been a private sector activity. However, a new phase may have begun with the discovery, in 1974, of tin dredging potential in the deep waters of West Malaysia's continental shelf. To late 1975, the Malaysian Government had granted offshore exploration rights only to a subsidiary of the state-owned company, Pernas.

Mineral Exactions in Thailand. Thailand, like Malaysia, has thus far placed primary reliance for mineral development on the private sector, although there are now signs that a government-owned entity may move actively into the exploration field. The government's main source of mining revenue is an ad valorem royalty tax on output. Mining enterprises are, however, also subject to income taxes on the same general basis as other businesses, with deductibility of the royalty from the income tax base. There are, additionally, a business tax and some minor taxes on mining income.

In 1973 the Thai mining industry contributed about 1.7 percent of Gross Domestic Product. Mineral exports, in 1972, were valued at about U.S. $105 million and amounted to 10.4 percent of total national exports. Of total mineral exports, tin metal contributed 80 percent, tungsten concentrates 11 percent, flourite 7 percent, with the remainder accounted for by seventeen other minerals. A significant zinc deposit is being brought into production in the northeast of the country. The chief nonexported mineral is lignite.

---

[f]Since M$10,000 is an exclusion, and not a deduction, there is also a "notch" provision that phases the tax in, gradually, for excess profits over M$10,000. Full weight of the tax is not felt until "excess profits" exceed M$11,111.

Thailand's market share of world tin output has been falling in the last few years (from 13.7 percent of world output in 1966 to 11.3 percent in 1974) despite the fact that Thailand is a relatively low-cost producer. Reportedly the reasons are a lack of investment capital for the capital-intensive offshore dredging sector and a lack of technological improvement in land-mining techniques.[4] About 40 percent of Thai tungsten production is a byproduct of tin mining.

About 65 percent of Thai tin output comes from onshore alluvial and eluvial deposits, about 20 percent from offshore sea dredging, and about 15 percent from mining hard rock tin or tin-tungsten lodes. The last are all small, primitive operations, with access by simple adits.

Dividing up tin output another way, by size and nationality of the controlling group, about 20 percent of output (the offshore dredging sector) is controlled by two foreign groups, the TASK group of four companies (controlled by London Tin Corporation), and a Union Carbide-Billiton joint venture. Three large Thai-controlled companies account for another 23 percent of output. The remainder (57 percent) are small, locally owned enterprises, many of which face problems similar to those of the "small mining sector" of Bolivia.

The Thailand Smelting and Refining Company (THAISARCO), a joint venture of Union Carbide and Billiton, commenced tin smelting in July 1965. By law, all tin concentrates must be sold to this smelter. Tin ore is smelted to a 99.92 percent purity.

State participation in mining up to 1975 has been limited to operation of two lignite mines through the Electricity Generating Authority, and a 16 percent participation in one of the London group's dredging companies.[5] Apart from the sea-dredging operations and the smelting company, the only significant foreign ownership is in one large fluorite mine.

Exploration and Exploitation. The Thai Government claims exclusive ownership of all mineral rights in Thailand. About 40 years ago, in order to protect rich farming lands from destructive placer and hydraulic mining, the government proscribed mining rights grants north of the eleventh parallel. About 1960 this policy was modified to allow mining rights in this same area only to Thai nationals or companies controlled by them. There have, apparently, been some exceptions granted to this rule.

A license to prospect a potential mineral area carries no right to exclusivity nor to a sequential mining lease. However, the Minister of Mines may award a prospecting license that is exclusive. In that case the licensee also has the exclusive right to apply for subsequent lease (subject to equal right by the owner of surface lands involved). Permissions to mine are also of two types. Temporary mining permits are valid only for 180 days. Mining leases confer a right to mine that is valid for a 52-year maximum (or shorter renewable term).

Since 1971, the government has actively pursued the granting of prospecting licenses, opening up areas for international bidding. The response was disappointing, with only one or two bids.[6]

Taxation

Output Taxes

• Production Royalty. All minerals are subject to an ad valorem tax that is deductible against the income tax base, and based on a presumptive price determined in a specific market for each commodity. The relevant price for tin is the price for tin metal in the Penang market. Once established, a posted price endures until the Penang price changes by over fifty *baht* per *picul.*[g] Given that price, the tin royalty (as is the business tax—for which see below) is collected on delivery of concentrates to the smelter, which acts as the tax collecting agency. The royalty is assessed according to assay, or, alternatively, it may be based on a presumed 72 percent tin content of the concentrate.

The presumptive price for tungstic oxide is that in the New York market, and is altered when the latter changes by 25 *baht* per *picul.* The taxable price for gold is that prevailing in the Bangkok market, as is that for other minerals.

Given the metallic content of the concentrates and the posted price, the royalty on tin ore is set by law as 25 percent of the excess of posted price over a floor price of 1,000 *baht* per *picul* (equivalent to U.S. 37.5¢ per lb.). Thus the royalty on tin is not quite proportional, but mildly progressive, given the exclusion feature. At a price for tin of $3.08 per fine pound the production royalty would amount to U.S. $0.645, or about 21.8 percent of the Penang export price. At a price of $4.00 per lb., the Thai royalty would be 22.6 percent ad valorem. By way of comparison, the Bolivian royalty plus export tax and earmarked taxes for the same grade of concentrate would be about 27.42 percent ad valorem (on the L.M.E. price) at a price of $3.08 and 31.72 percent at a price of $4.00.

The rates for all other minerals are proportional and lower than those on tin. For minerals other than tin, rates are established from time to time by regulation. However, by law, rates on tungstic oxide

[g]The Thai monetary unit, the *baht*, was worth U.S. $0.05 in mid-1975. The Thai unit of weight used for tin, the *picul*, is equivalent to 133.1/3 lbs.

may not exceed 20 percent and those on other minerals may not exceed 15 percent. Some of the relevant rates currently in force are:

Tungsten	10%	Fluorite or iron ore	4.5%
Gold	3.57%	Other metal ores	2.5-3.0%
Antimony, lead, zinc copper, in ore	5%	Coal and gypsum	1.5%
Above, in metal	2.5%	Other nonmetallics	0.5%

• Business Tax. This is a sales tax on gross receipts received by businesses generally. As regards minerals, metallics are subject to business tax assessed at a rate of four percent on gross receipts only if exported. The converse applies to non-metallic minerals (i.e., they are taxed only if sold locally).

Income Taxes

Mining enterprises are taxed on the same basis as other industries. Corporation tax rates are 15 percent on the first 500,000 *baht* (U.S. $25,000) of profit, 20 percent on the next 500,000 *baht* and 25 percent on the excess. Individuals and family corporations pay tax at progressive rates ranging from 10 percent to 50 percent at the margin.

Asset depreciation is permitted at the same rate as is used in the enterprise's own accounts, subject to a limit set in the depreciation schedule. The limits may be regarded as generous; for example, the limit on machinery is 20 percent of cost. Further, mining companies may take cost depletion on concession and exploration costs over a ten-year period.

Other Taxes

Other taxes on mining firms in Thailand include a municipal surcharge of 0.4 percent added to the 4.0 percent business (gross sales) tax, and a 25 percent tax on remittances abroad.

Final Comment on Thailand. By way of final summary and comment it would be useful to quote some of the conclusions reached in a recent study of the Thai mining sector. The report's authors are highly critical of the output royalty, and particularly its disproportionate impact on tin. They observe that the tax encourages high-grading of potential ore. They cite one major potential investor who wanted to work a low-grade ore but was deterred by the tax structure. The authors believe that the royalty structure reflects a conscious incentive policy of diversification away from tin. They approve of a government plan—still very much in the discussion stage—of creating a state-owned Mineral Resources Development Corporation, as an active explorer, and they recommend the Quebec Government's SOQUEM as a model.[7]

Mineral Exactions in Indonesia. The Indonesian mining sector is dominated by large-scale enterprise, both state-owned and foreign. Some small-scale tin mining activity is carried on in Sumatra, but makes no significant contribution to overall output. Manganese is, however, mined exclusively by small-scale, labor-intensive enterprises, primarily in West Java. Major mineral exports include tin (1974 output of about 24,000 tons of contained metal), bauxite, nickel ore, copper concentrates and manganese. The mining sector inclusive of petroleum and natural gas accounted for 9.6 percent of GDP in 1973-74. The nonhydrocarbon mining sector may have accounted for as much as 2.5 percent of GDP in 1973-74, although no separate figures for nonhydrocarbons are provided in the national accounts. For 1973, minerals exports exclusive of hydrocarbons were 5.4 percent of total exports. Tax revenues from minerals other than oil and gas amounted to slightly less than one percent of total government tax collections in both 1974 and 1975. (Tax revenues from oil and gas accounted for 54 percent and 57 percent of total government revenues in the same period.)

The large state enterprise P.T. Timah is responsible for nearly all tin production, principally on Bangka and neighboring islands near the Malay Peninsula. Another state enterprise, P.T. Aneka Tambang, is active in the mining of nickel and a variety of metals, and is the only enterprise now active in mining bauxite. Still another state enterprise, P.T. Batubara, concentrates on coal production, largely on the island of Sumatra.

Since the establishment of a "New Order" administration in 1966, the government has actively sought foreign investment in the minerals sector. The first major contract was signed in 1967 with Freeport Gulf Sulphur, for exploitation of a large, high-grade copper deposit in Irian Jaya (Indonesian New Guinea). Other foreign mining companies that subsequently signed "contracts of work" include Pacific Nickel, Kennecott, Alcoa and INCO. All "second-generation" contracts (i.e., those signed after the Freeport agreement) are basically similar as regards terms of land and tenure, stipulated land rents, taxes and royalties. They are not governed by a "mining code" as such. However, agreements signed in 1976 and beyond will be of a new "third generation" variety, which is expected to bring the income tax treatment of mining companies closer to that applicable to firms in the industrial sector of the economy, and which will likely involve important changes in output-related taxes on mining.

Exploration and Exploitation. The "second generation" agreements spell out specific contract areas within which the contracting firm may make a "general survey" over a two-year period.[8] By the end of the first year 25 percent of the area must be relinquished, and (with some well defined exceptions) by the end of the second year, 50 percent.

The exploration period lasts for a further 36 months beyond the general survey period. At the end of that interval the contract area must be reduced to 25 percent of the original grant, except in certain cases where the company can document particularly heavy and continuing exploration outlays, in which case the exploration period may be extended.

After completion of detailed exploration, a further 12 months are allowed for feasibility studies. A construction period of unspecified length follows, after which the firm may commence mining operations.

Taxation

• Land Rents. These are geared to the phases of the contract, noted above. During the general survey period land rent is U.S. $0.005 per hectare per annum, for all land remaining under the firm's control. From the exploration through the construction periods, land rent is U.S. $0.10 per hectare per annum. During the operating period, land rent rises to U.S. $2.00 per hectare per annum, except for lateritic deposits (bauxite and nickel) and other extensive surface deposits, for which land rent is $1.00 per hectare per annum.

Output Taxes, Private Sector Firms

Indonesian royalties for significant minerals are given in Table 5-1. These royalties have remained unchanged since 1967-68. Three features of the royalty system should be especially noted:

1. Some royalties are expressed in terms of "per kilogram element in mineral or ore." That is, they are expressed in terms of the metallic content in each kilo of concentrates or metal exported.
2. Some royalties, especially for non-metallic minerals but also for tin, bauxite, etc., are expressed in terms of "per metric ton of washed and unwashed mineral." For example, the royalty for tin (when the international price is U.S. $3,000/metric ton) is

Table 5-1. Indonesian Royalties, Selected Minerals (as of January, 1976)

Mineral	*(1) Royalty per kg. element in mineral or ore U.S. $*	*(2) Royalty per ton of washed & unwashed mineral U.S. $[a]*	*(3) International price of metal per metric ton U.S. $[b]*
Bauxite	–	0.25	500
Copper	0.025	–	700
Lead	0.015	–	300
Zinc	0.010	–	300
Tin	–	210.00/ton	3,000
Manganese	–	0.35	–
Gold	11.0	–	1,100,000
Silver	0.25	–	25,000
Sulphur	0.002	–	–
Asbestos	0.50	–	–
Magnesite	0.50	–	–
All sorts of clay	0.05	–	–
Iron in iron ore	–	0.25	80
Iron in complex ore	0.0005	–	–

[a]At metals prices as given in column (3).

[b]When world market prices differ from prices as given in column (3), the royalties are increased or decreased in direct proportion to the difference.

$210.00 per metric ton. Since the metallic content of all concentrates is presumed to be 70 percent, the $210.00 per ton royalty is equivalent to an ad valorem royalty of 10 percent. (The 70 percent presumptive provision is intended to penalize export of low-grade concentrates or low-grade ores, and to provide incentives for further beneficiation of ore.)

3. The royalties as indicated in the table are those applicable at an international *metal* price as given in column (3). When actual world prices exceed column (3) prices, the royalty is increased in direct proportion to the excess of actual *metal* prices over column (3) prices.[h] By the same token, when actual world prices are *less* than column (3) prices, the royalty is reduced accordingly. In effect, then, the royalty system is of an ad valorem nature, for those *metals* having "international prices."

[h]Example: The tin royalty (applicable only to private firms) is expressed at U.S. $210.00 per metric ton of mineral, when the international price is U.S. $3,000 per ton. If the prevailing tin price is instead $6,000 per ton, the royalty is $420.00 per metric ton.

Corporation Taxes, Private Sector Firms

In general, *private* firms operating under "second-generation" contracts pay corporate tax rates that differ according to the minerals won, as indicated below.

Minerals	Tax Rate on Taxable Income	
	First 10 yrs. of operation	*After 10 yrs.*
Category (i)		
Copper. lead, zinc, iron titanium, asbestos, sulphur, etc.	35%	42%
Category (ii)		
Nickel, cobalt, bauxite	37.5%	45%
Category (iii)		
Tin	40%	48%

Also, a *tax credit* is allowed to the companies once their total investment surpasses U.S. $75 million. The credit is eight percent of the investment by the company, subject to certain limits.

Taxes and Royalties on Public Sector Firms (State Enterprises)

These enterprises are subject to much the same tax regimen as the private firms, with two significant exceptions: in practice, state enterprises pay no royalties; however, they do pay a 10 percent export tax that is not paid by private firms (Table 5-2). Further, state enterprises apparently do not pay land rents.

The 10 percent export tax is levied on the f.o.b. value of the metallic content of all "unprocessed" minerals exported by state enterprises. (Metallic content is, in Indonesia, presumed to be 70 percent of the volume of tin concentrates exported.) However, the definition of what is and what is not "processed" is somewhat vague. In 1974, 45 percent of Indonesian tin was exported as concentrates. Such concentrates are not deemed to be processed, and therefore are subject to the 10 percent export tax. (It should be noted that by mid-1976 all of Indonesia's tin exports will be exported as tin metal.)

Further, state enterprises are subjected to the full 45 percent rate of company tax (*pajak perseroan*), while private firms have, as noted, received rate reductions for the first 5-10 years of production. In addition, in 1973 and 1974 state enterprises paid (for the first time) a "dividend" to the government equal to 25 percent of after-tax profits.

Table 5-2. Output and/or Export Taxes on Tin as Percent of Export Value: Bolivia, Indonesia, Malaysia, and Thailand (concentrates with 35 percent or more of metal content)

L.M.E. cash price per fine pound[a]	*Ad valorem equivalents of output and export taxes in each country*			
	Bolivia[c]	*Malaysia*[d]	*Thailand*[e]	*Indonesia*[f]
$3.08[b]	27.42%	16.84%	21.8%	10%
$4.00	31.72%	16.88%	22.6%	10%

[a]Taxes in Malaysia and Thailand are levied on the Penang Price *ex-smelter.* In 1974 the average Penang price was 95.9 percent of the L.M.E. cash price per fine pound of tin. (Source: *Metals Week*, various 1974 issues.)

[b]$3.08 was the average price per fine pound in the third week of August 1975.

[c]Export tax plus regalia (Bolivia), plus 0.42 percent for three small earmarked taxes: 0.6 percent for universities; 0.18 percent for departments; 0.18 percent for regional development taxes.

[d]Export tax plus export duty surcharge (Malaysia), as a percent of Penang metal value. The typical Malaysian concentrate is 75 percent tin metal.

[e]Production royalty (does not include 4 percent business tax on gross receipts), as a percent of Penang price.

[f]Indonesian export tax is 10 percent when concentrates are exported by state enterprises. State enterprises do not pay the 10 percent export royalty if they have paid the export tax. In 1974, the state enterprise P.N. Timah accounted for about 94 percent of total production of tin-in-concentrates and 100 percent of tin metal exports.

South American Countries

Mineral Exactions in Chile. In 1973 and 1974 the mineral industries of Chile contributed over 11 percent of the nation's Gross Domestic Product and about 86 to 89 percent of the value of its exports. By far the most important mineral produced is copper.

Chilean legislation distinguishes among three categories of mine, *gran minería* (producing 75,000 metric tons or more of copper annually), *mediana minería* (mines whose capital stock is 70 or more times the minimum annual wage payment to mine workers), and *pequeña minería* (the residual of small mines). The *gran minería* sector, which accounts for about 80 percent of total Chilean copper production, was partially nationalized between 1964 and 1969, and completely nationalized on July 10, 1971. In 1966-70, the government's share of income (taxes and equity return) from the large mining sector contributed 15.1 percent of total government revenue.

The present discussion is mainly restricted to copper mining and to the period of partial nationalization, before 1971.[i]

[i]The five large copper mines nationalized in 1971 (names of original owners in brackets) were: Chuquicamata (Anaconda), El Teniente (Kennecott), El Salvador (Anaconda), Andina (Cerro), and Exotica (Anaconda). In the case of

Exploration and Exploitation. The present situation is unclear, although it is known that a number of companies from the United States, Germany, Japan and Canada have been actively reviewing exploration projects in Chile, following overthrow of the Allendé Government on September 11, 1973, and the subsequent financial settlements made with Anaconda and Kennecott. In 1974 the only exploration done was by two government mining organizations, CODELCO and ENAMI.

Under the Mining Code of Chile the state owns all minerals, whether or not it owns the surface. As in Bolivia, deposits of radioactive materials may be exploited only by the National Commission on Nuclear Energy or by state enterprise. The mining code permits unrestricted prospecting for all other minerals, with the proviso that permission of the occupant is required where applicable.

Exploration concessions are granted for a maximum term of two years. Active exploration must commence within six months, or the concession may be forfeited. After discovery of minerals, mineralized portions of the exploration concession may be converted to an exploitation or mining concession.[9]

Taxation

Output Taxes. The only output tax is a gross royalty charge of two percent on the sales proceeds of small mining enterprises. This tax substitutes for all other forms of taxation on income originating in the small mining sector. There are no export taxes on output in Chile.

Income Taxes. The income tax treatment of mining enterprises varies according to type of firm, with the large firms bearing heavier tax responsibilities. In general, the medium mining sector is subject to the same income tax as other enterprises operating in Chile. Such firms are subject to a "first category" income tax that is 35 percent on domestic corporations, and 17 percent on all other commercial and industrial enterprises. All current costs of production are deductible, as is depreciation, calculated on the straight-line method (5 to 30 percent per annum depending on the category). In addition, for medium mining firms with foreign equity participation, a 40 percent rate is applied to dividends remitted to nonresidents and to the whole income of branches of nonresident companies.[j]

---

iron ore, which is of growing importance in Chile, the major producer, Bethlehem Chile Iron Mines Company, was purchased by the Chilean Government on March 31, 1971. For a thorough discussion of the role of copper in the Chilean economy, see Ricardo French-Davies and Ernesto Tironi (eds.), *El Cobre en el Desarollo Nacional* (Santiago: Ediciones Nueva Universidad), August, 1974.

[j]Thus a domestic corporation that remits its entire after-tax profit to nonresident owners would be taxed at 61 percent total rate. If its entire after-tax profit were reinvested in Chile, its rate would be 35 percent. The branch of a foreign corporation would be taxed at 57 percent in any event.

Special tax provisions also apply to medium mines in addition to the non-income-tax benefits discussed below. They may negotiate accelerated depreciation and may be granted guarantees of "no change in taxes" for periods of ten (or more) years, where the output of a new project is exported.

For the firms classified as *gran minería*, taxation takes several forms: government participation in ownership, income taxes on the private return, dividend taxes, an "overprice" scheme, and a reinvestment requirement. The first two are typically negotiated on an ad hoc basis, and vary from mine to mine.

In what follows, more attention is paid to the former Kennecott El Teniente mine, on which more information is available.

Government Participation. In the case of the El Teniente operations, the government accepted in 1965 Kennecott's offer to sell a 51 percent ownership interest, on condition that Kennecott's tax rates be renegotiated. Proceeds of the sale were used to expand the mine.[10]

Two new mines, Exotica and Andina, were initiated in 1964 as joint ventures, with 25 percent ownership to the Chilean Government and 75 percent to Anaconda and Cerro Corporation, respectively—with special tax treatment. Cerro's interest in Andina was later reduced to 70 percent. For Chuquicamata and El Salvador, a 51 percent interest was transferred to the Chilean Government as of January 1, 1970, with provision for the government to purchase Anaconda's remaining 49 percent interest by the end of 1981.

Income Taxes, Dividend Taxes, and the Reinvestment Requirement. El Teniente paid a basic profits tax of 20 percent. Andina and Exotica paid 15 percent. Chuquicamata and El Salvador paid rates "slightly higher" than El Teniente.[11] Permitted depreciation rates also varied. El Teniente was given a 15-year amortization period, Andina and Exotica a 10-year period. In addition, dividends remitted by the large mines to their foreign owners bore a tax of 30 percent. Further, all of the companies were required to reinvest 4 percent of their profit share in Chile.

The "Overprice" Scheme. In June 1969, prior agreements were revised to give the government a progressively higher share of profits as copper prices rose. The government was to receive 54 percent of each United States cent per pound when the price of copper exceeded 40 cents per pound. Similarly, it was to receive 62 percent of the excess over 50 cents per pound and 70 percent of the excess over 70 cents per pound. The overprice scheme has some of the earmarks of a progressive royalty. In terms of economic incentives, the scheme is not related to true profits, but to world prices.

However, the requirement could also be viewed as a change in the division of profit of the joint venture. For example, when the price of copper was 40 U.S. cents, the government's share in the profit of El Teniente was 51 percent. But at a price of 50 cents, the government's share would be 51.6 percent of the profit.

Other Taxes

Mining firms are also required to pay a variety of other taxes and levies, including those earmarked for housing and for social security.

Large and medium mines (like other enterprises) are required to pay a housing tax of 7 percent of profit, which amount is deductible in computing income subject to the income tax. Alternatively, 5 percent of profit may be invested directly in employee housing.

Payroll Taxes

Social Security and related levies total about 44 percent of remuneration to wage employees and can reach 50 percent for salaries employees. Contributions for the social security fund alone are 40.4 percent in the first instance and 44.4 percent in the second. These figures are thus somewhat higher than the 28.3 percent social security levy paid in early 1975 on remunerations by medium mining firms in Bolivia (apart from the 3.5 percent paid by the employee).

In addition, there are contributions for the Labor Accidents and Occupational Disease Fund (1 to 4 percent of remuneration), educational funds and a small labor service tax, levied at rates of 0.75 percent and 0.46 percent of wages and salaries.

Tax Incentives

The only significant tax incentive that appears available to mining firms is the privilege of importing investment goods duty-free, provided such goods are used in export production.

Final Comments on Chile. Three classes of mine exist in Chile. The large mines were subject to individual arrangements involving government participation, income taxes and dividend taxes. These arrangements ended with the nationalization of the large mines in 1971. Medium mines are not taxed in a manner different from the taxation of other commercial or industrial enterprises. They must pay income taxes, dividend taxes and a host of social security taxes. Small mines contribute only a single 2 percent gross royalty.

Mineral Exactions in Peru. The major mining products of Peru are copper (of which 210,000 tons were produced in 1973) and iron ore (of which 9.5 million tons in the form of ore, concentrates or pellets were produced in 1973). Other metallic minerals, chiefly silver, zinc, lead and tungsten, are mostly byproducts. Peru's mining laws are of special interest due to six features. First, the royalty that is levied on exported output is really a prepayment of profit-tax liability. Second, the Peruvian approach to mine profit taxation and metal marketing contains compelling incentives for government participation, with guarantees (financial and marketing) substituting for an explicit governmental contribution to equity. Third, the Peruvian tax on mine profits is partially progressive in structure, with the added feature that marginal rates vary directly with the ratio of cash flow to invested capital. Fourth, Peruvian mining land rents and area rents are relatively high in comparison to those in such nations as Bolivia and Chile. Fifth, the very smallest mining enterprises are exempt from both royalty and profit taxes.

The sixth distinctive feature of the 1970 Peruvian mining code is a guarantee, under the relevant normative law, that no new taxes will be added to those in effect on April 14, 1970. The *rates* of existing taxes may, however, be changed. Further, contracts drawn up for new mines or mine expansions may now include protection against rate changes until such time as the original investment has been recovered (twice that time for mixed, i.e., private/state, enterprises). There is, however, no guarantee against nationalization. The mines of the Cerro de Pasco Corporation were nationalized, with effect from January 1, 1974, with payment subject to negotiation. Also, in July 1975, the Peruvian Government announced the nationalization, without compensation, of Marcona Mining Company, the country's major iron ore producer.

A government-owned corporation, Minero Peru, generally holds the government's participation interests.[12] A provision of the 1970 code is that all new copper refineries must be owned by the government company (the first such refinery should come on stream in 1975), although existing refineries were not affected. More significantly, the state company took over all marketing activity. This means, for example, that if (as is generally the case) a mine development financing scheme requires long-term delivery commitments, such commitments require a government guarantee, which the government will only undertake in return for an equity position.

Exploration and Exploitation. Minerals belong to the state, except as removal rights are granted. Exploration concessions of up to 1,000 hectares are valid for five years. An annual land rent of about 2.7 *soles* per hectare is levied in advance on each exploration concession.[k] In addition, firms must pay a contribution for "work conditions" that rise from 200 *soles* per hectare in the first year to 500 *soles* per hectare in the fifth year for large claims (small claims pay one-tenth of these amounts).

At present, the most active explorer is the government-owned Minero Peru. However, a few foreign companies, especially Mitsui Mining, have also been active recently. But private sector exploration, whether foreign or domestic, has in general declined substantially in the past few years, partly due to political conditions and partly to the stringency of the "work schedule" requirements.

Exploitation concessions do not require prior exploration concessions. Larger concessionaires pay a one-time fee of 220 *soles* per claim and 3 *soles* per hectare plus an annual land tax per hectare: 7.5 *soles* for gold or coal, 9.0 *soles* for metallic concessions, and 30 *soles* for nonmetallic concessions. (Smaller concessionaires pay exploitation fees about one-third as high.) The law also requires that if mining concessions are not developed expeditiously, according to approved "work schedules," they revert to the state without compensation.

Taxation

Area Rental. In addition to the land rent on exploitation concessions, mine operators in Peru pay annual area rental of 250 *soles* per hectare (about U.S. $6.00) on the exploitation area.

Output Taxes. A proportional, ad valorem gross sales tax, levied at the export stage, is imposed on the f.o.b. value of all mineral exports in excess of 800,000 *soles* per annum (approximately U.S. $18,600 at mid-1975 exchange rates). Rates applied are 4 percent on metals and 2 percent on nonmetallics. This is really not a distinct tax but a prepayment of profits tax, because export taxes paid are creditable against profit tax liability.

Taxes on Income. Prior to 1970 mining profits were subject to special taxation, at progressive rates. These rates were differentiated between domestic-owned and foreign-owned firms. The lowest marginal rate was 20 percent, the highest 68.5 percent. Incentive benefits included 100 percent depreciation of new investment.

Taxes on mining profits were substantially altered in 1970, when a new mining code was issued. The basic rate structure and preference to domestic ownership (and state participation) were retained. Major

[k]Translations into U.S. currency, in this section, are based on the exchange value of the Peruvian *sol* in mid-1975, approximately: 1 *sol* = U.S. $0.023. In May and June of 1976, Peru made two major devaluations, which were followed by a sequence of minor devaluations. As of November 18, 1976, the official exchange rate was 1 *sol* = $0.0148.

innovations were guarantees of "no change in taxes" and "recovery of investment," linking the tax rate to cash flow return, and changes in write-off and tax credit provisions.

The tax base is now defined as actual revenue less actual cost. Costs include depreciation on plant and equipment at 20 percent per year and 10 percent on other capitalized mining assets. Exploration expenditure may either be expensed against mining income or amortized over the life of a mine.

For profits of up to 100 million *soles* per year, tax rates were unchanged from previous law. The rate structure is progressive in four steps, rising from 20 percent to 40 percent up to 100 million *soles*, so that the average rate on 100 million *soles* (about U.S. $2.3 million) is 37.465 percent.

On profits above 100 million *soles*, there is a single proportional rate. However, that rate will vary, year by year and company by company, depending on the ratio of cash flow (net sales less direct production costs, excluding interest and depreciation) to book value of fixed assets (total investment, written down five percent yearly). There are 23 step-wise alternative ratios that are applicable, ranging from a cash flow return of 10 percent and under to 100 percent and over. For each such ratio, there is a corresponding applicable incremental tax rate ranging from 29 percent at the lowest cash flow return to 55 percent at the highest.[13]

There is an additional tax on domestically owned mining corporations. For these firms, a 25 percent tax is levied on the amount of dividends paid, but this amount may be credited by *local* shareholders against their personal income tax liability.

An additional tax, amounting to 30 percent of profits (net of the basic tax) must also be paid by mining firms wholly owned by foreigners. Hence for foreign-owned mines the tax rate is $0.3 + 0.7b$, where (b) is the applicable basic rate. Specifically, for foreign firms, the rate on the first 100 million *soles* of profits becomes about 56.2 percent and the rate on profits over that amount varies from 50.3 percent to 68.5 percent, depending on the ratio of cash flow to asset value. Thus the maximum rate is the same as under the prior law, but instead of being reached in progressive stages, that rate applies to the whole of the increment of profit over U.S. $2.3 million, but only when cash flow is as great as asset value.

There is a reduction of the basic tax rate for mixed enterprises. When the Peruvian state has at least 25 percent participation in a foreign-controlled mine, the government may grant (subject to negotiation) a remission of the basic tax of up to one-third. This reduced basic rate is then subject to a 30 percent additional tax.

Thus, where (b) is the basic rate the minimum tax rate payable by a foreign-controlled firm could (under these circumstances) be as low as $0.3 + 0.467b$. It follows that for such a firm, the rate on the first 100 million *soles* of profits is about 47.5 percent to 55.7 percent, depending on the ratio of cash flow to asset value.[1]

Finally, Peruvian law allows a reinvestment credit. Up to 40 percent of pre-tax profits, if reinvested in the mine (or in another mining venture) is exempt from profit tax. The limit on this exemption is 300 million *soles* (approximately U.S. $7 million) per annum, and projects must receive government approval.

Final Comment on Peru. Permutation of the Peruvian mine profit tax structure is a mathematician's delight. The system is geared to the assumption that most mining investment will be large scale and of foreign source. It is interesting to note that the higher the "domestic rate," the smaller the relative penalty imposed on foreign-owned firms.

The Peruvian system does appear to accomplish a number of the purposes of mineral tax policy. Although no numbers are available, the system does seem suited to obtaining a large revenue share for the state without adverse effect on the marginal cost structure of firms; it also actively encourages state participation in ownership. Its most novel feature is that tax rates escalate with rising metal prices, thus securing a large measure of windfall gains for the state, in addition to a portion of natural resource rents. The method is also sensitive to downward movements of metal prices.

The Peruvian tax system, however, requires a considerable measure of sophistication and accurate record-keeping on the part of the taxpayer. It would be difficult to adapt to an environment of small-scale mining and low tax compliance. However, it should be noted that the creditable export tax probably suffices to get some contribution from the small-scale end of the industry (although not from the very smallest mining firms, which are exempt).

Mineral Exactions in Colombia. The minerals sector in Colombia is relatively small compared to those of Bolivia, Peru, Chile and the other nations surveyed in detail in this chapter. Even with petroleum included, the mining sector has accounted for only about two

---

[1]It can be shown through algebraic manipulations that if the average basic tax is greater than 50 percent, the foreign firm would be better off, so far as its net retention goes, to hand over a 25 percent interest to the government, for no consideration, *provided* that it then obtained the full one-third remission of basic tax. It would be in that position if its taxable profit exceeded U.S. $14.5 million *and* if its ratio of cash flow to national asset value exceeded 90 percent.

percent of GDP in recent years. Exclusive of petroleum, the sector accounts for less than one percent of GDP. Until 1975, mining activity other than petroleum has been traditionally heavily concentrated in the production and export of precious metals and emeralds, although the country has produced iron ore for a number of years and, with the Hanna agreement of 1970, expects to export substantial volumes of nickel in 1976-1980. Registered export values for the sector reached about U.S. $42 million in 1972 (about five percent of total recorded exports in that year).

The mining sector in Colombia has been subject to the generally applicable income tax for a number of years, along with departmental (subnational) royalties and small stamp taxes levied by the central government. Prior to the general tax reform of October 1974, the sector enjoyed fairly liberal allowances for depletion under the income tax.[14] Before October 1974, mining enterprises were entitled to take advantage of four separate types of depletion-related deductions, two of which were available only in cases where legally established reinvestment requirements were met. Depletion as defined under these four concepts took the form of percentage depletion, but the taxpayer could also opt for cost depletion instead.[m] Income of all corporations, including corporations in the mining sector, was taxed at a rate ranging from 12 percent to 36 percent, plus several surcharges (for excess profits, housing, etc.) which could result in tax rates as high as 72 percent on taxable income. However, few mining enterprises ever paid tax rates that approached this level, given the benefits of the percentage depletion allowances available to all firms prior to October of 1974.

The tax reform of 1974 provided for a 40 percent flat rate of income tax on business income of all corporations, with rates half as high for all other forms of business organizations (e.g., limited-ability companies and partnerships). (However, for tax purposes, after-tax income of limited-liability companies and partnerships is imputed to the owners and must be declared on their *personal* income tax returns.) As far as the mining sector is concerned, the other major change enacted was the abolition of the right to percentage depletion for all enterprises except for taxpayers having either (a) concessions

[m]Under cost depletion the original cost of a nondepreciable asset such as a mineral deposit is divided by the number of tons of mineral to be produced, and this amount is deducted as a cost of each ton mined, up to the point where all such costs are amortized. Under percentage depletion, the depletion deduction is expressed as a specified fraction of the gross value of production, and typically may continue to be taken long after the original cost of the deposit is fully amortized. For example, statutory depletion rates for certain minerals in the United States are: lead, zinc and bauxite (22 percent); gold, silver, copper (15 percent); and coal (ten percent).

in force before October 28, 1974 (the effective date of the new law); (b) mining contracts for which the mineral subsoil has been recognized as private property. Since the overwhelming proportion of mineral extraction activities is carried out on lands belonging to the national government, the effect of the reform was, for all practical purposes, to deny henceforth availability of percentage depletion to virtually all undertakings begun after October 1974.

For firms engaged in mining activities on government-owned land, and for which contracts were not in force prior to October 1974, cost depletion is required, over periods set by the Ministry of Mines and Energy for each type of mineral. That is, tax-free recovery of investment in depletable mineral assets is allowed only up to the extent of costs incurred in acquiring nondepreciable mining assets. Such outlays may not be expensed, but must be amortized over a period set by the Ministry of Mines. Thus, the depletion deduction available to the mining sector is almost identical to that provided for all amortizable assets (regardless of the economic sector where they are used) under the basic income tax law now in force.

Finally, for any one firm, all exploration expenses may be amortized against mining income from other exploitations in Colombia at an annual rate of not more than 10 percent, up to the initiation of exploitation activity. Once production begins, the unamortized balance of exploration costs may also be included in the base for depletion of affiliates or subsidiaries operating in Colombia.

Final Comments on Colombia. In summary, insofar as the mining sector was concerned, the Colombian tax reform in 1974 resulted in a substantial tightening, on balance, of tax benefits available to the mining sector, relative to other productive sectors of the economy. The mining sector exclusive of petroleum has been subject to generally applicable income taxes for over two decades, and the 1974 reform, with its virtual abandonment of the system of percentage depletion, served to bring the tax treatment of the mining sector closely in line with nonmining activity.

Developed Countries

Recent Developments in Canadian Mine Taxation. The purpose of this section is to indicate the nature of the very substantial changes recently enacted in the Canadian taxation of mining revenues. The pace of recent change has been truly staggering. By way of example, the federal government presented three budgets in the 14-month period from May 1974 to June 1975, and each of them involved sharp changes in the method of taxing mine (and oil and gas) profits. Partly as cause and partly as effect of the federal moves, provincial mineral taxes have been altered even more drastically. In this connection it is interesting to note that while a strong "demonstration effect" is evident among the provinces, the period 1973-76 has also witnessed an unprecedented divergence of mineral tax practice among the provinces. One by one, it seems, they have discovered the concept of "mineral rent," but each has devised its own formula for rent extraction.

This section does not attempt to trace all of the recent peregrinations in Canadian mine taxation. Rather, it attempts to sketch the "latest" federal and provincial positions, placing them in the context of pre-1972 taxation.

In comparing Canadian mineral tax experience to that of other countries, one factor must constantly be borne in mind: the dominance of the federal-province issue. This dominance applies to all public finance matters, but nowhere more so than with respect to mineral levies. The provinces generally play a more independent role in the total public sector than do, for example, the states of the United States. With regard to minerals, the Canadian constitution vested ownership of the public lands in the provincial governments, and special provincial levies on mineral output predate the federal income tax. Whether levied in the form of royalties, mine profit or acreage taxes, provincial levies have been regarded as fees paid for the privilege of private exploitation of natural resources originally in the public (i.e., provincial) domain. Federal imposts are confined to the income tax which, until recently, bore very lightly indeed on the mineral industry.

The traditional form of mineral levies has been severely jolted in the 1970s. Three major causes may be identified as having contributed to the present "disequilibrium." The first is the decade of debate over tax reform in Canada, which culminated in the enactment of a new federal Income Tax Act with effect from January 1, 1972. The federal reform itself was not cataclysmic so far as minerals were concerned, but the debate did lead to a reexamination of a number of propositions with respect to mineral taxation. Moreover, the mining industry was able to deflect a more onerous reform of its federal corporation tax only by invoking the aid of the provincial governments—thereby confirming the industry as a dependency of the provinces. However, given the provinces' mounting revenue requirements, postponement of heavier taxes on mining was bound to be short-lived.

The second causative factor was the demonstration effect stemming from the OPEC-induced escalation of oil levies in 1973. This was paralleled in Canada by a tripling of oil royalty rates in the province of Alberta between 1972 and 1974. An increase in royalties and taxes on other minerals in Ireland, Jamaica, Australia, Guyana and elsewhere followed quickly.

Finally, the third cause operating in Canada was the election of NDP (socialist) governments in several provinces (British Columbia, Saskatchewan and Manitoba) between 1969 and 1972. The NDP Government of British Columbia introduced new mineral royalties legislation in February 1974 and Ontario raised its mining taxes in April 1974. The federal budget of May 6, 1974, then moved to protect what that level of government regarded as the erosion of its corporate tax base.

As other provinces, one by one, emulated the leaders, an industry that quite recently was the most lightly taxed in the country now faced what it regarded as confiscatory taxation. In late 1975 and early 1976 there were signs that tax pressure on the mining industry had begun to abate. But the tax burden on the industry will remain much higher than in the pre-1972 period.

The Corporation Income Tax on Mining, before 1972. The corporate tax field was (and still is) occupied by both the federal and provincial governments. Two of the provinces, Ontario and Quebec, have their own collection apparatus. The federal government acts as collector on behalf of the other provinces, whose taxing statutes, like those of Ontario and Quebec (in most respects) follow federal practice. To make room for these collateral provincial taxes the federal government abates ten percentage points of a corporation's taxable income earned in a province (i.e., the taxpayer's nominal federal tax liability is reduced by 10 percent of his taxable income). Provincial rates ranged from 10 to 13 percent. The total rate applied, including the provincial rate, was in the 22 percent range on the first $35,000 of income and in the 52 percent range on the excess.[n]

Mining income was subject to the general rules governing the corporate income tax and the personal income tax, but in defining income, this industry was given the benefit of four special kinds of provisions. The first of these permitted a corporation whose principal business was in the nonfuel mineral field to expense currently all costs of Canadian mineral exploration and development, excluding land costs and treatment plants. Shaft sinking and tunneling, after the mine came into production, were also subject to 100 percent

[n]These rates are only approximate. They varied from province to province, and were subject to periodic surtaxes and reductions, for stabilization purposes.

depreciation. Mining structures not subject to the expensing provision were depreciated for tax purposes on a diminishing balance basis—similar to other industry—but at rates that may be considered favorable (30 percent for mining machinery, buildings and equipment).

The second distinctive provision was the outright exemption from income tax of the first three years' profit from a new mine. The exemption applied to income derived in Canada from the proceeds of a mine through the "prime metal stage" (e.g., to the stage of refined nickel or pig iron), so long as the output remained the property of the exempted corporation.

The third group of mining tax aids took the form of three types of depletion allowances, which reduced the base on which tax was calculated. The important one was the operator's allowance (generally a reduction of 33 1/3 percent) applied to net mineral production profits (again to the prime metal stage) after all other deductions. Nonoperators (i.e., royalty holders) received a 25 percent allowance on their gross mineral incomes. Finally, shareholders in mining corporations were given a personal depletion allowance of 10 to 20 percent against mineral-source dividends.

The fourth special provision was that income received by prospectors and their financial backers (grubstakers) from the sale of mineral property was excluded from taxable income.

Provincial Mining Taxes before 1974. As noted, the provinces have imposed taxes or royalties on mineral production since well before the end of the nineteenth century. A constitutional difficulty, however, arises with respect to specific or ad valorem royalties on output. Constitutionally, the provincial taxing power is restricted to direct taxation, and the courts have ruled that a royalty on general mineral production—where privately owned mineral rights may be involved—is an indirect tax. Consequently, the form that mining taxes have taken in the provinces is that of a "profits tax" (which, to confuse the issue, is sometimes termed a "royalty" in the taxing statute).

Although there were significant differences among the provinces, all of their statutes had in common an attempt to apply the tax to the "pit's mouth" value of the ore, less mining costs. The general approach was to start with a mining company's revenue from its total extraction and processing activities and then to deduct costs of processing and marketing, depreciation of processing plant and a "processing allowance" (typically eight percent of original cost of depreciable property used in processing the ore). A common provision was that the processing allowance was to be not less than 15 percent nor more than 65 percent of the profit on the integrated enterprise. Finally, direct costs and depreciation on the mine itself were deductible. Further deductions for exploration and development varied from province to province.

In the more important mining provinces, the statutory rate of tax applied had risen over time to a 15 percent figure, but the effective rate, after processing allowances, was considerably lower. (Newfoundland, however, applied a general 5 percent rate.)

It is to be noted that provincial mining taxes were, before the federal tax reform, recognized as expensible deductions (royalties) for determining income subject to federal (and provincial) income taxes.

The Reformed Federal Income Tax Act of December 1971. Ten years of debate produced some extension and rather more restriction of tax advantages to the mining industry. For corporations whose principal business was in the mineral field, the writeoff of Canadian exploration and development costs against income from any source was retained and, indeed, extended (costs of mining land were added to expensible items). Other taxpayers (i.e., individuals and non-principal-business corporations) could now write off the equivalent expenditures to the full extent of current income from Canadian mineral properties (including royalties), or, alternatively, depreciate them at 20 percent (on a diminishing balance basis) against income from any source. For the first time, foreign exploration and development could be charged to Canadian income (depreciable at 10 percent of the unclaimed balance).

On the other hand, the three-year tax exemption for new mines was terminated effective January 1, 1974. To offset this change, all depreciable assets related to a new mine (including, e.g., a refinery and a town site) could now be written off as fast as income from the mine permitted, and in no case at less than 30 percent of the unclaimed balance. Similar provisions apply to a mine expansion involving an increase in mill capacity by at least 25 percent.

Shareholders' depletion and the prospectors' and grubstakers' allowance ended in 1971. However, gains on mining property sales are now taxed at preferred capital gains rates. Nonoperators' depletion (25 percent) was to end in 1976.

The existing operators' depletion (33 1/3 percent) was to be unchanged until the end of 1976. Thereafter, this allowance was to remain, with the proviso that it would have to be "earned." Depletion was to become, in effect, an incentive to expand mining facilities. Specifically, the form of the old allowance would be retained. But, commencing in 1977, one-third of mineral income after all other eligible deductions would become a taxpayer's *maximum* allowance in a year. The *actual* allowance would depend on his spending on exploration, development, and capital assets related to a new mine, a major expansion, or mineral processing facilities. Each \$3.00 of such expenditure would "earn" the taxpayer \$1.00 in depletion. Expenditures made between November 7, 1969 and December 31, 1976 would build up a bank of earned depletion that might be claimed in 1977 and afterward.

As a concession to the provinces when the new Act became fully effective in 1977, the former deductibility of provincial mining taxes from income subject to the corporation income tax would be replaced by a more generous tax credit. Mining profits (but not oil and gas) would be "abated" by 15 percentage points of the federal corporation tax to "make room" for the mining taxes.

Recent Provincial Tax Developments

Income Tax. Provincial income tax changes have generally followed the foregoing (and subsequent) revisions to the federal statute. One exception is the Province of Ontario, which refused to adopt the earned depletion concept and will continue the automatic 33 1/3 percent depletion allowance (Ontario retained the three-year exemption for new mine income until 1974, but then replaced it with accelerated depreciation). Quebec has retained the prospectors' and grubstakers' exemption.

Other Taxes on Mining

British Columbia. We noted above that in 1974 British Columbia was the first province to depart from the traditional form of provincial mine profit taxes in favor of innovative and far more onerous mineral levies. It is also the latest province to be heard from, closing out the cycle of provincial experiment by a return to something very like the earlier mode of taxation. A December 1975 election, that brought a change of government, was the immediate cause of this policy reversal. But it is likely that even if the ruling party had not changed, substantial modifications would have been made in the 1974 legislation.[o] There is little doubt that the British Columbia mining industry was badly hurt by the 1974 tax measures. Exploration expenditures fell from \$27 million in 1973 to \$22.4 million in 1974 and to a reported \$5.6 million in 1975.[15]

[o]The government that was subsequently defeated had appointed a commission to review mining taxation, in November 1975.

The 1974 legislation had introduced two new mineral royalties, in addition to the existing 15 percent mining profit tax. A basic royalty of 5 percent (2 1/2 percent for 1974 only) applied to the "net value" of "designated minerals" produced from Crown lands. "Net value" was determined as the international metal price less smelting, refining and transportation costs. The applicable rate might be reduced if the mineral were smelted or refined in the province and if there were a decline in the selling price of the mineral.

The more contentious royalty was the second one, the so-called "super-royalty," very similar to a tax measure adopted in the Philippines in 1973. This tax took one-half of the amount by which the "gross value" of thirteen "designated minerals" ("net value" as above, plus transportation costs) exceeded 120 percent of the "basic value" of the mineral.[p] "Basic value" initially depended on an average of prices from 1969 through 1973. In both 1975 and 1976 basic value was escalated to allow for half the percentage increase shown in the Canadian wholesale price index.[q]

The latest legislation has repealed the superroyalty retroactive to April 1, 1976. The basic royalty will also disappear in 1977. Royalty payments made in 1976 will be creditable against a new profit tax, gradually, over a period of years. That new profit tax, taking effect in 1977, will be much like the earlier one (which continues for coal and certain industrial minerals). Henceforth the net income from mining the metallic minerals, asbestos and gypsum will be taxed at a 17 1/2 percent rate. That rate is reducible by a proportional processing allowance of 15 percent (i.e., the net rate of tax can be reduced to 14.875 percent). British Columbia also introduced the possibility of reducing the mining tax through spending on exploration and mine expansion ("earned depletion," following the example of Quebec, for which see below).

Ontario. New legislation in the mining province of Ontario was comparatively more favorable to the mining industry. Ontario enacted a revision of the existing mining profit tax, designed to double the yield in 1974. The existing 15 percent flat rate on mine profit was replaced by a graduated rate schedule, ranging from 15 percent (on company profit between $100,000 and $1 million) to 40 percent (on profit over $40 million). Perhaps more significantly, the proces-

[p]The Act made no provision for deducting the new royalties from the base of the continuing provincial mine profit tax.

[q]Among the 13 designated minerals (for superroyalty purposes), basic values for 1976, applicable to the province's six most important, were: copper, $0.67 per lb.; zinc, $0.14 per lb.; lead, $0.15 per lb.; molybdenum, $2.03 per lb.; asbestos, $311.86 per ton; and iron ore, $14.21 per ton.

sing allowance was converted into an extremely potent incentive to process in Canada, and in northern Ontario in particular. Henceforth, processing facilities outside Canada will be denied the 8 percent allowance (which, as noted above, is a notional deduction in arriving at "pit-head profit").[r] On the other hand, processing within Canada will carry varying rates of allowance that rise with the stage of processing, and are weighted in favor of northern Ontario. The lowest rate is 8 percent (on concentrator facilities anywhere in Canada); the highest is 35 percent (on fabricating facilities, beyond the refiner, if located in northern Ontario).

As a spur to new mining ventures, the Ontario Act provides that, despite its "progressive" rate structure, a new mine, regardless of its ownership, may elect treatment as a separate taxpayer.

Quebec. Quebec followed Ontario's lead in introducing a graduated rate schedule: 15 percent on profit between $150,000 and $3.15 million, rising to 30 percent on profit over $20.15 million. This change was accompanied by increases in depreciation and processing allowances, the latter being confined to Quebec assets. These allowances could be as much as 15 percent of original cost, still subject to the existing 65 percent maximum of total profit. The most interesting Quebec innovation, however, is an extension to the mining tax of the principle of "earned depletion," derived from the federal income tax. An investment allowance of $1.00 will be granted for every $3.00 of exploration and of investment in new processing facilities within the province. The total of such allowances is limited to one-third of mine profit.

Manitoba. Manitoba has devoted more study to the subject of mineral taxes than has any other province. A hastily designed bill, intended to tax mining profits in excess of a "base," at 50 percent, was introduced and withdrawn in mid-1974. Instead, in 1975 the province opted for a milder form of "excess profit" taxation, whose amount depends on the real return realized by producers.

The Manitoba mine profit tax (termed a "royalty") now involves a two-stage calculation. The whole of a mine's profit is taxed at 15 percent. Determination of that "profit" largely follows conventional

[r]As originally written, in defining taxable mineral profit, the revised Ontario Act would have denied the deductibility not only of the eight percent capital allowance on processing facilities outside Canada, but would also have disallowed all non-Canadian direct processing costs. The last restriction would have impacted with particular severity on Falconbridge Nickel Mines Ltd., which processes its Ontario nickel ore in Norway. In December 1975 Ontario retroactively rescinded the denial of foreign expenses for five years; a review is promised at that time.

Canadian practice, except that the tax base now includes processing as well as mining operations.[s]

The innovative feature is that each producer computes an annual "investment base," which consists of the notional current value of its plant and equipment; i.e., the investment base contains a built-in inflation factor applied to original asset cost and to the depreciation which is netted out. The firm's "profit base" is then calculated as 18 percent of its "investment base." To the extent that its "conventional profit" exceeds the "profit base," that excess is taxed at an additional 20 percent.

Saskatchewan. Production from Saskatchewan's fabulously rich potash endowment began only in 1962. However, by 1968 Canada (i.e., Saskatchewan) had attained second place (behind the U.S.S.R.) in world potash output, and first place in exports. The province is estimated to hold more than 40 percent of the world's known reserves of potash.

Industry/government relations began to deteriorate during the 1970-74 period, when market prorationing of potash was imposed.[t] Bad feeling intensified with the imposition, effective July 1, 1974, of a new potash reserves tax, designed to raise about five times the former royalty revenue. In principle, this tax applies to millrate to the value of mine, plant and ore reserves. But implementation follows a complicated formula that makes the levy vary with the price of potash, the capacity of the individual mine, its ore reserves and its invested capital.[16] One of the twelve potash mining firms refused outright to pay the reserves tax. Most other companies began withholding tax payments as the industry mounted a legal challenge to the legislation. The government responded in December 1975 with two bills that grant it power to nationalize the industry at what industry spokesmen call "bargain basement prices."[17]

In Spring 1976 both sides adopted a more conciliatory tone. The province opened discussions with several potash producers with a view to negotiating rather than compelling their takeover. The provincial Premier intimated that he would be satisfied with public ownership of half the province's industry.[18]

[s]The Manitoba tax now extends to processing profit on Manitoba ore treated outside the province. Current exploration may be expensed. Depreciation may be claimed at 10 to 20 percent of historic cost, diminishing balance.

[t]In May 1975 a court ruled the prorationing system unconstitutional and awarded a complainant company $1.5 million in damages. Pending appeal, the government has amended this legislation retroactively. Saskatchewan potash mines still pay a "prorationing fee"—really an output royalty—of $1.20 a ton.

Unlike its approach to the potash industry, the Saskatchewan Government made a concerted effort to accommodate industry opinion when it launched a new plan for taxing its uranium producers, in the spring of 1976.[u] The modified Saskatchewan uranium tax proposal as features reminiscent of both the new Manitoba scheme (see above) and the renegotiated Bougainville Copper contract in Papua New Guinea (see Table 5-3). Uranium producers will pay two taxes. A universal 3 percent output royalty will replace the existing 12.5 percent mine profit tax. A progressive excess profit tax, with marginal rates between 15 and 50 percent, will be applied to operating profit when the ratio of operating profit to capital investment exceeds 15 percent. However, no mine will be liable for the excess profit tax until accumulated operating profit exceeds the amount of invested capital.[v]

Nova Scotia. Under its previous legislation, Nova Scotia imposed profit taxes at low graduated rates (for profit of up to $10 million the rate was 6 percent and rose by one percent for each additional $5 million). In March 1975, the province increased the tax to the greater of 15 percent of profit or 2 percent of the net selling price minus transportation costs. The new Act also empowers the Cabinet to impose up to three times the normal royalty if minerals are shipped out of the province in an unrefined state.

Newfoundland. A provincial Royal Commission report inspired the 1975 increase in Newfoundland's mineral exactions. That Commission's main concern had been a problem peculiar to Newfoundland: much of the province's mineral endowment (mainly iron ore) has been parcelled out to private concessionaires who assign mining rights to operating companies and themselves receive royalty income. Payment of private royalties was an expense to the operators and reduced the liability for provincial mining taxes. The Commission recommended that private royalty payments become subject to the mining tax and be withheld and remitted by the operator. The Commission also advocated that, for future mines, an ad valorem output royalty be negotiated for each mine on an ad hoc basis, before a mining lease was granted. These royalties, when applicable, would replace the profit tax and would be designed to capture all

---

[u]A not unreasonable speculation is that, well endowed though Saskatchewan may be with uranium, relatively the province is in a more monopolistic position with respect to potash potential.

[v]The invested capital "base" is the undepreciated original capital investment in a mine and mill. In two concessions to the industry the province agreed to include prior exploration and property preparation spending, with interest imputed to the time of startup, as well as costs of social overhead capital.

projected returns in excess of a "normal" return on the investment. Another notable concern by the Commission was with respect to the usually ignored negative spillover costs imposed by a mining industry (adverse effect on persons, on infrastructure and an environment). The Report spoke favorably—without explicit recommendation—of a "rehabilitation tax."

The amended provincial mining levy now provides for taxing private royalty revenue at 20 percent. Remission is to be made by the operator and blended with his own liability in such a way that the total profit derived from a mine (gross of private royalties) is taxed at not less than 16 percent.[w] (The previous rate was 5 percent on profit net of private royalties.)

The Latest Federal Position. It would be tedious to trace all the federal budgetary twists with respect to mineral income. In brief, the federal government's response to provincial initiatives was to hasten the demise of unearned depletion and of the deductibility of provincial mining taxes (extended now to oil and gas royalties, as well) from the income tax base. No mineral levies by a province in whatever form, including compulsory provincial marketing boards, have been permitted as deductions since 1974. Also in 1974, "earned" depletion replaced "automatic" depletion three years ahead of schedule. Moreover, the maximum that may be "earned" was reduced to 25 percent of production profits rather than 33 1/3 percent.

Further, development expenditures may no longer be expensed currently. Rather, *development* costs are now recoverable against taxable income at a 30 percent rate. *Exploration* expenditures remain subject to current writeoffs. Curiously, in the light of the harsher treatment of development expenditures, depreciable mining (but not petroleum) assets are still recoverable as fast as income from the mine permits.

Beginning in 1976, the general rate of federal corporate income tax will be 46 percent.[x] Given the 10 percent "abatement" in favor of the provinces, federal and provincial corporation tax rates will range from 46 to 49 percent. In 1976 (but not 1974 or 1975, when

[w]The statute sets the rate on the operator's profit net of private royalty payments at 15 percent and the rate on royalties at 20 percent. But it also presumes that private royalties are not less than 20 percent of total mine profit. Hence the total rate on profit gross of royalties is 16 percent when actual royalties are 20 percent or less of profit.

[x]The favored position formerly occupied by mining in tax legislation seems to have been assumed by manufacturing, where for explicit manufacturing activities (rather narrowly defined) profit is taxed at a 40 percent federal rate.

the rate for mineral profit was lower), this general rate will apply to mineral producers.

However, beginning in 1976 there will be a new "resource allowance" in arriving at taxable income "in recognition of provincial resource levies." When calculating taxable income, the firm must first deduct operating expenses and depreciation from revenue. This intermediate amount, called "resource production profits," is then reduced by 25 percent (the resource allowance). From the residue the firm then deducts interest and permitted exploration and development charges, as well as earned depletion. The net difference is the firm's taxable profit. It may be noted that under this system, a dollar of incremental exploration (or interest)—even apart from "earning" depletion—carries a higher "tax credit" rate than does a dollar of operating expense. The latest overhaul of British Columbian taxes disallows the deduction of a resource allowance in computing the provincial portion of the jointly collected corporation income tax.

As noted at the start, this latest federal position may not be "final." The entire area of mineral exactions in Canada is still in a highly fluid state. The Canadian mineral industry itself, after a long history of relative calm on the tax front, is highly agitated, to say the least.

Mineral Exactions in Eire (Ireland). The case of Ireland is unique, because although it is a very "old" country, with a long history of mineral production in the nonmetallics (cement, coal, gypsum), it is a very "new" country in terms of important production of metallic minerals. Value of metallic minerals produced has risen from virtually zero in the early 1960s to about $75 million in 1973. In this growth very generous tax inducements played no small part. The chief metallic minerals now produced are zinc, copper, and lead (in concentrates).

However, the recent history of Irish mineral exactions is not unique, but parallels that of the less developed countries. Starting with extremely generous tax legislation passed in the early 1960s, designed to induce the inflow of foreign capital, and the subsequent discovery of several base metal deposits, Irish tax legislation is now in a state of transition to high taxes on mining and government equity participation.

The initial tax inducement offered around 1960 was a four-year exemption from income tax, and a four-year period of half-taxation for the output from nonbedded mineral deposits. Following the 1961 discovery of the first major base metal deposit by Northgate

Exploration, this exemption was extended to 20 years. A mining consulting firm has described this legislation as "not a good move because it provided such a gap between the tax situation in Ireland and elsewhere that it invited strong criticism from opposition parties and the more nationalistic groups among the Irish public."[19]

In the event, following a change in government in March 1973, the new administration announced its rejection of the existing agreements, and a review of mining tax legislation. By late 1975, comprehensive new legislation had not yet been enacted. In the meantime, a major discovery of lead-zinc ore was made at Navan, Ireland, by Tara Exploration Co., Ltd. (31 percent owned by Cominco Ltd., and associates, and 20 percent by Noranda Mines Ltd.). The government refused to allow Tara to proceed to production under the existing legislation. After protracted negotiation, an agreement was worked out in February 1975, in effect giving the Irish Government about 66 percent of future profits. It is expected that production will reach, by spring 1977, the target rate of 7,500 tons a day, more than any other zinc producer in Europe.

The following description of taxation describes separately the existing general mine tax legislation, and the agreement with Tara. All the taxes discussed below are based on income. It is worth emphasizing that general mine-tax legislation may be changed within the course of 1976.

Exploration and Exploitation. Mining rights to about 2/3 of the republic are government-owned. Prospecting licenses are granted for a two-year period, renewable for a further two years. When minerals are found, leases may be granted subject to the payment of a dead rent which merges in royalties on minerals produced. In advance of production, the Minister fixes royalties on a mine-by-mine basis.

Taxation (excepting Tara)

"Royalties" on Metallic Mining. "Royalties" are based on actual revenue minus actual cost, excluding any allowance for depreciation or depletion (i.e., on operating profit). Rates are progressive over time, negotiated separately with each mine, but with the highest level at 10 percent.

Corporate Income Tax. In general, the rate levied on profit is 50 percent. Deduction of mine development costs is allowed on an accelerated basis. Depreciation may also be claimed by a mine, quarry or smelter, on the basis of the expected life of the extracting unit. However, special exemptions are available for producers of nonbedded deposits, such as copper, lead, zinc, pyrite, barite, gold and silver. Under existing legislation, such firms are exempted from corporate income tax (or personal income tax, if unincorporated) for a 20-year period, if production starts before 1986.

A special regimen exists for the Tara mine. The normal 50 percent corporate profit tax will apply to Tara, but there will also be an additional 4.5 percent "royalty" which will apply, as a flat tax on profit. Apparently, this "royalty" will be applied to the same base as the corporate profit tax. Finally, the Irish Government receives a 25 percent equity interest in Tara, free of charge.

Other Countries

Basic information on mineral taxation for the eight countries discussed in previous sections, plus nine other nations, is provided below in tabular fashion (Table 5-3). Aside from the eight countries treated in the text of this chapter, the table includes details of mining taxation for Gabon, Jamaica, Liberia, Mexico, New Caledonia, Papua New Guinea, the Philippines, Zaire and Zambia.

I. *BOLIVIA*

Share of Mining in GDP (1973): 12.3%, excluding hydrocarbons.

Share of Mining in Exports (1973): 67.6%, excluding hydrocarbons.

Major Minerals Produced: Tin, zinc, silver, tungsten, copper, bismuth, antimony.

(1) *Royalty, Production or Export Taxes*	(2) *Taxes on Income*	(3) *Other Taxes*	(4) *Government Participation and Other Comments*
A. *Salient Features* 1. *Royalty on exports:* Based on presumptive market price and (fixed) presumptive cost. 50% reduction for any mineral production in excess of average over previous 3 years. 2. *Export tax:* Based on gross value of product at time of export, and grade of mineral, after deduction of royalty. 50% reduction for excess production also applies. B. *Rates* 1. *Specific rates*, on progressive scale, by mineral, over designated price range. Proportional rates on presumptive net income over that range. For tin from privately owned mines with output over 4.5 tons per month, rates range from 2.5¢ US per fine lb. (at official price of $US 1.60) to 53.2¢ at official price of $US 2.90. At prices above that, rate is 38% of difference between presumptive price and $US 1.50. Publicly owned mines and small mines slightly lower. Other minerals have similar structure, but rates are lower. 2. *Proportional rates*, depending on type of mineral, purity of exports, and category of exporter. For publicly owned mines and larger private mines, rates range from 2% on zinc to 6.5% on low-content tin (under 35% fine metal content) and 7.5% on high-content tin. Small mines pay in the range of 1.5% to 5.5%, and cooperatives in the 1% to 5% range.	None on mining enterprises. But National Smelter and Mining Bank subject to 20% tax on profits.	Land rent (license fees) on mineral holdings. Educational tax and other earmarked taxes. Social security levies. Sales commission to government bank when minerals sold through this institution.	A. *Government Participation* COMIBOL, a government enterprise, is the largest mining company. ENAFBOL, another government company, operates a tin smelter. B. *Other Comments* COMIBOL nominally subject to taxes similar to those on the private sector (mining). Until recently, most such taxes allocated to its own educational and health expenditures and to its capital reserves.

*For sources see extensive bibliographical commentary following table.

Table 5-3 (cont.)

II. *CANADA – Federal and Provinces*

Share of Mining in GDP (1974): 8.4%

Share of Mining in Exports (1974): 19.4%

Major Minerals Produced: Copper, nickel, zinc, iron ore, lead, molybdenum, gold, silver, potash.

(1) *Royalty, Production or Export Taxes*	(2) *Taxes on Income*	(3) *Other Taxes*	(4) *Government Participation and Other Comments*
No royalties at Federal level. For structure and rates of provincial royalties, see below.	A. *Salient Features* 1. *Federal:* Corporate and personal income tax, as other income sources. All exploration expensible. Accelerated depreciation (30% diminishing balance of mine development, including land costs). Other mine assets depreciated at 30% or more, as income allows. Up to 25% reduction in tax base for earned depletion ($US 1.00 reduction for each $US 3.00 of new investment). Prior "resource allowance" is 25% of operating costs less depreciation. 2. *Provinces*: Mine profits tax, based on notional profit at pit-head. Base derived by deducting processing allowance, and mine expense and depreciation from total revenue. B. *Rates* 1. 46%, with abatement of 10% for provencial taxes. 2. Vary, but generally 15% or progressive from that rate.	Minor license fees and land rents in provinces.	A. *Government Participation* 1. *Federal*: Government-owned Canada Development Corporation is major shareholder in one major mining company. 2. *Provinces*: Quebec government owns active exploration company. B. *Other Comments* 1. *Federal*: Provincial mine profit taxes and royalties not deductible from tax base. System here described is effective 1976. Canada has highest per capita mineral output in world. 2. *Provinces*: Additional or alternative taxes in some provinces; see below. Minimum exploration spending is required to hold claims.

IIa. *CANADA – British Columbia*

Major Minerals Produced: Copper, lead, molybdenum.

(1) *Royalty, Production or Export Taxes*	(2) *Taxes on Income*	(3) *Other Taxes*	(4) *Government Participation and Other Comments*
A. *Salient Features* 1. *Ad valorem royalty* on presumptive revenue net of smelting and refining and transportation cost (to disappear in 1977). 2. From 1974-76, a "*super royalty*" on excess of presumptive revenue net of smelting and refining over a basic value per mineral, dependent on average 1969-1973 price, escalated by half the change in wholesale price index. B. *Rates* 1. 5% 2. 50%	New *profits tax* to take effect in 1977: A. *Salient Features* 1. Royalty payments made in 1976 creditable against this tax. 2. "Earned" depletion allowance as in Federal income tax. 3. 15% processing allowance. B. *Rates* Net income from mining metallic minerals, asbestos and gypsum taxed at 17.5%.		

IIb. *CANADA – Manitoba*

Major Minerals Produced: Nickel, copper.

(1)	(2)	(3)	(4)
	A. *Salient Features* 1. *Tax on profit* from mining including processing and extending to ore treatment outside province. Deductions include current exploration, and 10-20% depreciation. 2. *Surcharge on profit* when it exceeds an 18% return on notional current value of plant and equipment. Latter is depreciated, inflation-adjusted, historical cost. B. *Rates* 1. 15% of basic profit. 2. 20% of excess profit.		This tax replaces general mine profits tax but not income tax.

Table 5-3 (cont.)

IIc. *CANADA – Nova Scotia*

Major Minerals Produced: Coal, Gypsum.

(1) *Royalty, Production or Export Taxes*	(2) *Taxes on Income*	(3) *Other Taxes*	(4) *Government Participation and Other Comments*
A. *Salient Features* *Alternative*: Ad valorem royalty selling net of transportation costs. B. *Rate*: 2% (if higher than profit tax alternative)	A. *Salient Features* *Alternative*: Mine profits tax (see above, Part II, col. #2, A-2). B. *Rate*: 15% (if higher than royalty alternative)		Higher of two tax alternatives must be paid.

IId. *CANADA – Saskatchewan*

Major Mineral Produced: Potash, Uranium

(1)	(2)	(3)	(4)
Uranium: 3% output royalty replaced (1976) 12.5% mine profit tax.	Uranium: progressive excess profit tax with rates ranging from 15 to 50%.	Property tax on net value of remaining ore in ground, as assessed (potash only).	As of 1974, property tax replaced royalty.

III. *CHILE*

Share of Mining in GDP (1973): 11%.

Share of Mining in Exports: 84% (1973); 89% (1974).

Major Minerals Produced: Copper, natural nitrates.

(1)	(2)	(3)	(4)
A. *Salient Features* Gross royalty charge on sales proceeds of small mining enterprises only. No other taxes on them. B. *Rate*: 2%.	A. *Salient Features* 1. Medium mines subject to same tax as other enterprises. 2. Large mining sector (up to nationalization) had special income tax. B. *Rates*: 1. 35% on domestic corporations; 17% on other enterprises. 40% surtax on income of non-resident companies. 2. Rates escalated with price of copper.	Housing tax Payroll tax Educational fund Labor service tax	*Government Participation:* Five large mines partially nationalized in 1964-1969. Completely nationalized in 1971.

Table 5-3 (cont.)

IV. *COLOMBIA*

Share of Mining in GDP (1973): Under 1%, exclusive of hydrocarbons.

Share of Mining in Exports (1972): 5%.

Major Mineral Produced: Iron ore.

(1) Royalty, Production or Export Taxes	(2) Taxes on Income	(3) Other Taxes	(4) Government Participation and Other Comments
Departmental (subnational) royalties; no Federal royalties.	A. *Salient Features* General personal and corporation income taxes apply. On mining concessions granted after October 28, 1974, deductions are substantially the same as for other sectors, including cost amortization. Before production, exploration chargeable (10% a year) against other mining income. Earlier law permitting percentage depletion still applies to mining concessions held before October 1974 and where mineral rights are privately owned.	Stamp taxes.	A. *Government Participation* Government has ownership and management participation in Hanna Company nickel project. B. *Other Comments* Percentage depletion now unavailable for all contracts signed after October 1974, except for claims on privately owned subsoil.

(1)	(2)	(3)	(4)
	B. *Rates:* 40% flat-rate corporation tax. Limited liability companies and partnerships taxed at half corporate rate, with after-tax income imputable to owners' personal income.		

V. *EIRE*

Share of Mining in GDP (1971): 2.3%

Share of Mining in Exports (1970): 5.5%.

Major Minerals Produced: Lead, zinc.

(1)	(2)	(3)	(4)
	A. *Salient Features* 1. Royalty on operating profit (revenue less operating costs). 2. Mines were exempted from corporate tax for 20 years. 3. Tara Mine to pay corporate tax and additional "royalty" on profit. B. *Rates* 1. Individually negotiated, progressive over time. Highest: 10%. 2. 50% plus 4.5%.	Dead rent on leases, merging with royalties.	Former tax favors will be removed, but replacement not yet announced. Tara is a separate recent arrangement.

Table 5-3 (cont.)

VI. *GABON*

Share of Mining in GDP (1972): 15.8%, excluding hydrocarbons.

Share of Mining in Exports (1973): 17.6%, excluding hydrocarbons.

Major Minerals Produced: Manganese, uranium.

(1) *Royalty, Production or Export Taxes*	(2) *Taxes on Income*	(3) *Other Taxes*	(4) *Government Participation and Other Comments*
A. *Salient Features* 1. Ad valorem royalty on F.O.B. value of production. 2. Manganese only: Duty and surtax on F.O.B. value of exports. B. *Rates* 1. Manganese: 5%; uranium: 3%. 2. 3% plus 2%.	A. *Salient Features* 1. Basic profit tax, on net profit, including allowances for depreciation and depletion. 2. "Complementary royalty." On manganese, equals difference between 20% of taxable income and the 5% royalty. On uranium, equals difference between distributed profit and 21% of value of production.	Dividend withholding tax is 13% (versus 18% on other industries).	Guarantee of stability of income and dividends taxes for 25 years.

VII. *INDONESIA*

Share of Mining in GDP (1974): 11.5%, inclusive of hydrocarbons.

Share of Mining in Exports (1973): 5.4%.

Major Minerals Produced: Tin, bauxite, nickel, copper, manganese.

(1)	(2)	(3)	(4)
A. *Salient Features:* Royalties (private firms only). Formally expressed at specific rates, but effect is that of ad valorem levy on presumptive market prices. On some minerals rate applied to weight of mineral content; on others (including tin and bauxite) to weight of output regardless of purity. Deductible from income tax base.	A. *Salient Features:* Corporation income tax paid by all enterprises (including state-owned). Private mining firms pay on a separate rate schedule determined by kind of mineral. These firms are allowed a 1/6 reduction from applicable rates during first 10 years of operation. Also a tax credit (8% of mining investment) is permitted when total investment exceeds $US 75 million.	Land rent paid by private firms on contract areas. Rates range from $US 0.005 per hectare per annum in initial survey period to $2.00 per hectare in operating period.	A. *Government Participation* Three wholly owned mining enterprises, P.N. Timah (mainly tin), P.N. Aneka Tambang (other metals), P.N. Batubara (coal). Some government participation in privately held concessions. B. *Other Comments* Private mining companies operate under negotiated contracts, which differ in detail, and of which 17 concluded from 1967 to 1972.

VII. *INDONESIA* [continued]

(1) Royalty, Production or Export Taxes	(2) Taxes on Income	(3) Other Taxes	(4) Government Participation and Other Comments
A2. In lieu of royalties, state-owned mining enterprises pay an export tax when minerals are exported in unprocessed form (e.g., tin concentrates). B. *Rates* 1. Effective ad valorem royalty on metallic content of a 70% tin concentrate is 10%. On tin ingot effective rate is 7%. On other metals, effective ad valorem rates range from 1% (silver, gold) of metal value to 5% (lead). 2. 10% of F.O.B. value.	B. *Rates* General rate (nonmining and state-owned mining firms) is 45%. Privately owned tin mines pay 48%; nickel, cobalt, and bauxite pay 45%; all other mining companies pay 42%. (prior to 1977)		No new mining contracts since 1972. Future direction may be indicated by petroleum concessions, where foreign companies operate as contractors to state company, with production sharing.

VIII. *JAMAICA*

Share of Mining in GDP (1973): 28.7%.

Share of Mining in Exports (1973): 76.1%.

Major Minerals Produced: Bauxite, alumina.

(1)	(2)	(3)	(4)
A. *Salient Features* Ad valorem royalty on primary output, based on presumptive price of refined product (aluminum ingot). Tax becomes lump sum if output falls below 90% of capacity. B. *Rates* 7.5% of ingot value, 1974-1975; 8.5%, 1976.		Land taxes.	*Government Participation* Government recently bought (at less than market value) 51% interest in three mining firms. Discussions continuing with others.

Table 5-3 (cont.)

IX. *LIBERIA*

Share of Mining in GDP (1972): 49.6%.

Share of Mining in Exports (1974): 73%.

Major Mineral Produced: Iron ore.

(1) *Royalty, Production or Export Taxes*	(2) *Taxes on Income*	(3) *Other Taxes*	(4) *Government Participation and Other Comments*
A. *Salient Features* Original concessionaire, LMC, paid specific export tax, with ad valorem surtax on rising prices. B. *Rates* $US 0.05 per ton plus 1% of excess over 115% of 10-year average price.	A. *Salient Features* Two companies pay income tax. B. *Rate* Nominal 50%.	Land taxes.	*Government Participation* Government holds shares in two companies and has 50% profit share in a third. Profit share is after interest, repayment of principal, and reinvested earnings.

X. *MALAYSIA*

Share of Mining in GDP (1972): 6.0%.

Share of Mining in Exports (1974): 23.2%.

Major Mineral Produced: Tin.

(1)	(2)	(3)	(4)
A. *Salient Features* 1. Export tax progressive, specific, deductible from income base. 2. Surcharge on tin exports imposed in 1974. Progressive, ad valorem. B. *Rates* 1. Maximum 16% of export value. 2. 1% to 5%.	A. *Salient Features* 1. Tax on undistributed profits imposed on all companies. Writeoff of mine prospecting. Initial depreciation allowances. 2. Development tax (income tax surcharge) on all business income. B. *Rates* 1. 40%. 2. 5%. 3. 10% on excess over "standard."		

XI. *MEXICO*

Share of Mining in GDP (1974): 1.6%, excluding hydrocarbons.

Share of Mining in Exports (1972): 6.0%, excluding hydrocarbons

Major Minerals Produced: Copper, lead, zinc.

(1) *Royalty, Production or Export Taxes*	(2) *Taxes on Income*	(3) *Other Taxes*	(4) *Government Participation and Other Comments*
A. *Salient Features* 1. Production tax on minerals at specific or ad valorem rates (based on presumptive price). Rates are higher the less processed the mineral. Reduction of 50-75% (depending on size) for Mexican-controlled firms. 2. Export tax, ad valorem (some cases progressive), based on presumptive price. Rates are higher the less processed the mineral. Reduction of 50-75% for Mexican-controlled firms. B. *Rates* 1. 6% range for copper, 3.5% range for lead, 2% range for zinc, 20% range for gold. 2. 20-35%.	A. *Salient Features* Business subject to graduated income tax. Cost depletion and depreciation of assets allowed. Current exploration currently deductible. B. *Rates:* From 5% to 42%. New mines eligible for 40% reduction over 12 years, if Mexican controlled.	Surface tax on mining concessions. Dividend withholding tax (foreign and domestic). Federal and state transactions taxes.	A. *Government Participation* Heavy direct government investment began in early 1970s. B. *Other Comments* Tax and land-concessions laws create compelling incentive to Mexican control.

XII. *NEW CALEDONIA*

Share of Mining in GDP (1970): 72%.

Share of Mining in Exports (1970): 98%.

Major Mineral Produced: Nickel.

(1)	(2)	(3)	(4)
Ad valorem tax on exports.	No income taxes as of 1974.	Import taxes.	A. *Government Participation* None as such, but emphasis on participation by French nationals. B. *Other Comments:* Revision of tax structure now being planned, with 50% tax on profits to be introduced gradually over 5 years and to replace export tax.

Table 5-3 (cont.)

XIII. *PAPUA NEW GUINEA* (Bougainville Copper)

(1) *Royalty, Production or Export Taxes*	(2) *Taxes on Income*	(3) *Other Taxes*	(4) *Government Participation and Other Comments*
No royalties or export taxes.	A. *Salient Features:* A proportional tax on net profit with a surcharge on profit over an amount representing a 15% return on assets (debt and equity), with a built-in inflation factor. B. *Rates*: On basic tax, 33 1/3%. On excess over 15% return, 70%.	15% withholding tax on dividends.	A. *Government Participation* Government and other domestic investors hold 20% of equity, for which they paid. B. *Other Comments:* Arrangements here are after renegotiation of 1974. Original terms were generous with 3 years tax holiday, and tax rate rising from 25% by 1% a year to 60% maximum.

XIV. *PERU*

Share of Mining in GDP (1972): 5.6% (including hydrocarbons)

Share of Mining in Exports (1972): 50.7%.

Major Minerals Produced: Copper, iron ore.

(1)	(2)	(3)	(4)
A. *Salient Features:* Pro forma royalty levied on F.O.B. value of exports in excess of US$ 18,600. Export tax payments are fully creditable against profit tax liability. B. *Rates:* 4% on metals; 2% on nonmetallics.	A. *Salient Features:* Separate tax on mining profits, with rates dependent on cash flow return to investment and on extent of government participation. Depreciation of fixed assets at 10% for buildings and 20% for plant and equipment. Exploration may be expensed or amortized. On approved mining projects, reinvestment credit of up to 40% of pretax profit (maximum: US$ 7 million). B. *Rates:* Basic rate schedule in two parts: 1. Progressive, from 20% to 40%, on first US$ 2.3 million of taxable profit.	High schedule of land rents, and of "work requirements" on exploration concessions. 25% tax on dividends paid by domestically owned mining corporation, creditable against personal income tax.	A. *Government Participation* Strong incentives for government participation. Government's interests held through two wholly owned companies, one of which has a monopoly on mineral marketing and on new refineries. B. *Other Comments:* 1970 mining code guarantees that no new taxes will be imposed on mining (but no guarantee as to rates).

XIV. *PERU* [continued]

(1) Royalty, Production or Export Taxes	(2) Taxes on Income	(3) Other Taxes	(4) Government Participation and Other Comments
	2. On profit over that amount, single proportional rate, dependent on ratio of cash flow to book assets. Rates vary from 29% to 55%. Foreign-owned firms pay additional 30% on profit net of basic tax. But basic tax is reducible by up to 1/3 where state has over 25% participation.		

XV. *PHILIPPINES*

Share of Copper (alone) in GDP (1973): 2.2%.

Share of Copper (alone) in Exports (1973): 15.3%.

Major Mineral Produced: Copper.

(1)	(2)	(3)	(4)
A. *Salient Features* 1. Gross output royalty, ad valorem on market value. 2. Export duty, ad valorem on market value (removed March 1975). 3. Premium duty on selected export products including copper, gold, silver. Imposed on excess over a declared "basic export price" (90¢/lb. for copper). 5-year exemption from all nonincome taxes for new mines. B. *Rates* 1. On gold, 1.5%. All other minerals, 2%. 2. 6% in 1974, 2% in early 1975. 3. 30% of excess.	A. *Salient Features:* Corporation and personal income taxes. 23% depletion allowance for metals (being phased out). B. *Rates* Corporations: 25-35%. Personal: 3-70%.	Residence tax. Mining occupation fees and rentals.	

Table 5-3 (cont.)

XVI. *THAILAND*

Share of Mining in GDP (1973): 1.7%.

Share of Mining in Exports (1972): 10.4%.

Major Minerals Produced: Tin, tungsten, fluorite.

(1) *Royalty, Production or Export Taxes*	(2) *Taxes on Income*	(3) *Other Taxes*	(4) *Government Participation and Other Comments*
A. *Salient Features* 1. Ad valorem royalty applied to presumptive market value of mineral content. Deductible from income tax base. On tin, rate applies to excess over a floor of US$ 0.375. On other minerals rate is proportional. 2. Business tax, applied to gross receipts of all businesses. Metallic minerals taxed only if sold locally. B. *Rates* 1. On tin, 25% of excess over floor; on tungsten, 10%. Other metals: 2.5-5%. 2. On metals, 4% of gross receipts.	A. *Salient Features:* Individual and corporation income taxes apply generally. Generous depreciation. Mining companies amortize exploration and concession costs over 10 years. B. *Rates:* Corporations: progressive rates from 15% to 25%. Individuals and family corporations: 10-50%.	Municipal surcharge: 0.4% added to 4% business tax. Remittances abroad taxed at 25%.	*Government Participation:* Very minor, involving a 16% interest in one dredge, and two lignite mines.

XVII. *ZAIRE*

Share of Mining in GDP (1969): 35.5%.

Share of Mining in Exports (1972): 85%.

Major Minerals Produced: Copper, cobalt, silver, diamonds, tin.

(1)	(2)	(3)	(4)
A. *Salient Features* 1. Export tax on minerals (as well as other commodities). Surtax on copper exports, based on F.O.B. price two months prior. 2. General turnover tax also applies to exports. B. *Rates* 1. Basic rate is 28.6% on copper, 20% on diamonds, lower on other minerals. Surtaxes on copper are progressive from 5% to 40%	A. *Salient Features:* As other corporations. Mines allowed deduction for prospecting and depreciation. B. *Rates:* 40%. But one concessionaire pays 50% in lieu of export taxes.	20% withholding tax. 5% tax on wages paid to foreign employees.	A. *Government Participation* Government expropriated major copper producer, whose mines are operated for a fee by Belgian group. Also has share (in return for concession) in other mines, and half ownership in major diamond producer. B. *Other Comments*: Reduction of 50% of profit tax on profits reinvested.

XVIII. *ZAMBIA*

Share of Mining in GDP (1972): 42.4%.

Share of Mining in Exports (1971): 97%.

Major Mineral Produced: Copper.

(1) *Royalty, Production or Export Taxes*	(2) *Taxes on Income*	(3) *Other Taxes*	(4) *Government Participation and Other Comments*
None now. Before 1970 had both output royalty and export taxes.	A. *Salient Features:* Mines pay mineral profit tax (deductible over a mine's first 3 years if profit after all taxes less than 12% of equity). Ordinary company profit tax paid on net income after deducting mineral tax. All capital expenditures (including exploration) currently deductible. B. *Rates:* Mineral profit tax on copper: 51%. On lead and zinc: 20%. On other metals: 10-15%. Company profit tax: 45% of residual. (Total tax on copper = 73.05%.)	No withholding tax on dividend remittances of (partially) nationalized companies.	A. *Government Participation* Two major companies 51% nationalized in 1970. Paid at book value, with bonds, redeemable out of future earnings. B. *Other Comment:* Prospecting and exploration licenses carry minimum expenditure requirement.

Notes on Bibliographical Sources for Table 5-3

The information concerning comparative mining taxes and mining shares in GDP and total exports in Table 5-3 was derived from a large number of published and unpublished sources, as well as from the specific experience of the authors in a number of the countries included. Limitations of space and the confidential nature of some of the documents made available to us preclude a complete listing of all sources employed for each of the eighteen countries and four Canadian provinces whose mineral tax systems are shown. Nevertheless, it is useful to cite the basic sources utilized for each country. In some instances, data and information for particular countries were taken from unpublished, restricted documents of government agencies and international institutions. In these cases, only a very general reference is indicated (i.e., "unpublished documents of the Government of Thailand").

The references as listed here are divided into two categories:

I. General source material, and
II. Country-specific sources.

Category I includes periodicals which from time to time publish up-to-date reports on changes in mining tax laws (i.e., World Mining) and books or documents which contain general survey information on mineral taxes in a number of important mining nations. Category II includes citations that specifically refer to mineral tax legislation in particular countries.

I. *General Source Material*

A. Jack N. Behrman, "Taxation of Extractive Industries in Latin America" in R.F. Mikesell (ed.), *Foreign Investment in Petroleum and Mineral Industries* (Baltimore: Johns Hopkins University Press/Resources for the Future, 1971).

B. Northcutt Ely, *Summary of Mining and Petroleum Laws of the World* (Washington, D.C.: United States Bureau of Mines, U.S. Government Printing Office). This compilation has been published in stages over the period 1970-74. References for specific regions and countries are supplied under category II.

C. *Engineering and Mining Journal* (New York: McGraw Hill). Various monthly issues from 1970-76.

D. *Mining Magazine* (Mining Journal Ltd., London). Various monthly issues, 1973-76.

E. United States Department of the Interior, Bureau of Mines, *Mineral Trade Notes.* Various monthly issues, 1973-76.

F. David N. Smith and Louis T. Wells, *Negotiating Third World Mineral Agreements* (Cambridge, Mass.: Ballinger Publishing Co., 1975).

G. *World Mining* (San Francisco: Miller-Freeman Publications). Various monthly issues, 1970-76.

II. *Country-Specific Sources (Selected References)*

A. *Bolivia* (see Chapter Six for complete bibliographical information).

B. *Canada*

Meyer Bucovetsky, *Tax Reform in Canada: The Case of the Mining Industry* (University of Toronto Ph.D. thesis, March 1972).

Budget Speeches (Federal Minister of Finance, November 18, 1974 and June 23, 1975).

Bulletin of Canadian Institute of Mining and Metallurgy. Various monthly issues.

Canadian Mining Journal. Various monthly issues.

Canadian Tax Foundation, *Report of Proceedings of the 26th Tax Conference*, Nov. 1974.

Statutes and Regulations of the Canadian provinces.

C. *Chile*

Northcutt Ely, *Summary of Mining and Petroleum Laws of the World*, Part I, *Western Hemisphere* (U.S. Bureau of Mines Information Circular 8482, 1970).

Arturo Alessandri Rodriguez (ed.), *Código de Minería* (Editorial Jurídica de Chile, 1970).

T.H. Moran, "Transnational Strategies of Protection and Defense by Multinational Corporations: Spreading the Risk and Raising the Cost for Nationalization in Natural Resources," *International Organization*, vol. 27, Spring 1973, pp. 273-87.

D. *Colombia*

Malcolm Gillis and Charles E. McLure, Jr., *La Reforma Tributaria Colombiana de 1974* (Bogota, Biblioteca Banco Popular, 1977).

E. *Eire*

Northcutt Ely, *Summary of Mining and Petroleum Laws of the World*, Part 5, *Europe* (1974).

Engineering and Mining Journal, vol. 176, no. 4, April 1975, p. 23.

F. *Gabon*

Northcutt Ely, *Summary of Mining and Petroleum Laws of the World*, Part 4, *Africa* (1974).

G. *Indonesia*

Departemen Pertambangan, *Buku Pertambangan Indonesia* (Jakarta, Indonesia; issues from 1973-75).
Biro Kerja Suma Luar Negri, Departemen Pertambangan, *Mining Legislation in Indonesia* (Jakarta, 1975).
Various production-sharing agreements concluded between the Government of Indonesia and foreign mining firms (1966-1972).

H. *Jamaica*

Malcolm Gillis and Charles E. McLure, Jr., "The Incidence of World Taxes on Natural Resources, with Special Reference to Bauxite," *American Economic Review*, vol. 65, no. 2, May 1975.
M. Manley, "Statement Before the Parliament of Jamaica on Current Bauxite Negotiations," May 15, 1974.

I. *Liberia*

Smith and Wells, *op. cit.*, Chapter Three.

J. *Malaysia*

Northcutt Ely, *Summary of Mining and Petroleum Laws of the World*, Part 2, *East Asia and the Pacific* (1970).
Republic ng Filipinas, Pambansang Sento ng Pananliksik sa Buwis, "Comparative Study of the Tax Treatment of Natural Resource Based Industries in Selected Asian Countries" (unpublished, Manila, August 1973), Appendix 1, Part 4, "Malaysia."

K. *Mexico*

Northcutt Ely, *Summary of Mining and Petroleum Laws of the World*, Part 1, *Western Hemisphere*, 1970.
Engineering and Mining Journal, vol. 177, no. 2, February 1976, p. 35.
Mexican Newsletter. Published monthly in Mexico D.F.

L. *New Caledonia*

Unpublished material of the International Monetary Fund.

M. *Papua/New Guinea*

M.L.O. Faber, "Bougainville Renegotiated: An Analysis of the New Fiscal Terms," *Mining Magazine*, December 1974, pp. 464-449.
"Heads of Agreement for Variation of the Agreement of 6th June 1967 between the Government of Papua New Guinea and Bougainville Copper Ltd." (Port Moresby, 1974).
Raymond F. Mikesell, *Foreign Investment in Copper Mining: Case Studies of Mines in Peru and Papua New Guinea* (Baltimore: Johns Hopkins University Press, 1975). Chapters 6-10.

N. *Peru*

Mikesell, *op. cit.*, Chapter Four.

Ministerio de Energía y Minas, *Ley General de Minería* (Lima: Oficina de Relaciones Públicas, 1973).

John Strasma, "Peru's New Approach to Mining Taxation," *Bulletin for International Fiscal Documentation*, vol. 24, June 1970, pp. 241-45.

O. *Philippines*

Republic ng Filipinas, *op. cit.*, Appendix 1, Part 2 (Philippines).

National Tax Research Center, *A Compilation of Taxes, Fees, Charges and Duties on Selected Philippine Export Products* (Manila: National Economics Development Authority, 1974).

P. *Thailand*

International Translations, "Mineral Royalty Rates Act b.E. 2509" (Bangkok, 1975).

A.J. Freyman, M. Tremblay and R. Rodger, "Thailand: Current Economic Position and Development Prospects in the Mining Sector" (mimeographed, July 1974).

Unpublished documents of the Government of Thailand.

Q. *Zaire*

Northcutt Ely, *Summary of Mining and Petroleum Laws of the World*, Part 4, *Africa* (1974).

Unpublished manuscripts of the IMF and IBRD.

R. *Zambia*

A.R.K. Bostock and Charles Harvey (eds.), *Economic Independence and Zambian Copper* (New York: Praeger, 1972).

Northcutt Ely, *Summary of Mining and Petroleum Laws of the World*, Part 4, *Africa* (1974).

NOTES TO CHAPTER FIVE

1. These and other reasons for the shift from royalties to income taxes are explained in Chapter Two of David N. Smith and Louis T. Wells, *Negotiating Third World Mineral Agreements: Promises as Prologue* (Cambridge, Mass.: Ballinger, 1975).

2. See Malcolm Gillis and Charles E. McLure, Jr., "The Incidence of World Taxes on Natural Resources," *American Economic Review*, May 1975, vol. 66, no. 2. There are other minor exceptions to the general move from royalties to income tax. It appears, for example, that Le Nickel, a French firm, still pays only royalties for the metal it extracts in New Caledonia, but that is hardly an independent developing country.

3. For a comparison of Malaysian *output* taxes on tin vis-à-vis those of Thailand, Indonesia and Bolivia, see Table 5-2.

4. See the recent World Bank report: A.J. Freyman, M. Tremblay and R. Rodger, "Thailand: Current Economic Position and Development Prospects in the Mining Sector," mimeographed, July 1974, pp. 10 and 14. The same report (at p. 15) also draws attention to high profit expectations in Thai mining (a 20 to 30 percent, or more, required return on invested capital), which may be a contributory inhibiting factor, quite apart from the political uncertainty the country has experienced in recent years, particularly since 1973.

5. Since the 1975 revocation of certain foreign-owned offshore mining concessions—mentioned in Chapter Four of this study—it appears that the Thai Government will shortly become a more direct participant in tin mining. See *Engineering and Mining Journal*, vol. 176, no. 7, July 1975, p. 152.

6. Freyman et al., p. 22.

7. *Ibid.*, p. 19 and Annex 3, p. 5.

8. For example, the contract area for the P.T. Baliem Valley Minerals contract of work was 1,916,000 hectares. See articles 4 and 13 of *Contract of Work Between the Government of Indonesia and P.T. Baliem Valley Minerals.*

9. For details, see Northcutt Ely, "Summary of Mining and Petroleum Laws of the World," Part 1, United States Bureau of Mines Information Circular 8482 (Washington, D.C.: U.S. Government Printing Office, 1970), pp. 119-121.

10. That Kennecott did not "suffer" from the 1964 nationalization is well documented in T.H. Moran, "Transnational Strategies of Protection and Defense by Multinational Corporations: Spreading the Risk and Raising the Cost for Nationalization in Natural Resources," *International Organization*, vol. 27 (Spring 1973), pp. 273-287. As one example, under the former tax regimen Kennecott could retain only about 20 percent of the mine's pretax profit. Under the 1964 agreement (after the government participation, income taxes and dividend taxes) Kennecott's share in the mine's profit would be 27.44 percent —on the assumption that Kennecott's share was entirely remitted as a dividend.

11. It is difficult to obtain a precise number on the rates paid by El Salvador and Chuquicamata. One source says that the government's total share in the profits of these mines—after participation, income taxes, and dividend taxes— was 75 to 76 percent, as contrasted with 72.56 percent in the case of El Teniente.

12. Presumably because of Minero Peru's overfull financial commitments, a new state-owned entity, Centromin, was organized to hold the government's new exclusive interest in the recently nationalized Cerro de Pasco's mines.

13. The full table of ratios and corresponding rates is reproduced in John Strasma, "Peru's New Approach to Mining Taxation," *Bulletin for International Fiscal Documentation*, vol. 24, June 1970, pp. 242-243.

14. See Malcolm Gillis and Charles E. McLure, Jr., *La Reforma Tributaria Colombiana de 1974* (Bogota, Biblioteca Banco Popular, 1977).

15. Data for 1973 and 1974 are from *Statistics Canada*; 1975 data are from the British Columbia and Yukon Chamber of Mines, as reported in *The Globe and Mail*, Toronto, April 29, 1976.

16. The reserves tax approximates a progressive royalty. When the price of potash is under $35.50 per short ton there is no reserves tax payable. By industry estimates, the effective "royalty rate" is about 38 percent when the selling price is $90 per ton. See *The Globe and Mail*, Toronto, June 21, 1975.

17. See *Engineering and Mining Journal*, vol. 177, no. 1, Jan. 1976, p. 145. The compensation formula, in the expropriation legislation, nets out the capitalized burden of the reserves tax.

18. *The Globe and Mail*, Toronto, Jan. 20, 1976, and March 5, 1976.

19. D.R. Derry, C.E. Michener, and J.K.B. Booth, "World Mining Exploration Trends," *World Mining*, vol. 27, March 1974, p. 49.

Chapter Six

The Structure of Bolivian Mining Taxes

Malcolm Gillis

GENERAL INTRODUCTION

Bolivian tax laws affecting the mining sector have not evolved in the same fashion as has legislation in other countries where the mining sector dominates economic life,[1] or for that matter in countries such as Malaysia, the Philippines or Colombia, where the mining sector accounts for far smaller shares of income, exports or fiscal resources. For example, Bolivia appears to be the only independent developing country with an economically significant mining sector in which income (profits) taxes are not applied to mining enterprises. In nearly all the developed countries, income taxes have been applied on the mining sector since before 1920. In the developing nations the shift to heavier reliance upon income taxes and less reliance on output taxes for the sector began in the 1950s and continued through the early seventies.

This chapter and Chapter Five also show that Bolivia's royalties and export taxes upon mining sector output are among the highest in the world. Virtually all of the countries with income taxes on the mining sector utilize output-related taxes as well, but at significantly lower rates than Bolivia. Given the large proportion of mining sector taxes in total Bolivian tax collections, and in the absence of income taxes on the sector, this is of course not surprising. However, if indeed output taxes affect mining enterprises' decisions in ways different from profits taxes, high tax rates on output may be resulting in substantial losses in the value of the national mining patrimony through high-grading of deposits and through their effects in distorting decision on concentration of minerals. Further, given the extreme difficulty, if not impossibility, of recognizing exploration and development expenses under output-type taxes (royalties and export taxes), exploration and development activity in Bolivia is surely curbed significantly more by tax factors than in other nations that utilize profits tax for the sector and recognize such outlays for income tax purposes.

A third major observation that arises from this chapter is that Bolivian land rents (*patentes*) imposed on mining concessions and/or claims are perhaps the lowest in the world. This is especially true for land rents on exploitation concessions for metallic minerals which, even for the largest concessions, are less than five percent of those charged in Peru and only fifteen percent of those charged in Indonesia. The adverse consequences of low rates for land rents have little to do with government revenues; an increase of 200 or 300 percent in land-rent rates would lead to only a negligible expansion in revenues from this source. Rather, very low rates for mining land rents allow and encourage claim holders to hold back promising mineralized areas from exploration and/or exploitation.

Fourth, it may be noted that Bolivia, like most countries where small-scale mining is a significant activity involving substantial numbers of workers (Thailand, Chile, Peru, Malaysia), provides for one or more forms of tax relief for the small mining sector (both *de facto* and *de jure*).

Finally, with the notable exceptions of Peru and Chile, most mining activity in the other countries reviewed in the previous chapter is carried on at altitudes of less than 1,000 meters (and even in Peru and Chile, at less than 3,000 meters). But in Bolivia, most mines are located at between 4,000 and 5,000 meters (with several major mines at altitudes above 5,000 meters). Work conditions at such altitudes, coupled with the obvious difficulties of installing and maintaining transport infrastructure and the narrow-vein nature of Bolivian ore bodies, all contribute to the high-cost structure of Bolivian mining, relative to most other countries.[3]

PERSPECTIVES ON THE OVERALL TAX STRUCTURE

Expenditures of the Bolivian public sector, defined broadly,[a] repre-

[a]A very broad definition of the public sector would include the central government budget, the INSS (National Social Security Administration), all public enterprises (e.g. COMIBOL, YPFB, the national airline, etc.) and the net budgetary outlays (exclusive of transfers) of subnational levels of government (departments, provinces, cities, etc.).

sent a sizable proportion of the nation's total output. According to a recent consolidation of public sector accounts, the public sector in the period 1971-73 accounted for approximately one third of gross domestic product (GDP), about equally divided between the central government, COMIBOL and the Bolivian Petroleum Corporation (YPFB).[4] But, *tax* revenue of the central government has never approached the proportions cited above. Rather, the ratio of central government taxes to GDP has rarely exceeded nine percent since 1968, ranging from a low of 8.4 percent in 1971 and 1972 to as high as 12.5 percent in 1974 (see Table 6-3). By way of comparison it might be noted that the average ratio for Bolivia (9.8 percent) for the period 1970-74 is just about equal to that of Colombia (9.4 percent) over the same period,[5] but well below the average for other South American nations, particularly those with substantial natural resource endowments such as Venezuela (19.6 percent) and Chile (29.9 percent). Nonetheless, the Bolivian ratio is not substantially lower than that of Thailand (12.8 percent) or Indonesia (12.1 percent).

The revenue system as a whole relies very heavily upon taxes on or relating to international trade (export taxes, royalties on mineral exports, customs duties, taxes on foreign exchange, etc.), with very minor dependence upon personal or business income taxes, sales taxes or property taxes. In fact, in both 1973 and 1974 total taxes paid by the mining and hydrocarbon sectors amounted to about 60 percent of total central government revenues in each year.[6] Virtually all of the taxes paid by these two sectors may be categorized as trade or trade-related taxes.

PERSPECTIVES ON MINING SECTOR TAXES

Table 6-1 contains a series, covering the period 1965-1974, of total *identifiable*[b] taxes collected from the mining sector *alone* (exclusive of hydrocarbons). This table furnishes the basis for the calculations in Table 6-2. This latter table shows that over the period 1968 through 1974, mining sector taxes accounted, on the average, for about 23.1 percent of total central government tax revenues, and for the most recent five-year period (1970-74), nearly 27 percent of the total. However, the sector's share in total taxes has been highly variable, ranging from as low as 11.4 percent in 1968 to as high as

[b]The term "identifiable" is used advisedly, since some tax collections known to be extracted from the mining sector (personal income taxes paid by private sector miners, taxes on asset revaluation, stamp taxes, etc.) are not given separately by sector in the revenue accounts. Thus, collections from these sources are not included in Table 6-1.

Table 6-1. Identifiable Taxes Paid by Mining Sector, 1965-1974—Excluding Social Security (in thousands of Bolivian pesos; includes COMIBOL royalties on a gross adjusted basis)

Name of Tax	*1965*	*1966*	*1967*
I. *National Government*	89,471.1	76,669.8	15,373.2
A. Customs Agency	88,365.6	71,461.7	114,051.2
1) *Regalías* (royalties)	88,365.6	71,461.7	97,251.2
2) Export taxes	–	–	–
3) Other customs revenues	–	–	16,800.0
B. Internal Revenue Agency	1,105.7	1,208.1	1,322.0
1) Mining property	(1.2)	(0.6)	(1.4)
2) *Patentes* (land rents)	(505.2)	(639.1)	(677.6)
3) Transfer taxes	(599.1)	(568.4)	(642.9)
4) Income taxes on mining agents	–	–	–
5) Gold and platinum taxes	–	–	–
C. Revaluation of Assets	–	–	–
II. *Earmarked Taxes*	3,644.5	4,271.8	6,914.5
A. Customs Agency	3,285.0	3,988.6	4,554.0
1) Subnational governments	(1,524.2)	(1,170.0)	(1,466.8)
2) Educational institutions	(1,760.8)	(2,818.6)	(2,236.2)
3) Departmental development corporations	–	–	(850.9)
4) Other earmarked taxes	–	–	–
B. Internal Revenue (universities, departments, municipalities)	359.5	283.2	317.0
C. ENAF (National Smelter)	–	–	2,043.5
GRAND TOTAL (I + II)	93,115.8	76,941.6	122,287.7
Total exclusive of earmarked taxes	89,471.3	72,669.9	117,416.7

Basic Sources: Ministry of Finance, Ministry of Mines, Republic of Bolivia. Data for royalties must be interpreted with care. Treasury data indicate that from 1968 through 1972, COMIBOL paid no royalties at all to the *central* government, and paid only B.$28.3 million in 1973. However, COMIBOL did indeed submit vouchers for taxes collected but not remitted, for the years 1970-1973. These amounted to B.$98.2 million in 1970, B.$60.4 million in 1971, B.$77.8 million in 1972 and B.$203.5 million in 1973. The Ministry of Finance advises that nearly all the value of these vouchers represents either the cost of functions performed by COMIBOL for the government or capitalization payments due COMIBOL from the government and should, for analytical purposes, be treated as taxes paid by the enterprise.

1968	1969	1970	1971	1972	1973	1974
97,023.7	164,916.1	334,034.0	150,849.8	232,228.0	797,538.9	1,493,927.5
95,345.4	162,486.3	331,982.5	140,495.1	228.771.5	789,205.9	1,481,469.1
86,441.4	153,048.1	274,969.3	108,948.8	121,949.1	392,172.7	881,747.2
–	–	–	–	98,362.4	316,324.0	481,359.0
8,904.0	9,438.2	57,013.2	31,546.3	8,388.0	80,709.2	110,362.9
1,678.3	2,429.8	2,051.5	10,354.6	3,456.6	8,332.0	12,457.4
(1.4)	(3.2)	(7.1)	(7.7)	(4.0)	(3.9)	(5.4)
(558.4)	(556.5)	(702.8)	(603.6)	(594.9)	(702.0)	(1,045.4)
(227.6)	(922.1)	(333.1)	(308.3)	(341.9)	(2,343.3)	(1,973.4)
–	–	–	(9,234.1)	(1,977.5)	(1,502.7)	(4,019.3)
(890.9)	(948.0)	(1,008.6)	(201.0)	(538.3)	(3,781.2)	(5,414.7)
–	–	–	–	–	–	–
8,375.1	13,747.4	24,439.1	13,635.3	18,166.0	47,148.9	222,897.9
5,062.3	8,122.8	18,239.4	8,156.1	8,403.2	31,565.3	192,209.3
(1,377.7)	(2,424.3)	(7,087.4)	(5,804.2)	(3,679.6)	(14,897.0)	(144,717.3)
(2,328.2)	(3,052.0)	(4,736.9)	(2,143.4)	(2,268.5)	(4,989.4)	(7,813.7)
(920.1)	(1,920.3)	(6,096.6)	–	(2,105.6)	(9,652.1)	(29,647.6)
(436.3)	(726.2)	(318.5)	(208.6)	(349.5)	(2,026.9)	(10,030.6)
244.7	700.7	724.7	972.5	1,146.2	1,176.3	1,205.4
3,068.1	4,924.0	5,456.9	4,506.6	8,616.7	14,407.3	29,483.2
105,398.8	178,663.6	358,413.1	164,485.0	250,394.1	844,687.9	1,716,825.4
99,200.9	168,892.1	338,482.3	155.155.3	240,844.4	811,946.3	1,523,410.5

Table 6-2. Total Identifiable Mining Sector Taxes Paid to Central Government and as Percent of Total Central Government Tax Revenues (Adjusted), 1968-74 (millions of Bolivian pesos)

	1968	*1969*	*1970*	*1971*	*1972*	*1973*	*1974*
A. Total identifiable mining sector taxes paid to the central government[a]	100.0	168.8	339.3	155.4	240.8	821.0	1,515.4
B. Total central government tax revenues	877.0	946.5	1,241.8	1,107.6	1,289.2	2,038.0	4,655.2
C. Mining sector taxes paid to central government as percent of total central government tax revenue [A + B]	11.4%	17.3%	27.3%	14.0%	18.7%	40.3%	32.6%

[a]From Table 6-1. Includes small earmarked tax paid to ENAF, the state smelter.

40.3 percent in 1973. As noted in Chapter Three, the extreme variability in tax collections from the mining sector has had considerable destabilizing effects upon the national economy. As long as the nation remains so heavily dependent upon tax revenues on commodity exports, this pattern will of course continue, barring any major international initiatives that may serve to stabilize mineral prices at levels acceptable to producing and consuming nations.

Table 6-2 includes only central government tax collections from the mining sector. The sector also is responsible for sizable earmarked taxes paid primarily to subnational levels of government (see Table 6-1). For both 1973 and 1974, such earmarked taxes amounted to about 40 and 13 percent, respectively, of total central government taxes collected from the mining sector. The role of mining taxes in the national economy may be placed in somewhat sharper focus by relating such taxes to GDP (Gross Domestic Product) since 1968. The share of mining taxes to GDP has displayed a rising tendency since 1968, going in somewhat erratic steps from 1.0 percent in 1968 to 4.6 percent in 1974 (see Table 6-3). Of even more interest, for purposes of later comparisons, is the ratio of total mining sector taxes to the Gross Sectoral Product of the mining sector (the share of total value-added generated by the mining sector). The ratio, like almost all other measures which incorporate mining sector values of any sort, has shown an extremely high degree of variability over the past decade (see Table 6-3), ranging from as low as 11 percent in 1968 and 1971 to nearly 38 percent in 1974.[c]

Current Tax Provisions

There are no less than 20 major types of taxes, plus over 30 low-yield "earmarked" levies, affecting the mining sector. Nearly all of these taxes are specific to the mining sector, with the major exception of the export tax, which applies to hydrocarbons and, nominally at least to certain agricultural exports. All of the major taxes applicable to mining are, for all practical purposes, assessed on gross value of exports, and are considered by the enterprises as equivalent to output taxes. Taxes geared to output account for over 90 percent of mineral sector tax collections and amount to a significant proportion of gross export values. As may be noted from Table 6-4, the effective rate of such taxes (total output-related taxes on mining as a percent of export values) has risen steadily since 1971, from 5.93 percent in 1971 to 20.71 percent in 1974.

[c]Given the low degree of reliability in the Bolivian national income accounts prior to 1977, care should be taken to avoid attributing too much importance to these or any other measures based on national income figures. In recognition of the many gaps in the national income accounts, the government began a thorough-going revision of the accounts in 1976.

Table 6-3. Total Mining Sector Taxes (Central Government Taxes Plus Earmarked Subnational Taxes) as % of GDP and as % of GDP Originating in the Mining Sector, 1968-1974

	1968	*1969*	*1970*	*1971*	*1972*	*1973*	*1974*
A. Total identifiable taxes paid by mining sector (from Table 6-1) in millions of pesos	105.4	178.7	358.4	164.5	250.3	853.7	1,708.8
B. Nominal gross domestic product (in millions of pesos)[a]	10,192.0	11,044.0	12,080.0	13,145.0	15,268.0	21,459.0	37,317.0
C. Nominal Gross Domestic Project originating in the mining sector (gross sectoral product in mining exclusive of hydrocarbons) in millions of pesos[a]	960.0	1,137.0	1,402.0	1,476.0	1,663.0	2,638.0	4,560.0
D. Total mining sector taxes as % of GDP (A ÷ B)	1.0%	1.6%	3.0%	1.3%	1.6%	4.0%	4.6%
E. Total mining sector taxes as % of GDP originating in the mining sector (A ÷ C)	11.0%	15.7%	25.6%	11.1%	15.1%	32.4%	37.5%
F. Total central government taxes as a % of nominal GDP (from Table 6-2 and Row C of this table)	8.6%	8.6%	10.3%	8.4%	8.4%	9.5%	12.5%

[a]Source: Consejo Nacional de Planificación, *Plan Operativo 1975* (La Paz, 1975), Anexo C.

Table 6-4. Effective Rates[a] of Output-Related Mining Taxes on Total Mineral Gross Export Values, 1971-74

	1971	*1972*	*1973*	*1974*
1. Central Government *Regalía* Collections as % of mineral export values	5.30%	5.19%	8.75%	11.53%
2. Mineral Export Tax Collections as % of mineral export values	–	4.18%	7.05%	6.24%
3. All Other[b] Output-Related Mining Taxes as % of mineral export values	0.63%	0.75%	1.11%	2.94%
4. Combined Output Related Taxes on Minerals as % of mineral export values (A + † + 3)	5.93%	10.12%	16.91%	20.71%

[a]Output-related mineral taxes as a percent of gross export values.

[b]Includes the following: 18 percent of total *regalía* collections earmarked for subnational government or their instruments, taxes on sulphur, earmarked tax paid to IIM, earmarked taxes for universities, taxes on gold.

THE REGALÍA: GENERAL OBSERVATIONS

Since the promulgation of the most recent mining code in 1965, the cornerstone of the structure of Bolivian mining taxes has been the *regalía* (royalty). The *regalía* was intended at that time to be the only tax applicable to the sector, as it was to replace a truly complex maze of low-yield taxes earmarked for almost every imaginable purpose. In turn, this jerry-built structure of taxes had evolved since 1957, when the country abandoned its five-year effort to collect income taxes from the mining sector. Over the past ten years the *regalía* has been responsible for 68 percent of total tax revenues collected from the mining sector, and about 27 percent of total tax revenues of the central government (1968-1974 only). For most of the decade since 1965, the *regalía* was the only major tax paid by mining firms, accounting for 86 percent of all mining sector taxes from 1965 through 1971. With the introduction of the export tax in 1972, the share of the *regalía* declined somewhat, but even from 1972 through 1974 amounted to 56 percent of total identifiable mining taxes. In 1974 this revenue source was 58.2 percent of total sectoral taxes and 19.0 percent of total central government taxes (see Table 6-5). The effective rate of the *regalía* on total exports (actual *regalía* collections as a percent of mineral export values) more than doubled from 1971 through 1974, rising from 5.3 percent to 11.5 percent. This occurred in spite of a reduction in *regalía* rates in February 1974.

Table 6-5. Particular Mining Sector Taxes, as % of Total Identifiable Mining Sector Taxes, and as % of Total Central Government Tax Revenues, 1974 (in millions of Pesos and %)

	As % of Identifiable Mining Sector Taxes	*As % of Total Central Govt. Revenues*
I. *Customs Agency*	97.2%	31.7%
1. *Regalía* (Royalty)	58.2%	19.0%
a) COMIBOL	35.1	11.4
b) Medium Mines	15.1	4.9
c) Small Mines and others	8.7	2.8
2. *Export Tax on Minerals*	31.8%	10.3%
a) COMIBOL	18.7	6.1
b) Medium Mines	7.1	2.3
c) Small Mines and others	6.0	2.0
3. *Import Duties and Exchange Taxes*	7.3%	2.4%
4. *Other Customs Revenue*	0.1%	–
II. *Internal Revenue Service*	0.8%	0.3%
1. *Patentes* (land rents)	0.1%	–
2. Income taxes on mining agents	0.3%	0.1%
3. Gold and platinum taxes	0.3%	0.1%
4. Transfer taxes	0.1%	–
III. *ENAFBOL's Share of Royalties*	2.0%	0.6%
Total	100.0	32.6%

(Numbers may not all add to totals due to rounding.)

The *regalía*, as it applies to most minerals, is nominally a tax on the "presumed income" of mining firms; but the tax operates in much the same fashion as royalties on mineral output in Indonesia, Thailand, and elsewhere. The nominal base of the tax for most minerals, including tin, is the difference between world mineral prices in specified markets (e.g., the London Metal Exchange) and a presumed cost set by the government. However, for certain mineral substances (bismuth, silver, sulphur, iron ore and 26 other specifically enumerated minerals), the nominal tax base is specifically gross value. Even for those minerals for which taxes are putatively assessed on presumed income (i.e., those for which a "presumptive cost" has been established), the *regalía* functions as a tax on the gross value *per unit* of output, with rates progressive over a range of prices between the level of presumptive cost and specified prices set for each mineral. Once the specified price is reached, the *regalía* rate becomes proportional with respect to increments in price and presumed income.

The *regalía* is normally expressed in terms of taxes due per "commercial unit" of the mineral in question. For such minerals as tin, copper, lead and zinc, the commercial unit is the "fine pound" (*libra fina*). This is intended to represent the actual metallic content per unit of the concentrate of metal exported. For other metals, such as antimony and wolfram, the commercial unit is the "long unit" (*unidad larga*), or 22.4 pounds of metallic content.

As noted, not all minerals have a presumptive cost for *regalía* purposes. For those that do, the level of presumptive costs has been changed infrequently since they were first established in 1965.[7] Presumptive costs for tin were changed only twice from May 1965 to January 1976, while those for wolfram, antimony and copper were increased three times over the same period (see Table 6-6). For those important minerals for which presumptive income is the putative base for assessing *regalía* rates, the *regalía* scale differs according to the mineral, with nominal rates higher for tin, tungsten and antimony, and with lower rates on zinc, copper and lead. For those

Table 6-6. Evolution of Presumptive Costs under the *Regalía*: Major Minerals, 1965-1976 (US $ per commercial unit)[a]

Mineral	*May 1965*	*December 1965*	*February 1970*	*February 1974*[b]	*January 1976*[c]
1. Tin	$1.10	$ 1.10	$ 1.10	$ 1.50	$ 1.80
2. Antimony	–	$ 4.00	$ 5.00	$ 7.50	$ 9.00
3. Wolfram	–	$20.00	$24.00	$30.00	$36.00
4. Copper	–	$ 0.30	$ 0.36	$ 0.59	$ 0.70
5. Lead	–	$ 0.12	$ 0.12	$ 0.15	$ 0.17
6. Zinc	–	$ 0.15	$ 0.15	$ 0.19	$ 0.215

[a]Commercial units for each metal are as follows: Tin, copper, lead, zinc: fine pound. Antimony and wolfram: "long unit" (22.4 fine lbs.).

[b]The same level of presumptive costs applied to all producers of particular minerals from 1965 *until* February 1974. In February 1974, differential cost scales were established for medium mines, COMIBOL and small miners, with the medium mines assigned lower costs (and thus higher taxes) per unit, and the small mines the highest presumptive costs. For example, in the case of tin, presumed cost for medium mines was $1.50 per unit in 1974, while that for small mines was $1.60, and for COMIBOL was $1.55.

[c]Presumptive costs for medium miners. Presumptive costs for COMIBOL and the small miners are slightly higher.

minerals for which the *regalía* is geared explicitly to gross value rather than presumed income (bismuth, silver, iron ore, etc.), ad valorem rates range from a high of 7 percent (silver) to a lcw of 1 percent (bismuth, sulphur). The evolution of ad valorem *regalía* rates on such minerals is summarized in Table 6-7.

Tin accounts for by far the greatest share of total *regalía* collections (see Table 6-8). Over the period 1970-74, this mineral was responsible for an average of 62 percent of total *regalía* collections. Antimony remains the second most important source of *regalía* revenue, as its average share was 16 percent over the most recent five-year period. Wolfram[d] has consistently been the third most important source of *regalía* collections, averaging about 11 percent of total *regalía* collections since 1970, but wolfram's relative share has declined steadily from 18 percent in 1970 to less than 9 percent in 1974.

The effective rate of *regalía* on both export values and on presumed income varies considerably for different minerals. Tables 6-9 and 6-10, respectively, show the effective rate of *regalía* on total gross export value and presumed income for each mineral. It may be noted that the weighted average effective rate on export value for all minerals was 11.53 percent in 1974. The effective rates for both antimony and wolfram were markedly higher than the average for all

Table 6-7. Evolution of Ad-Valorem and Specific *Regalía* Rates on Gross Unit Values for Minerals *Not* Assigned Presumptive Costs

Mineral	*December 1965*	*October 1968*	*February 1974*	*January 1976*
1. Bismuth	3.0%	3.0%	3.0%	3.0%
2. Silver[a]	7.0%	7.0%	7.0%	7.0%
3. Iron ore	2.5%	2.5%	1.0%	1.0%
4. Sulphur	0	0	1.0%	1.0%
5. Various enumerated[b] minerals	0	0	1.0%	1.0%
6. Miscellaneous minerals not specifically exempted	2.5%	2.5%	1.0%	1.0%
7. Nonmetallic minerals (except sulphur)	0	0	1.0%	1.0%

[a]Lower rates prevailed for low-grade concentrates of this metal.
[b]Asbestos, Mica, talc, magnesite, cadmium, marble, etc.

[d]In this volume, the term wolfram is used interchangeably with the word tungsten.

Table 6-8. Relative Share of Various Minerals in Overall *Regalía* Collections 1970-1974

	Relative Share in *Regalía* Revenues				
Mineral	*1970*	*1971*	*1972*	*1973*	*1974*
1. Tin	37.8%	58.9%	79.3%	66.1%	67.1%
2. Antimony	35.6	6.1	4.8	12.9	12.3
3. Wolfram	17.1	19.8	7.8	5.3	8.7
4. Copper	5.8	3.5	2.3	4.9	3.7
5. Lead	0.3	6.1	0.3	1.9	1.9
6. Zinc	0.0	0.2	0.1	5.8	4.0
7. Bismuth	0.9	1.4	0.3	0.1	0.1
8. Silver	2.2	3.8	3.6	2.9	2.7
9. Gold	0.3	0.1	0.4	0.7	0.5
10. Others[a]	0.0	0.0	0.1	0.2	0.1
Totals	100.0%	100.0%	100.0%	100.0%	100.0%

Source: Departamento de Estudios Economicos, Ministerio de Minería y Metalurgia, May 1975.

[a]Sulphur, asbestos, mica, iron ore, etc.

[Note: Columns may not add to 100.0% due to rounding.]

minerals in 1974, while that for tin was somewhat above the average. The average effective rates for lead, copper, zinc, bismuth and silver were well below the average. This information does not necessarily imply that minerals such as lead, copper and zinc are "undertaxed" relative to, say, antimony and tungsten (wolfram). The effective rates of *regalía* in Table 6-9 are computed on *gross* export values, *not* net export values. It can be shown that concentrates with a low value-to-bulk ratio, such as lead and zinc, have significantly higher realization costs (transport, smelting and other handling costs) as a percent of gross export value than is the case with minerals such as tungsten and tin.[e] Therefore, a more appropriate measure of effective rates of *regalía* would be the ratio of royalties to *net* export values (gross export values minus realization costs). Unfortunately, reliable data for net export values for all important minerals are unavailable, so that as a measure for comparing the relative *regalía* burden among different minerals, the effective *regalía* rate on gross export values suffers from some serious limitations.

A somewhat more revealing indicator of the relative tax treatment of different minerals under the *regalía* is measured average effective

[e]Realization costs can range from as low as 20 percent of gross export values for tin concentrates to as high as 60 percent for many zinc concentrates.

Table 6-9. Effective Rates of *Regalías* on Gross Export Value, Various Minerals, 1970-1974

	Effective rate[a] of *Regalía* on Export Value (in %)				
Mineral	*1970*	*1971*	*1972*	*1973*	*1974*
1. Tin	9.87%	6.79	8.25	12.34	16.16
2. Antimony	30.60	8.18	6.26	18.37	23.51
3. Wolfram	25.85	17.70	8.96	11.75	23.03
4. Copper	12.44	5.15	3.17	9.11	12.91
5. Lead	1.17	0.12	0.52	5.51	9.10
6. Zinc	0.00	0.19	0.83	5.53	5.96
7. Bismuth	2.96	3.00	1.40	0.34	0.45
8. Silver	5.68	5.60	5.70	5.73	5.46
Average effective rate (weighted) for *regalías*	–	5.30	5.19	8.75	11.53

Source: Dept. de Estudios Económicos, Ministerio de Minería y Metalurgia.
[a]Effective rate of royalty: total royalties (per mineral), divided by total gross export values (per mineral).

rates of *regalía* on presumptive income. In Table 6-10 we may note that in 1974 the percentage of *regalía* collections to total "presumed" income was between 36 and 38 percent for tin, antimony, wolfram and copper, but only 21 percent for lead and 13 percent for zinc. This is not surprising, given that the statutory maximum rate of *regalía* may not exceed 38 percent of presumed income for the first four minerals, whereas the maximum statutory rates for zinc and lead are 14 and 16 percent, respectively.

Tin

As noted in Table 6-8, tin has been responsible for more than two-thirds of total *regalía* revenues in recent years. In 1974, the tin *regalía* alone accounted for nearly 13 percent of overall central government tax revenues. Precisely because of the preponderant share of this mineral in overall mining sector taxes and its relative importance in total taxes, successive governments have been extremely reluctant to make adjustments in the tin *regalía* that might involve some risk of substantial revenue loss. Thus, even in the face of a 400 percent increase in the internal price level from 1965 until 1976, the "presumed" cost of tin per unit (i.e., the price at which liability for *regalía* begins) was increased only twice over the same period, while those for antimony, wolfram and copper were increased three times.

Table 6-10. Effective Rates of *Regalía* on Presumed Income, Various Minerals, 1970-1974

	Effective Rate of *Regalía* on "Presumed Income,"[a] in (%)				
Mineral	*1970*	*1971*	*1972*	*1973*	*1974*
1. Tin	29.25	25.65	29.96	34.84	37.23
2. Antimony	37.45	22.41	21.68	31.63	36.29
3. Wolfram	38.25	32.18	21.30	28.48	36.70
4. Copper	28.28	19.98	13.40	17.71	36.22
5. Lead	9.21	0.33	4.30	15.24	20.56
6. Zinc	0.0	11.55	5.52	14.99	13.12

Source: Departmento de Estudios Económicos, Ministerio de Minería y Metalurgia.

[a]Presumed income of all producers is calculated by subtracting total presumed cost for each group of producers from total export value for each group of producers.

Further, the "presumed" cost of tin rose by only 36 percent from 1965 until 1976, while those for antimony and wolfram were allowed to increase by 88 percent and 50 percent respectively. However, a decree issued in September 1975, allowed further increases in "presumptive costs" for all major minerals (effective on January 1, 1976) of 20 percent each for tin, antimony, and wolfram, 19 percent for copper and 13 percent for lead and zinc, so that between 1965 and early 1976 presumptive costs for the three major minerals increased by 64 percent for tin, 125 percent for antimony and 80 percent for wolfram, still well below the rates of increase in the domestic price level over the same period.

As is the case for all Bolivian *regalías*, the structure of the principal tax on tin is quite simple. No royalties are assessed when metal prices fall below "presumed costs." For prices above this level, the royalty scale is given in terms of cents per U.S. dollar at different prices per unit, and makes no mention of "presumed income" per unit. Exports of minerals in metallic form are free of *regalía* (and export tax), the *regalía* having been levied at the point of sale to a domestic smelter. The 1975 royalty rate was more or less proportional with respect to price (when price exceeded "presumed cost" per unit) but was progressive with respect to presumed income (the difference between official prices and presumed costs) up to a price of U.S. $2.90. Thereafter, the *regalía* became proportional on presumed income, owing to a provision that restricted taxes to 38 percent of presumed income.

Figures 6-1, 6-2, and 6-3 portray, for both the *regalía* and the export tax (discussed subsequently), several noteworthy relation-

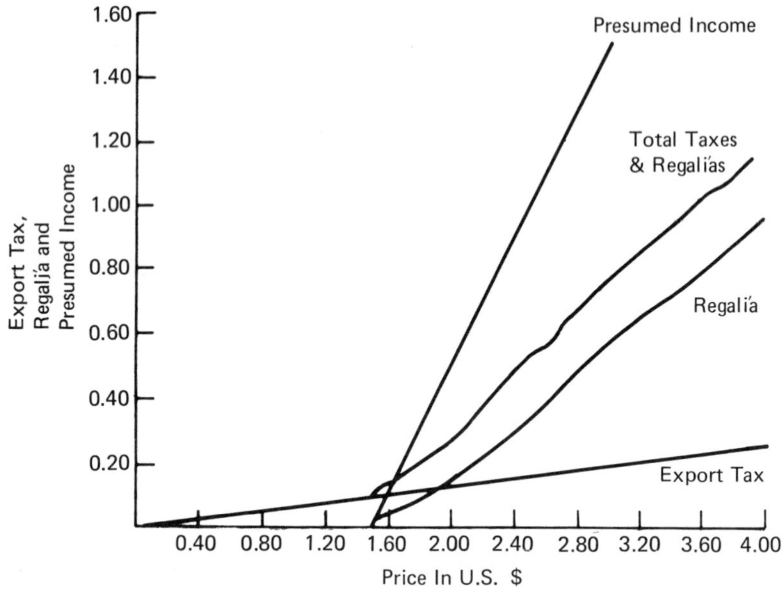

Figure 6-1. *Regalía* and Export Tax on Tin (Medium Miner Exporting a 33.3% Concentrate) Related to Price (1974 – 1975)

ships. Figure 6-1 shows the more or less proportional relationship between the 1975 *regalía* and unit export value (price) for all prices in excess of U.S. $1.50 (the presumed cost in 1975 for a medium mining firm). Figure 6-2 relates the *regalía* (and the export tax) to presumed income, while Figure 6-3 shows the *regalía* (and the export tax) as a *percent* of presumed income at various levels of the latter.[f]

Antimony and Wolfram

Collections from antimony accounted for 12 percent of total *regalía* collections in 1974. The antimony *regalía* is modeled on the same general pattern as that for tin. The tax as a percent of price rises in progressive fashion from zero at the presumed level of costs (U.S. $7.50 per long unit in 1975, U.S. $9.00 in 1976) to $18.50,

[f]Other than the slight differentiation established between medium miners, COMIBOL and the small miners, there is only one other exception to the generally applicable *regalía* on tin: microcrystallized or colloidal tin concentrate with a fine metal content of less than 20 percent is entitled to a reduction of 50 percent in *regalía* rates.

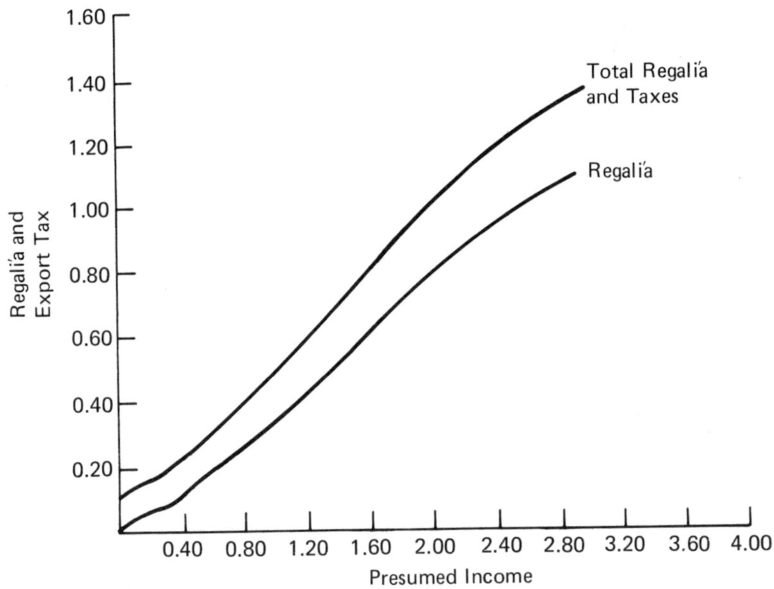

Figure 6-2. *Regalía* and Export Tax on Tin (Medium Miner Exporting a 33% Concentrate) Related to "Presumed Income" (1974 – 1975)

where it becomes roughly proportional. As in the case of tin, the law provides that the maximum rate of tax on "presumed income" may not exceed 38 percent. Thus, although the effective tax rate on presumed income rises quickly from 18 percent (at prices just above the level of presumed costs), it quickly reaches its peak (at U.S. $18.50 in 1975) and is constant thereafter.

The *regalía* on wolfram, responsible for nearly 9 percent of total *regalía* revenues, contains no special features which distinguish it from the levies on tin and antimony, as it displays much the same effective rate pattern, and is subject to the same 38 percent maximum tax on "presumed income."

Other Minerals

Lead and zinc are often found in the same complex minerals in Bolivia, and both receive the same treatment under the present *regalía* structure. Whereas the maximum rate of tax on "presumed income" for tin, antimony, copper and wolfram is 38 percent, the

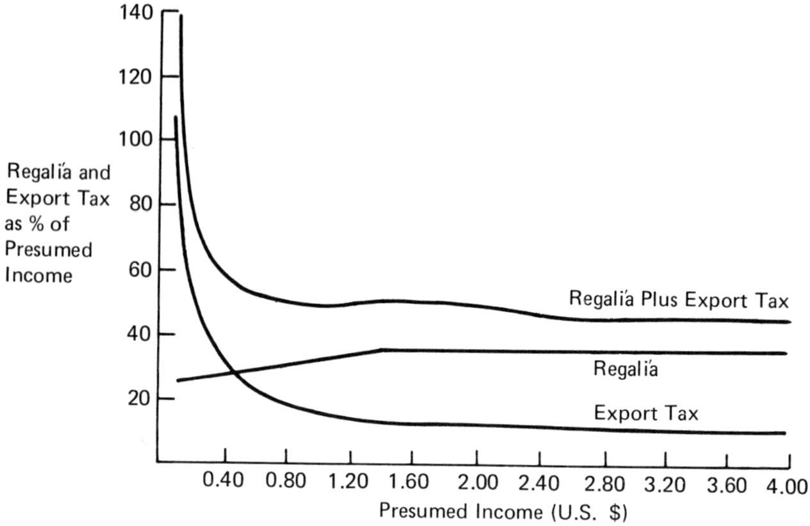

Figure 6-3. *Regalía* and Export Tax on Tin (Medium Miner Exporting a 33.3% Concentrate) as % of Presumed Income (1974 – 1975)

limit for zinc is 14 percent and that for lead is 16 percent. As a result, although lead accounts for 3 percent and zinc for 10 percent of export value, their relative shares in total 1974 *regalía* receipts were only 2 and 4 percent respectively.

Whereas lead and zinc mines are subject to lower limits on effective tax rates, copper mining benefits from tax relief provided across the entire *regalía* scale, for intermediate and low-grade exports of copper concentrate. Specifically, for all copper concentrates below 25 percent but above 15 percent in metallic content, royalties are reduced by five percentage points for each percentage point below 25 percent of metallic content. For all concentrates below 15 percent but greater than 10 percent, royalties are further reduced by 10 percentage points for each point below 15 percent of metallic content. Exports of concentrates of metal content of 10 percent or less are exempt from royalty.[g] Thus, the effective rate of the *regalía*

[g]Thus, *regalía* rates in 1975 for one fine pound of copper exported by a medium mining firm were (at a price of U.S. $0.70 per fine pound):

40% concentrate: $0.0374
20% concentrate: $0.0281
15% concentrate: $0.0187
10% concentrate: $0.0000

for "low" content (15 percent) copper is exactly half that for "high" content copper. As copper also benefits from lower *export* taxes for low-grade copper concentrates, the mineral is one of the most lightly taxed of all metallic exports.

Only the six substances discussed above are assigned presumptive costs. For all other minerals, the *regalía* is assessed strictly on the basis of gross export value. Since 1965, bismuth and silver have been subject to *regalía* rates of 3 and 7 percent respectively. For silver, however, the first 500 grams of content per metric ton are exempt from the *regalía*. From 1965 until 1974, iron ore exports were subject to a *regalía* of 2.5 percent of gross value, but this rate was reduced to 1.0 percent in February 1974, presumably to lighten the overall tax burden on prospective ore from SIDERSA's iron ore deposits in eastern Bolivia. Sulphur pays a normal *regalía* of one percent, but in addition in subject to an earmarked central government *regalía* of 5.85 percent of gross value. All other minerals, metallic or nonmetallic, pay a one percent *regalía* on gross values.

THE EXPORT TAX

This levy was created in October of 1972[8] as part of a broad stabilization package accompanying a devaluation of 70 percent in the peso-dollar rate. The export tax was originally intended to capture some of the "windfall" gains from the devaluation, and was to have expired in October of 1973. Budgetary constraints in 1973 led to the renewal of the tax in that year,[9] and like other "emergency" fiscal measures elsewhere in the world, it has apparently become a permanent fixture in the revenue structure. The tax initially applied to a wide variety of exports, including not only minerals and hydrocarbons, but cotton, meat and cattle exports and a variety of other products. Although cotton remains nominally subject to the export tax, virtually all other agricultural products have by now been exempted, so that in 1974 minerals and hydrocarbons provided over 98 percent of total export tax collection. The mining sector by itself has accounted for nearly 65 percent of total ad valorem export tax revenues since inception of the levy in 1972 (see Table 6-11). Over the same period, mining sector export tax collections alone have accounted for 11 percent of total central government taxes. Initially, the tax was 20 percent of net export values, but in October 1973 was converted (primarily for administrative reasons) to a levy on *gross* values, and the nominal general rate reduced to 7.5 percent or less on most mineral exports. The effective rate of mineral export taxes on total gross mineral exports was slightly over 7 percent in 1973 and about 6.25 percent in 1974.

Although the most common applicable rate of tax for minerals is

Table 6-11. Ad-Valorem Export Taxes: Relative Share of Mining Sector in Total Collections and Other Ratios

	1972	*1973*	*1974*
I. Total ad-valorem export tax collections (millions of pesos)	139.0	452.2	801.0
II. Mining sector: total export taxes paid (from Table 6-1)	98.4	316.3	481.4
III. Hydrocarbons sector: total export taxes paid	34.1	66.3	300.0
IV. Mining sector export taxes as percent of total	70.8%	70.0%	61.0%
V. Hydrocarbon export taxes as percent of total	24.5%	14.7%	37.5%
VI. Effective rate of export tax on gross mining sector exports	4.18%	7.05%	6.24%
VII. Effective rate of export tax on gross hydrocarbon exports	–	5.82%	7.77%
VIII. Relative share of COMIBOL, the medium mines and small mines in total mineral export taxes			
A. COMIBOL	74.2%	56.3%	58.8%
B. Medium mines	3.4	20.9	22.2
C. Small mines and others	22.4	22.8	19.0
	100.0%	100.0%	100.0%

Source: Ministry of Finance.

7.5 percent, there are numerous exceptions by type of producer and mineral, metallic content (*ley*) of mineral, and volume of mineral. Table 6-12 summarizes the rate structure applicable from October 1973 until 1976. As may be noted, the highest rates (8.5 percent) are levied on tin, antimony, wolfram and copper, but there is substantial rate differentiation by type of producer and grade of concentrate. For example, high-grade tin concentrates exported by COMIBOL and the medium mines carry a 7.5 percent rate, while low-grade tin concentrates coming from cooperatives are exempt from export tax. Most other important minerals (lead, zinc, bismuth, and silver) are taxed at substantially lower rates than tin, antimony, wolfram and copper. Here again, rates are highest on high-grade concentrates and on COMIBOL and the medium mining sector, and lowest on low-grade concentrates and the cooperatives, with small miners in between.

The export duty is highly regressive with respect to nominal

Table 6-12. Ad-Valorem Export Tax: Structure and Rates (October 1972 through December 1975)

Type of Mineral	*Export tax rate by type of producer and grade of concentrate (in percent)*								
	COMIBOL and medium mining: applicable tax rate (%)			*Small mining: applicable tax rate (%)*			*Cooperatives and special contracts: applicable tax rate (%)*		
	High grade	*Medium grade*	*Low grade*	*High grade*	*Medium grade*	*Low grade*	*High grade*	*Medium grade*	*Low grade*
1. Tin[a]	7.5	6.5	6.5	5.5	4.0	0.0	4.5	3.0	0.0
2. Antimony	7.5	7.5	7.5	5.5	5.5	5.5	4.5	4.5	4.5
3. Wolfram[b]	7.5	7.5	7.5	5.5	5.5	5.5	5.0	5.0	5.0
4. Copper[b]	7.5	7.5	7.5	5.5	5.5	4.5	4.5	4.5	3.0
5. Lead	3.0	3.0	3.0	2.0	2.0	2.0	1.5	1.5	1.5
6. Zinc	2.0	2.0	2.0	1.5	1.5	1.5	1.0	1.0	1.0
7. Bismuth	3.0	3.0	3.0	2.0	2.0	2.0	1.5	1.5	1.5
8. Silver	3.0	3.0	3.0	2.0	2.0	2.0	1.5	1.5	1.5
9. Iron ore	1.5	1.5	1.5	1.5	1.5	1.5	1.5	1.5	1.5
10. Sulphur[c]	6.5	6.5	6.5	1.0	1.0	1.0	1.0	1.0	1.0
11. Other minerals	1.5	1.5	1.5	1.5	1.5	1.5	1.5	1.5	1.5

[a]For export tax purposes, "high" grade tin is concentrates of 35% or more of metallic content, "medium" is 20%-35%, and "low" is less than 20%. Any mineral exported in metallic form (effectively tin, antimony and bismuth) is free of both export tax and royalty, both having been paid upon sales to domestic smelters.

[b]"High" grade copper concentrates are those with 35% or more of metallic content. For purposes of the *export* tax (but not the *regalia*), "low" content copper is that of 13% or less.

[c]Rates for sulphur exports by small mines and cooperatives refer only to sulphur *ore* exported through the National Association of Sulphur Producers (ANAPH). Refined ANAPH sulphur pays a 2% rate.

"presumed income" as calculated under the *regalía*. This may be noted for tin in Figure 6-3, which indicates that at very low world mineral prices, the export tax as a percent of presumed income exceeds 100 percent, but rapidly falls to 30 percent at slightly higher prices and thence to 10 percent of "presumed income" at the highest possible mineral prices.

The combined burden of the *regalía* and the export on both minerals prices and presumed income is shown for several minerals in Table 6-13. The prices used in the example range from the lowest price at which presumed income is positive (at which point *regalía* liability begins) to the highest prices reached by the particular minerals in recent years. A fairly clear pattern of combined effective

Table 6-13. Pattern of Combined Burden of Export Tax and *Regalía* on Price and "Presumed Income" per Unit of Various Minerals (1974-1975 rates)

Mineral	*Presumed income per commercial unit (US $) (medium mining enterprise)*[a]	*Higher Grades*		*Lower Grades*	
		Regalía *and export tax as % of per unit price*	Regalía *and export tax as % of per unit presumed income*	Regalía *and export tax as % of per unit price*	Regalía *and export tax as % of per unit presumed income*
1. Tin	$ 0.10	9.1	145.0	8.1	129.0
	0.50	14.8	49.0	13.8	55.0
	1.00	21.1	52.8	20.1	50.3
	1.50	26.5	53.0	25.5	51.0
	2.00	29.2	51.2	28.2	49.4
	2.50	31.3	50.0	30.3	48.4
2. Antimony	$ 1.00	9.6	81.8		
	5.00	17.9	44.8		
	10.00	28.1	49.1		
	20.00	35.1	48.3		
3. Wolfram	$ 2.00	8.6	138.0		
	30.00	23.5	47.0		
	50.00	31.2	50.0		
	100.00	36.7	47.8		
4. Copper	$ 0.01	7.9	474.0	7.7	462.0
	0.05	9.7	124.0	8.6	110.0
	0.10	12.3	84.8	9.9	68.3
	0.20	17.1	67.7	12.3	48.7
	0.30	20.3	60.3	13.9	41.3
5. Lead	$ 1.00	3.1	50.0		
	5.00	5.5	22.0		
	10.00	9.4	23.5		
	20.00	12.1	21.3		
6. Zinc	$ 0.01	2.1	42.0		
	0.05	4.1	19.6		
	0.10	6.8	19.8		
	0.20	9.2	17.9		

[a]The lowest price indicated in the table is that just above presumed cost levels.

rates is evident: at very low minerals prices (low levels of presumed income) the effective rate on price is moderate (from 2.1 percent in the case of zinc to 9.6 percent in the case of antimony). But at low world prices the two taxes account for an extremely high proportion of "presumed income," from 474 percent in the case of high-grade copper exported by a medium mining firm to only 42.0 percent in the case of lead. At high minerals prices (high levels of "presumed income"), the two taxes account for a substantially higher proportion of price (from 9.2 percent for zinc to 35.1 percent for antimony), but much lower percentages of presumed income than at lower prices (60.3 percent for copper, 17.9 percent for zinc). Thus, the two levies together behave quite progressively with respect to price, but quite regressively with respect to presumed income. The *regalía* is responsible for all of the progressivity on prices, while the regressivity on income is entirely due to the export tax.

Other observations in previous sections of this chapter have indicated that the fiscal system tends to favor lead, zinc and copper, relative to tin, antimony and wolfram. Except for copper, this statement is easily borne out by the information in Table 6-13. As for copper, the system does not especially favor this mineral when exported by a medium mining firm or by COMIBOL, but copper exports by small miners and cooperatives benefit substantially from both *regalía* and export tax relief measures.

THE INCENTIVE FOR EXCESS PRODUCTION

The incentive for excess production (*sobreproducción*) involves preferential treatment under both the *regalía* and the export tax, and is the principal tax incentive available to the mining sector. Its original aim was that of helping to rectify the general stagnation in mineral exploration and production discussed elswhere in this volume. The beginnings of this incentive date only from 1973.[10] The incentive originally provided for a 100 percent exemption from the *export tax only*, for all production in any three-month period in excess of average production over the previous two years, but this measure was never implemented. A few months later, provision was made for an exemption of 50 percent, from both *regalías* and export taxes, on the excess of 1974 (and 1975 and 1976) production by any single firm over the average of mineral production in 1971 and 1972. The most recent adjustment in the incentive, made in April 1975, attempted to restrict some of the revenue loss from very large incentive benefits that would have accrued to *new* firms having no production in the base years of 1971 and 1972. Thus, the incentive has had a brief but already checkered history.

As the measure now stands, the benefits of the incentive are available to all firms. However, the volume of "excess production" enjoying the 50 percent reduction in *regalías* and export taxes is defined differently for pre-existing enterprises and those established after 1971, and for medium mining as opposed to small mining and "other exporters." The provisions and safeguards are quite complex.[h] For 1974 the incentive was available for only the last 10 months, but several medium and small mining firms moved quickly to secure relief under the measure in that year. COMIBOL, although nominally eligible for benefits under the incentive, either had insufficient production increases in 1974 or decided to forego the benefits of the incentive in the case of all minerals. The former explanation is the more plausible one.

For tin, eight of the 28 medium mining firms (and no small mines) filed for export tax-*regalía* reductions in 1974. For tin producers among these particular firms, an average of 56 percent of their production benefitted from the incentive provisions.[12] This amounted to 28 percent of the tin exports of medium mining firms in 1974, or 4 percent of total Bolivian tin exports in 1974.[i] The incentive also affected a small proportion of total wolfram and antimony exports, and an extremely large share of total zinc exports (see Table 6-14), particularly that of the medium mining sector. Fully 39 percent of the 1974 zinc exports of the medium mining sector enjoyed a 50 percent reduction in *regalías* and export taxes under the incentive, in a year in which total zinc exports declined by nearly two percent. It is possible to argue that zinc exports would have declined even further in the absence of the incentive, but such an argument is barely credible, since all affected firms had begun to expand

[h]For medium mining firms established before 1971-1972, a minimum level ("base") of annual mineral production, below which firms are ineligible for the 50 percent reduction, has been established for each mineral. (This "minimum base" of annual output is set at 66 fine metric tons for tin and wolfram, 240 tons for antimony and zinc, 180 tons for lead and copper concentrates, 120 tons for copper in complex mineral concentrates, and 60 tons for silver.) Firms established *during* the 1971-72 period are entitled to select a slightly lower minimum base, depending on the number of months of operation of the firm in 1971-72, while firms established *after* 1972 may take as a "base" for each mineral a level of production equal to one-half that for firms in operation prior to 1971. In all cases described in foregoing *and* subsequent sections, if the average of *actual* annual production in 1971-72 exceeds the "minimum base," the enterprise is required to use actual averages in lieu of the "minimum base."

[i]In one firm, 81 percent of total tin production enjoyed the 50 percent reduction in *regalías* and export taxes provided under the law as then in effect.

Table 6-14. Incentive for Excess Production: Distribution of Benefits, 1974 (export figures in thousands of fine metric tons)

				Medium Mining Exports		*Small Mining Exports*	
Mineral	*Total 1973 exports (1)*	*Total 1974 exports (2)*	*Percent of total exports benefiting from incentive (3)*	*Total in 1974 (4a)*	*Percent benefiting from incentive (4b)*	*Total in 1974 (5a)*	*Percent benefiting from incentive (5b)*
1. Tin	28,409.6	28,953.8	3.4	3,546.9	28.7%	–	–
2. Wolfram	2,768.5	2,582.7	0.8	1,258.3	2.0%[a]	–	–
3. Antimony	14,799.9	13,060.1	5.5	7,140.5	4.0%[b]	3,003.9	13.0%[c]
4. Zinc	49,568.3	48,599.9	21.5	11,785.4	88.5%[d]	3,850.1	10.0%

Source: Departamento de Estudios Económicos, Ministerio de Minería y Metalurgia.

[a]All benefits went to one medium producer, who received relief on 61% of his 1974 output of wolfram.

[b]Most benefits went to two producers, who received relief on 100% and 59% respectively of their 1974 output of antimony.

[c]Only four small mining firms enjoyed relief on 60% of their antimony output.

[d]All benefits went to two producers of zinc.

productive capacity in 1969-1970, several years in advance of the first version of the incentive in November of 1973.

The revenue cost, in terms of export taxes and *regalías* foregone, of the incentive for excess production was B. $30.4 million (U.S. $1.521 million) in 1974 alone. This amounted to about 2.2 percent of collections of the two taxes in 1974.

PATENTES (LAND RENTS)

A *patente*, or land rent, is paid by all holders of exploration or exploitation claims in Bolivia. Land rents have never accounted for more than a very small fraction of government revenues, and in 1974 represented less than one tenth of one percent of total mining sector taxes paid to the central government. The importance of the device lies not in its potential for mobilizing revenue, but in its effects upon exploration and exploitation activity in the mining sector. The present level of land rents was established in the 1965 Mining Code, in fixed peso terms, and has not been altered since. The rates are quite low relative to levels prevailing in other important mining countries,[13] and are so minimal as to allow a flourishing business in mine-hunting (*cazaminas*), wherein speculators hold large mineralized tracts idle until some growing mineral enterprise is willing to pay enough (sometimes as much as $1 million just for survey rights) to convert the claim into an active one. In this way, some claims have remained in private hands for two decades without any exploration or exploitation activity at all.[j] This situation has contributed materially to the stagnation in exploration discussed in earlier chapters. Further, virtually all pecuniary benefits of holding these claims off the markets have accrued not to the state, but to a relatively small number of families.

The present land-rent structure is progressive with respect to claim size, with annual exploration land rents set at half those for exploitation. Annual obligations for each type of claim are as follows:

	Annual Land-Rent Obligations per Hectare	
No. of Hectares	*Exploration Claims*	*Exploitation Claims*
0-5,000	B.$0.60 ha.	B.$1.20 ha.
5-10,000	1.20 ha.	2.40 ha.
10-15,000	1.80 ha.	3.60 ha.
15-20,000	2.40 ha.	4.80 ha.
20,000 and above	3.00 ha.	6.00 ha.

Note: In 1975 the peso-dollar exchange rate was B$20:US$1

CUSTOMS DUTIES, EXCHANGE TAXES AND SALES TAXES

This group of taxes bearing on the sector accounted for about 7.3 percent of total central government taxes collected from the mining sector in 1974, or about 2.4 percent of total central government tax revenues in the same year (see Table 6-5). The average level of nominal rates in the overall structure of Bolivian import duties is low relative to that of most other Latin American countries, and has been reduced significantly in recent years, as the average unweighted theoretical tariff declined from 30.6 percent in 1971 to 19.1 percent in 1974.[14] Over the same period, the ratio of total duties actually collected to total import values declined from 19.4 to 13.3 percent. Imports of the mining sector are subject to even lower nominal and

[j]The Mining Code requires that such claims will be revoked in the absence of activity. *Exploration* claims may be nullified if after six months no exploration has taken place. *Exploitation* claims may be nullified if after two years (or in any two-year period) no production has taken place. However, holders of idle claims can easily circumvent these requirements by purchasing minerals from nearby mines and vending them as output of the idle claim.

"effective" rates of duty than for imports in general. Nominal duties on imports of material and equipment by the mining sector range from 2 to 8 percent, with most such imports subject to rates of 2 and 6 percent. However, some mining sector imports are and will be subject to rates of duty of 70 percent or more, owing to Bolivia's responsibilities under the Andean Pact for Regional Economic Integration.[k] A duty of 70 percent was established in 1975 on imported perforating equipment used in mining. Under the Andean Pact, Bolivia was assigned this particular industry, and as soon as the firms producing this equipment began operations in 1975, all member nations were obliged to levy a 70 percent protective duty. Other such measures appear to be on the horizon, since Bolivia's obligations under the Andean Pact require that she move to the common *minimum* external tariff by 1980 and the common overall external tariff by 1985. (See Chapter Nine.)

Domestic sales taxes do not apply to minerals sold within the internal market. Further, the import sales tax does not apply to *direct* imports of materials and equipment by mining enterprises, although the tax applies at a 2 percent rate to imports of such materials by wholesalers and other distributors. A tax on purchases of foreign exchange (*impuesto sobre divisas*) applies at a rate of 1.8 percent on purchase of exchange by private sector firms (including mining firms) and at a 1.6 percent rate on such purchases by state enterprises such as COMIBOL.

INCOME TAXES

As noted in previous chapters, *enterprises* engaged in mineral exploitation are not subject to generally applicable income taxes. However, income from certain activities in the mining-metallurgical sector is taxable, as well as all distributions of profit by mining firms to private persons.

Since October 1969, all state enterprises, including those active in mining exploitation, metallurgy or "finance," have been liable for a special 20 percent tax on net profits. COMIBOL, however, was later exempted from this tax, even after it had accumulated reserves for income tax liability.[15] Other state enterprises within the mining sector, including ENAFBOL and BAMIN, are still subject to this provision, although neither firm has ever paid any income tax. Because for tax purposes smelting and refining activities are not deemed to be a part of the "mining" sector, private smelters are

[k] For further details regarding Bolivia's role in the Andean Pact, see Chapter Nine.

subject to entity-level income taxes. But there is no record of these firms ever having paid income taxes of any kind.

Although the private medium mining enterprises are not subject to income tax at the entity level, profit distributions by such firms are taxable under the personal income tax. For this reason they have been required since 1960 to file informational income tax returns with Renta Interna. The agency has apparently encountered severe difficulties in enforcing tax liabilities on such distributions, and evasion is reportedly substantial. Since most small mining firms are not organized as juridicial entities, their entire net income, after deduction of certain costs, is nominally subject to personal income tax. For small mining operations that do not keep adequate books, special methods of income presumption apply.[16] Under these procedures the taxpayer first computes the net value (i.e., net of "realization" costs) of total sales to BAMIN, COMIBOL or mineral purchasing agents. From this figure, certain allowable costs are deducted (including *regalías* and export taxes paid, land rents, wages, commissions and interest paid to BAMIN), and a percentage rate of income presumption is applied to the balance. The rate of income presumption ranges from 10 percent on balances up to B. $650 thousand to a maximum of 25 percent on balances in excess of B. $1.7 million. Then, the income so presumed is included in total income, and subjected to generally applicable income tax rates (after deduction of personal exemptions and other generally available tax reliefs).[l]

Comercializadores (mineral purchasing agents) also pay an income tax geared to presumed income.[m] In this case, profits are presumed to be 40 percent of the *regalías* collected by these firms and remitted to the Treasury, on all sales to them (thus reinforcing incentives to understate *regalía* collections). Such firms pay a tax of 30 percent of the income so presumed.

OTHER CENTRAL GOVERNMENT TAXES AFFECTING MINING

This group of taxes is of very limited revenue importance, largely because of substantial evasion and avoidance. It includes a 30 percent withholding tax on dividends paid abroad, a 15 percent tax on

[l]These provisions apply to owner-operated mines. In the case of leased claims, the lessor pays a 15 percent tax on rental income from the claim.

[m]Three or four foreign enterprises dominate this trade. Such firms serve much the same function as BAMIN, collecting minerals from small (and sometimes medium) firms for later export.

interest payments and a 1972 tax on asset revaluation. All these levies are easily avoided at present—in fact, several medium mining firms flatly refused, during 1972-75, to pay the latter tax (5 percent on the revalued basis of assets, nominally payable within 30 months from June of 1973), on grounds that unlike firms subject to income tax, they received no offsetting benefits from asset revaluation. The 30 percent withholding tax has affected few mining enterprises in recent years, largely because of the very limited foreign equity participation in the sector up through 1974. But, with growing foreign interest in joint venture arrangements in the medium mining sector, the tax may soon become a focus of controversy. As such taxes go, a 30 percent rate is on the high side (although Thailand's tax on remittances abroad is, at 25 percent, almost as high); but considering that as of yet there are not taxes on income at the entity level in the mining sector, a 30 percent rate presents no real problem for the moment. But should Bolivia shift to an income tax in the future, the rate and structure of this tax will have to be seriously reconsidered.

OTHER INCENTIVE LEGISLATION AFFECTING THE MINING SECTOR

The most important tax incentive offered at the moment, as far as mineral exploitation activity is concerned, is that for "excess production," discussed in a previous section. In addition, the generally applicable incentive system established under the Investment Law of 1971[n] does contain some additional provisions relevant for mining. Approved investment in mining and metallurgical complexes is entitled to the full range of incentives offered in the Investment Law. These include 100 percent exemption from import duties on machinery and equipment and most raw materials, "drawbacks" (refunds) of all duties incorporated in exported products, exemption from taxes on transfer of capital (provided it is invested in approved projects), full exemption from all taxes on manufactured exports, and other minor incentive benefits. Mineral extraction activities not combined with metallurgical investment (other than beneficiation) enjoy some, but not all the benefits cited above. Most importantly, such undertakings do not receive import duty exemptions for raw materials, nor are they eligible for customs "drawbacks" or exemptions from taxes on manufactured exports.

[n]This decree (Decree 10045 of December 1971) is commonly called the Investment Law (*Ley de Inversiones*).

EARMARKED TAXES PAID TO SUBNATIONAL GOVERNMENTS[o]

The mining sector is subject to a wide variety of taxes imposed by or shared by subnational government units and their dependencies. The most important of these "*rentas destinadas*" is that portion of the *regalías* distributed to departments, provinces, Regional Development Corporations, and cities. Although the pattern of distribution within subnational government entities differs from region to region, the combined subnational share going to all jurisdictions in any one department is always 18 percent of total *regalía* collections. Such payments have been sizable in recent years, as earmarked *regalías* for subnational government are the third most important tax paid by the mining sector (see Table 6-1).

Although over thirty earmarked taxes (not including central government taxes) are assessed upon the mining sector, there are only two other types of earmarked taxes of any major significance: those for universities and for financing the Mining and Metallurgical Institute (IIM). The Customs Agency collects a tax of 6 per mill on all mineral exports, the proceeds of which are earmarked for university finance. The Internal Revenue Service collects seven other earmarked taxes for the benefit of particular universities. Finally, the medium mining firms pay an ad valorem tax of 1 percent on mineral exports (other than those channeled through BAMIN), the proceeds of which are earmarked for supporting the budget of the IMM.

OTHER TAXES AND LEVIES AFFECTING THE MINING SECTOR

Special tax arrangements exist for radioactive minerals and for a large concession contract for gold placers that dates from the latter fifties. In addition, social security levies paid by employers and employees in the mining sector are sizable: combined employer-employee contributions exceed 34 percent of payrolls in the medium mining sector, and reach 20 percent of payrolls in the small mining sector. COMIBOL's social security contributions are so large that the government allows the enterprise to retain 10 percent of total royalties due on exports to finance outlays of this type. Normally the practice elsewhere is not to count such payments as taxes, but as contributions to insurance-type pension and health schemes. How-

[o]At one point in recent (1956) Bolivian history, over 104 separate earmarked taxes were applied to the mining sector by the central government alone. Gomez d'Angelo, *op. cit.*, p. 151.

ever, if the social security scheme is used to commandeer resources to finance general development projects (not the present case in Bolivia) or if covered workers discount future social security benefits heavily (as is apparently the case in Bolivia), then some portion of social security contributions must be considered as a "tax," the incidence of which might be compared to any other levy used to finance general government activity.[17]

It is clear that private sector miners apply an extremely heavy discount to the value of future benefits to be financed by their own as well as employers' contributions. This is due to the fact that few private sector workers can easily collect their entitled health or pension benefits, and those that do succeed only after a long lag (2-4 years after retirement, for pension benefits). By all accounts, administration of the National Social Security System (CNSS) is extremely deficient, characterized by high administrative costs, lack of accumulation of reserves for retirement and death benefits, and failure to meet obligations to covered workers, especially private sector workers.

BRIEF SKETCH OF TAX TREATMENT OF OTHER SECTORS

Taxation outside the mining sector has been reviewed in great detail by a recent Fiscal Reform Commission, headed by Professor Richard A. Musgrave.[18] Thus, this section will only highlight some of the basic differences between the tax treatment of mining and that of the other sectors.

The Hydrocarbons Sector

Bolivia is not, by any standard, a major producer of petroleum or natural gas; 1974 production of petroleum averaged less than 60,000 bbls./day, and was held to about 40,000 bbls./day through most of 1975. Even so, the nation produces at present more than twice as much as necessary to satisfy domestic requirements, and exports the excess to neighboring Latin American countries and to North America. Notwithstanding its small size relative to that of major producing nations such as Venezuela, the hydrocarbons sector was in 1974 second only to the mining sector as a source of foreign exchange and fiscal resources, accounting for 27.8 percent of the former and 30.0 percent of the latter. Foreign oil firms active in Bolivia under the Hydrocarbons Law of 1972 operate under contracts with YPFB (Yacimientos Petrolíferas de Bolivia), the Bolivian state corporation, which oversees all aspects of activity in the sector, from exploration through marketing.

Like the mining sector, the hydrocarbons sector is exempt from income taxes. Rather, YPFB pays both a production tax and an export tax to the central government and a production tax to the departmental government under whose jurisdiction production takes place. The central government production tax is levied at a rate of 19 percent on the value of production at the well-head. The 20 percent export tax applies not to gross export value, but to YPFB's net export earnings, after payment of central and subnational taxes and certain other deductions. The effective rate of the export tax only, on gross export values, was about 5.8 percent in 1973 and 7.8 percent in 1974 (see Table 6-13). The departmental tax is levied on well-head values as well, and applies at a rate of 11 percent. For crude oil consumed in the national market, the 19 percent central government tax and the 11 percent departmental tax apply. Crude oil used for domestic consumption was in effect valued at slightly more than U.S. $3.00/bbl. in 1974-75, as opposed to the more than U.S. $12.00/bbl. fetched by Bolivian crude in international markets in the same period.[p]

Taxes paid to the central government account for an even higher fraction of sectoral GDP in hyrdocarbons than is the case for the mining sector. Although it is worth repeating that considerable skepticism is advisable in interpreting any Bolivian national income accounts data prepared before 1976, it is clear that taxes paid in 1974 by the hydrocarbons sector represented more than 50 percent of GDP originating in the sector. The national income accounts data suggest that central government taxes alone, as a percent of GDP originating in the hydrocarbons sector, have risen from 2 percent in 1970 to over 74 percent in 1974 (see Table 6-15). The higher relative burden on hydrocarbons relative to nonfuel minerals is neither surprising nor out of line with the situation in other countries. Although both sectors are similar in the sense that continued high levels of exploration investment are normally required to sustain production over long periods, a petroleum enterprise will typically have much lower marginal costs in exploitation activities than is the case for mining. Thus the margin of surplus that can be appropriated by the state in the form of taxes will normally be higher for hydrocarbons than for mining.

---

[p]There is therefore a very substantial subsidy upon domestic consumption of crude oil. Two important manifestations of this subsidy are very low domestic prices for gasoline (U.S. $0.20 per gallon as of October 31, 1975) and for industrial fuels. The price for crude oil used in the domestic market was raised to $10.00 per barrel in November 1975, while the export price of Bolivian crude moved up to U.S. $13.45 per bbl.

Table 6-15. Central Government Taxes as Percent of GDP Originating in Mineral, Hydrocarbons and Other Sectors (Taxes as % of Gross Sectoral Product) 1970-1974

I. Aggregate GDP and Sectoral GDP (B $ millions)

Year	*Aggregate nominal GDP (1)*	*Nominal GDP originating in mining sector (exclusive of hydrocarbons) (2)*	*Nominal GDP originating in hydrocarbons sector (3)*	*Nominal GDP originating outside nonrenewable resources sectors 1 – (2+3) = 4 (4)*
1970	12,080	1,402	268	10,410
1971	13,145	1,476	403	11,266
1972	15,268	1,663	501	13,104
1973[a]	21,459	2,638	966	17,855
1974[b]	37,071	4,560	1,742	30,679

II. Ratio of Central Government Taxes to Overall GDP, Gross Sectoral Product of Mining and Hydrocarbons Sector and Nonrenewable Resource Sector (i.e., GDP outside Mining and Oil)

Year	*Central government taxes as % aggregate nominal GDP (1)*	*Mining sector: ratio of central govt. taxes paid to sectoral GDP in mining[c] (2)*	*Hydrocarbons sector: ratio of central govt. taxes to sectoral GDP in hydrocarbons[c] (3)*	*All other sectors ratio of central govt. taxes to GDP originating outside mining and hydrocarbons (4)*
1970	10.3%	24.2%	1.9%	8.6%
1971	8.4%	10.5%	13.8%	8.0%
1972	8.4%	14.5%	35.5%	6.6%
1973	9.5%	31.1%	41.6%	4.6%
1974	12.5%	33.2%	74.3%	6.0%

[a]Preliminary figures of the National Planning Office (CONEPLAN).

[b]CONEPLAN estimate.

[c]Includes *only* central govt. taxes paid by each sector. If taxes paid by mining and oil sector to subnational governments are also included, the ratio for both sectors is higher (see Table 6-3).

Tax Treatment of Other Productive Sectors

Compared to the nonrenewable natural resources sectors in Bolivia, and indeed relative to the situation prevailing in other countries at a similar stage of development, the remainder of the economy bears a relatively small tax burden, however the concept is defined. This is particularly true in the case of the agricultural sector, especially commercial agriculture, and only slightly less so for industry, the services and transport sectors. Productive sectors outside mining and hydrocarbons have consistently accounted for between 81 and 85 percent of GDP in each year since 1970, but over the same period these same sectors have been responsible for less than 58 percent of total central government tax revenues. For example, in 1973 and 1974 these sectors together accounted for 83 percent of GDP, but only 40 percent of total central government tax revenues. This state of affairs is, of course, not necessarily undesirable, provided the state is able to meet its legitimate revenue requirements without recourse to inflationary methods of finance or without strangulation of the natural resources sectors, especially when world minerals prices fall to low levels, as in 1975.

The relative tax burden as between the natural resources sectors and the rest of the economy can be placed in sharper focus by a comparison between the "tax ratio" for the minerals and hydrocarbons sectors and the "other" sectors (see Table 6-15). In 1970, the ratio of central government taxes to GDP originating outside the natural resources sectors stood at 8.6 percent, as compared to 24.2 percent for mining and about 2 percent for hydrocarbons. Between 1970 and 1973, the ratio for "all other" sectors fell steadily to 4.6 percent, while that for mining and hydrocarbons rose to 31.1 and 41.6 percent respectively. As a result of installation of some new taxes and improved enforcement of others in 1974, the apparent ratio for all other sectors combined rose slightly in that year, to 6.0 percent.

None of the discussion above should be taken to mean that greater "tax effort" in the agricultural, industrial and other sectors is *required*; rather, the discussion merely underscores the point that ample scope exists for more effective taxation of these activities, should that be deemed advisable, whether for equity, resource allocation, economic stability or revenue reasons.[q]

Corporate entities outside of mining were, prior to 1974, subject to profits tax rates of 20 percent. However, from January 1 of 1974 until late 1976 the generally applicable rate of business income tax was 30 percent, and firms reinvesting a substantial portion of profits benefit from large reductions in tax rates, down to as low as 9 percent for enterprises reinvesting 100 percent of profits.

Sales taxes and custom duties bear somewhat more heavily on

[q]The virtual absence of taxes of any kind on commercial agriculture in eastern and northeastern Bolivia has undoubtedly been one of the principal reasons for the heavy investment by *mining* firms in this line of activity.

productive sectors other than mining and hydrocarbons. Although detailed breakdowns of nominal rates of duty on imports of these other sectors are unavailable, we may deduce that given the low rates applicable to mining and hydrocarbons, nominal rates for "other" sectors in 1973 and 1974 must have been substantially higher than the 18 percent average for *all* imports.

Evasion of all taxes in the agricultural, industrial, services and other sectors is reportedly quite high: the Finance Ministry cites figures ranging from 30 to 40 percent for both the income tax and the sales tax, and evasion is patently much higher for all taxes affecting commercial agriculture.

NOTES TO CHAPTER SIX

1. See Chapters Two and Five.
2. See Chapter Seven.
3. See "Tin Mining Today: Placer and Lode Mining and Beneficiation," *World Mining*, vol. 30, no. 2, February 1977, pp. 42-46.
4. These figures should be interpreted with care, for the operating and investment expenditures included in the figures cited have been measured on a gross basis, and may provide a misleading impression of the actual extent of public sector command over national economic resources. For example, the current revenue of the consolidated public sector during the same period was less than 25 percent of GDP, resulting in large internal public sector deficits.
5. See Malcolm Gillis and Charles E. McLure, Jr., *La Reforma Tributaria Colombiana de 1974* (Bogotá: Biblioteca Banco Popular, 1977), Chapter One.
6. See Chapter Two.
7. Decree Law No. 07148 of May 7, 1965.
8. Decree Law No. 10550 of October 27, 1972.
9. Decree Law 11143 of October 26, 1973.
10. Decree Law 11143 of October 1973.
11. Biministerial Resolution No. 304 of April 10, 1975.
12. Source: Departamento de Estudios Económicos, Ministerio de Minería y Metalurgia, Memo, May 20, 1975.
13. See Chapter Five.
14. Ministry of Finance, Memo of July 1975, p. 4.
15. Decree Law No. 11333 of February 8, 1974, Article 9.
16. See Supreme Decree No. 090833 of February 2, 1970, as amended by Supreme Decree No. 11512 of June 5, 1974. Cooperatives are subject to special legislation in this regard (see Decree No. 11154 of October 26, 1973).
17. For a good summary of the issues involved in such situations, see E.M. Kassalow (ed.), *The Role of Social Security in Economic Development* (U.S. Department of Health, Education and Welfare, Research Report No. 27, November 1967), especially Chapters 3-7.
18. Richard A. Musgrave et al., *Fiscal Reform for Bolivia* (Forthcoming 1977).

Chapter Seven

Evaluation of the Present System of Mining Taxation

Malcolm Gillis, Glenn Jenkins, Louis T. Wells and Brian Wright

INTRODUCTION: METHODOLOGY FOR EVALUATING TAXES ON MINING

Any study of the allocative or distributional impact of an existing tax structure involves choice of an alternative structure to serve as a benchmark for purposes of comparison. A no-tax, no-government expenditure situation is not a realistic benchmark. In a study of mining taxation, the best approach is to assume that real government expenditure is fixed at the current level, and to compare the effects of the existing tax structure with an alternative reference tax of equal real yield.

For present purposes, we have chosen to rely upon the most commonly chosen benchmark, a proportional income tax, in evaluating mining tax systems. Other, alternative reference taxes, such as a uniform-rate value-added tax, could also be used as a basis for comparison.[1] But in this study, the proportional income tax has been selected, not only because value-added data by sectors is often unavailable in less developed countries, but because this tax most closely resembles the two principal options we offer for reform of mining taxes in the case under review. However, it should be noted that even the available data often do not permit detailed and precise comparisons between the effects of the present tax system and the

[a]The authors wish to acknowledge the painstaking efforts made by a number of persons who commented on an earlier draft of this chapter. These include Ralph Beals, Michael Roemer, Carl M. Shoup and Joseph J. Stern. As always, the commentators are absolved from responsibility for remaining errors.

benchmark system. Rather, the reference tax will provide a general background against which evaluation of components of the existing revenue system can be undertaken.

The first step in constructing the benchmark system is the determination of income that would be subject to an income tax. We use the case of Bolivia as an example of this procedure. From Table 7-1, we know that *presumed* income in the mining sector in 1973 and 1974 was U.S. $88.1 million and U.S. $178.0 million respectively. However, these figures undoubtedly overstate the probable base of a true income tax in both years. Presumptive costs for tin, lead, and zinc were unchanged between 1965 and February 1974, and for antimony, wolfram and copper had been constant since early 1970, while actual per unit costs of domestic and foreign inputs rose by no less than 12 to 15 percent over the nine-year period through 1973. Presumptive costs under the *regalía* were increased in early 1974, and there is some justification for the assumption that the level fixed at that time was more or less accurate for January 1974. However, costs of all inputs rose dramatically *during* (and subsequent to) 1974, especially materials costs (see Chapter Two and Appendix B). Thus, the figures in Table 7-1 likely overstate the base for the reference tax. An attempt to adjust the base to reflect actual cost developments is made in Table 7-2. The adjusted data suggest that the

Table 7-1. ***Presumed*** **Income, Bolivian Mining Sector 1973 and 1974**

	Presumed Income as Computed for Regalía *Purposes*[a] *(US $ thousands)*	
Mineral	*1973*	*1974*
1. Tin	$46,421	$ 99,513
2. Antimony	10,077	18,329
3. Wolfram	4,591	13,121
4. Copper	6,909	5,702
5. Lead	3,015	5,082
6. Zinc	9,572	16,965
7. Others	7,486	19,316
Total	$88,071	$178,028

Source: Data provided by Departamento de Estudios Económicos, Ministerio de Minería.

[a]For minerals 1-6, presumed income was computed directly from the difference between actual gross revenues and presumed costs established for each mineral under the *regalía.* Other minerals (bismuth, silver, iron ore, sulphur, etc.) are not assigned presumptive costs for *regalía* purposes. For these minerals, it was assumed that their share in total export value matched their share in total presumed income (8.5% in 1973, 10.85% in 1974).

Table 7-2. Estimate of Probable Income Tax Base, 1973-1974 (US $ thousands)

Mineral	*1973*	*1974*[c]
1. Tin	40,151[a]	90,162
2. Antimony	9,895[b]	17,027
3. Wolfram	3,888[b]	12,149
4. Copper	6,386[b]	4,415
5. Lead	2,375[a]	4,282
6. Zinc	7,605[a]	14,420
7. Others	6,737[a]	16,902
Totals	77,037	159,357

[a]For these minerals (tin, lead, zinc, others) recorded 1973 presumptive costs were increased by 12.5% to arrive at probable actual costs in 1973. For all these minerals, no presumptive cost adjustments had been made since 1965.

[b]For these minerals (antimony, wolfram, copper) recorded 1973 presumptive costs were increased by 8% to arrive at probable actual costs in 1973. For all three minerals, presumptive costs had been changed in Feb. 1970, by between 15% and 20%.

[c]For 1974, it was assumed that the assigned levels of presumptive costs put into effect in February 1974 were approximately accurate indicators of actual cost levels on Jan. 1, 1974. Although costs per unit for materials, labor and other inputs increased by much more than 12.5% from early 1974 to end 1974, it is assumed here that for the year as a whole average actual costs were only 12.5% higher than on Jan. 1.

probable base of a true income tax was about U.S. $77 million in 1973 and U.S. $159 million in 1974.

From these figures, together with data on actual revenue collections for 1973 and 1974, a rate of income tax for each year may be derived, such that if this rate *had been* applied on the probable income tax base, it would have generated an amount of revenue equivalent to actual *regalía* revenues plus export tax collections in 1972 and 1973. The results given in Table 7-3 show that a perfectly administered income tax levied at a rate of about 44 to 45 percent would have generated the same mineral tax revenue yield for the *central* government that was actually generated by the *regalía* and the export tax. In order to also replace *regalías* collected for both the central and the subnational governments, the required income tax rate would have been between about 50 and 51 percent for both 1973 and 1974. For purposes of evaluation, a 50 percent tax rate has been adopted as the rate for the reference tax, on the assumption that under any reform alternative, subnational governments would continue to receive their proportionate share of the principal tax levied on the mining sector (i.e., 18 percent of the present *regalía*, or

Table 7-3. Computation of Income Tax Rate Required to Generate Revenue Yield Equal to Actual *Regalía* and Export Tax Collections 1973 and 1974 (monetary units in US $ thousands)

	1973	*1974*
I. Probable tax base (from Table 7-2)	$77,037	$159,357
II. Actual *regalía* plus export tax collections, mining sector (central govt. only)	35,424	68,555
A. *Regalía*	(19,608)	(44,487)
B. Export taxes	(15,816)	(24,068)
III. Income tax rate required to generate same central government yield as *regalía* and export taxes (II/I)	46.0%[a]	43.0%[a]

[a]Note, however, that subnational governments receive an 18% share of gross *regalía* collections. Thus, total *regalias* collected in 1973 and 1974 were US $23,912 and US $54,252 thousand respectively. To replace gross (national plus subnational) *regalías* plus export tax revenues the reference tax rate required would have been:

1973 – 52.6% ($39,728/$77,037)
1974 – 49.5% ($78,860/$159,357)

about 12 percent of total *regalia* plus export tax collections). Nevertheless, it should be noted that if the income tax would have induced more production than under the *regalía*, the required tax rate under the reference tax would be lower than 50 percent.

MAJOR PROBLEMS OF AN OUTPUT-BASED TAX SYSTEM

An output-based tax system, such as the existing revenue system on the mining sector in Bolivia, is not without its merits. The principal taxes levied upon the sector in Bolivia (*regalía* and export tax) are relatively simple in terms of both administration and compliance, although both possess an excessive number of rates, which detract from ease of administration. Further, although revenues from the sector have fluctuated substantially from year to year, the system as it now stands provides greater short-run stability of tax receipts than any feasible alternative. However, the system leaves much to be desired on a number of counts, in terms of design, implementation and ultimate effects. The major drawbacks of the present system, relative to a proportional income tax of equal yield are as follows:

1. Certain adverse incentives for efficiency and for conservation in exploitation and concentration.
2. Failure to recognize, as deductions from taxable income, outlays on exploration, mine development and mining health and safety, with consequent underinvestment in such activities.
3. Failure to provide for adequate state participation in total returns to the sector in years of favorable market conditions (e.g., in 1974).
4. Overtaxation of the sector in lean years (e.g., in 1975), with adverse consequences for the amount and composition of total investment in the sector.
5. Inability of the system to achieve even "rough justice" in taxation of firms having comparable net income.
6. Because the tax structure is not sensitive to economic fluctuations, arbitrary adjustments in the structure must be made as economic conditions change. Lack of predictability and continuity in application of tax laws and regulations can result in greater uncertainty for producers, with consequent negative effects on investment.

Effects on Incentives for Efficiency and Conservation in Exploitation and Beneficiation

As discussed in Chapter Four, mining taxes geared to gross output involve, relative to a proportional income tax, greater "excess burden" in the sense that they lead to greater wastes in extraction and beneficiation. In particular, a *specific* tax on mining output can cause substantial "high-grading" of ore bodies, a serious social concern especially when an ore body lends itself to selective mining. In such cases, the royalty encourages bypassing of low-tenor ore that would otherwise have yielded a return in excess of direct costs of extraction. In Bolivia, this is commonly called "killing of deposits" or "robbing of ore bodies" (*jukeo*). Such a royalty can also adversely affect the optimal choice of technology and/or process intensity in beneficiation, leading to "creaming" of concentrates and consequent waste of contained metal. *Ad valorem* royalties also involve distortions in operators' calculations of various decisions, but to a somewhat lesser degree than specific royalties. Observers of the Bolivian mining scene have devoted substantial attention to the effects of taxes on the "high-grading" question as far as ore bodies are concerned.[2] Indeed, the economic significance of this problem may have been overstated in the Bolivian context. However, comparatively little analysis has been directed to the effects of taxes on "creaming" of concentrates, a problem that may be at least as significant as the "high-grading" of ore bodies.

Taxes and "High-Grading" of Ore Bodies. In discussing high grading, it is very useful to draw a clear distinction between two very useful polar "ideal types" of deposits: those with fixed investments which determine output capacity already in place, and those to which no substantial investment commitment has been made.

We consider first those deposits with exploitation investments already in place. For such cases the effects on short-run exploitation decisions of output taxes and income taxes may be very different, even if the two taxes yield exactly the same amount of revenue on a given volume of production.

If a rational profit-maximizing mine operator expects prices to remain constant, he makes his exploitation decisions as follows: If the incremental cost of extracting a unit of mineral from a deposit is less than its mine-mouth value, then the unit is extracted; otherwise it is left in the ground. Mining taxes calculated on volume of output, such as the Bolivian *regalía* and the export tax, are viewed by mine operators as equivalent to reductions in the price of each unit of mineral extracted. Such ad valorem taxes can cause a quantity of mineral which would have been extracted if there were no taxes to remain in the ground after the end of exploitation, because the net of tax mine-mouth value is less than extraction cost. When this occurs the deposit is said to be "high-graded." The term is misleadingly narrow, as it is often applied to any ore left unextracted because of a tax, regardless of the relative grade.[b]

The cause and effects of high-grading are illustrated for the very simplest case in Figure 7-1. We assume that mine-mouth value depends linearly on grade, given constant prices, and that grade is a linear function of quantity prodoced. The mine-mouth value of the marginal unit of ore is represented by the curve V_o. In this short-run analysis, capital costs are fixed. Short-run extraction cost is assumed constant per ton of ore, regardless of grade. Curve C_o represents extraction cost per unit. Quantity produced is Q_o, the quantity which maximizes "profits," the latter given by the area CAB. These profits can be considered as short-run surplus, which may be offset, wholly or partially, by fixed capital costs.

Now, if an ad valorem tax is imposed at rate $\frac{BF}{BO}$, then net of tax marginal mine-mouth value is given by curve V_N. Quantity produced falls to OQ_T. Profits are reduced to GEB. Tax revenue under the ad valorem tax is $CDEG$.

An income tax, on the other hand, would allow deduction of

[b]Ore may also be high-graded on the basis of cost-related factors such as accessibility, width of vein, etc., as well as grade.

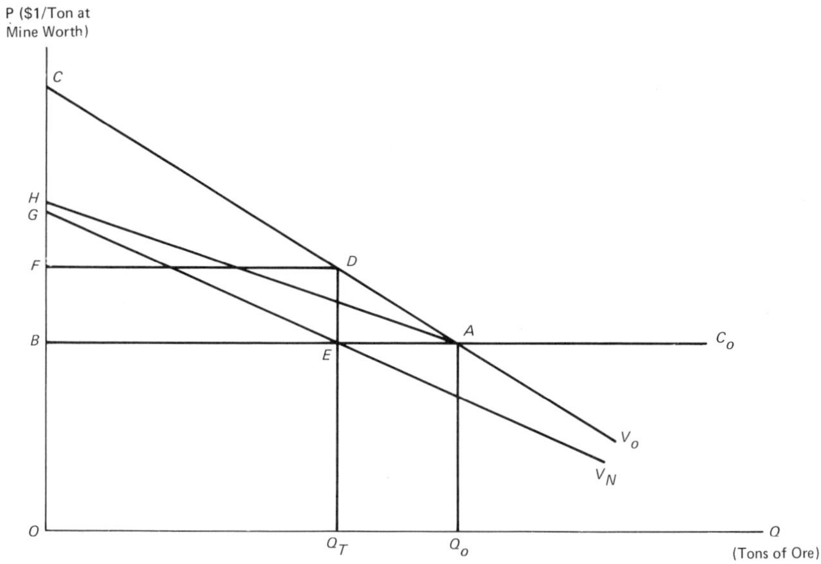

Figure 7-1. A Simple Illustration of Causes and Effects of "High-Grading"

current costs given by curve C_o. Therefore the tax base would be CAB. We assume that a tax rate $\frac{CH}{HB}$ is such that income tax revenue, area HAB, equals area $CDEG$. Output stays at the pretax level, OQ_o, and after-tax profits are CAH.

Under the income tax, profits plus tax revenue are CAB, whereas under the output tax they sum to area $CDEB$. The difference, area DAE, is the social cost of the "high-grading" induced by the output tax. To help keep relative magnitudes in perspective, note that the cost of high-grading is much less than the sale value of the loss in output. Specifically, in the example given, this cost is half the value of the tax revenue lost by the output reduction (This would only be true to a first order approximation if curve V_o were nonlinear).

In the above example, we assume that all production takes place at once, so that there are no problems of timing production or discounting future costs and revenues. In mining, it is more realistic to assume in the short run that the output rate is determined by the stock of capital (e.g., mine shafts, beneficiation plants, etc.) As long as capital is fixed, high grading is manifested by one or both of the following:

1. a decrease in the life of the mine, with sequence of exploitation unaltered
2. an alteration of the sequence of extraction, advancing the exploitation of rich deposits, by passing over poorer prospects which would previously have been taken first.

The first effect occurs in mines which have no technical constraints on the sequence of extraction of ores of different qualities. In such mines a profit-maximizing operator will extract ores in order according to their quality, taking the ores of highest net value first.[c] This is a direct consequence of the fact that if prices are constant the opportunity cost of leaving a unit of ore in the ground for one period is equal to its extraction value times the rate of return on capital, that is, the return which could be had by extracting and selling the ore, and placing the proceeds in the best available investment opportunity. So other things being equal, the more valuable the ore, the less profitable it is to delay its extraction.

The extraction sequency is followed till ores are reached for which current costs equal extraction value, net of taxes. At this point extraction ceases. The effect of an output tax is to raise the "cutoff" ore value. Ores which have net social value less than the rate of tax are left in the ground.

The loss to society of an output tax, relative to the reference proportional income tax of equal yield, is the social value of all ores which will be left unexploited because of the tax. This is calculated at the time they would have been extracted, and discounted back to the present, at the social rate of discount. In the simple case illustrated in Figure 7-1, where discounting is unnecessary since all production takes place in one period, the social cost is measured by the area of triangle *DAE*. In the multiperiod case, this loss occurs in the future, so that its significance relative to taxes and profits is reduced by discounting.

The second case, in which the extraction sequence is altered, is more complicated, but often more realistic. Technical factors often prevent the operator from sequencing according to ore quality. The choice is often between taking lower-grade ore now, or extracting it later at much higher cost. For example, when using block-caving methods, the technology is such that it is generally economically infeasible to return to recover lower-grade ore which has been bypassed. Another example typically discussed by mine operators is the choice of the width of cut when following a vein which has a rich center and either:

---

[c]The highest *value* ores in this sense may not have the highest *grade* if high-grade ores have higher extractions costs.

1. sharp and unpredictable variation of the vein
2. gradually diminishing grade related to distance from the center.

In both cases, the average grade of the ore extracted depends upon the width of the cut. Mine operators typically claim that it is often infeasible, or at least much more expensive, to go back and widen the sides of a narrow cut. So instead of taking just the rich center of the vein first and then coming back to take the sides, they take a wider cut of lower average grade.[d] Hence under the reference income tax, the optimal extraction sequence might call for the taking of some low-grade ore before taking other higher grades. The imposition of an output tax might cause more of the lower-grade ore to be bypassed, thus advancing the extraction of some higher-grade ores.

In analyzing the social cost of high-grading in this case, it is necessary to consider what happens to the ore which is bypassed because of the tax. It appears at first that there are two alternative cases, with different social costs:

1. the bypassed ore can be recovered later
2. the bypassed ore is never recovered.

In fact, given constant prices and costs, the bypassed ore will not normally be recovered later, as the following simple example demonstrates.

Consider the choice between two ore bodies, A and B, each of volume equal to one period of mine output, with constant gross value of extraction of $V(A)$ and $V(B)$ respectively. A is of lower grade than B so $V(A)$ is less than $V(B)$. In the reference income tax situation, A is extracted immediately before B. This implies that:

$$V(A) - C_1(A) - \text{MAX}\ [(V(A) - C_2(A))/(1 + r), 0]$$

$$\geqslant V(B) - C_1(B) - (V(B) - C_2(B))/(1 + r), \qquad (7\text{-}1)$$

where $C_1(A)$

is the cost of extracting (A) in period 1; $C_2(A)$ is

the cost if extracted after (B), in the second period.

$C_1(B)$ and $C_2(B)$ are similarly defined.

Introduction of an output tax of 100k percent of gross value will not alter the extraction sequence if:

---

[d]We are not referring here to the situation where the vein width is below the minimum feasible cut width. In that case some low-value ore must be extracted to allow access for men and machinery.

$$(1-k)V(A)-C_1(A)-\text{MAX}[((1-k)V(A)-C_2(A))/(1+r),0]$$
$$\geqslant (1-k)V(B)-C_1(B)-\text{MAX}[((1-k)V(B)$$
$$-D_2(B))/(1+r),0] \quad (7\text{-}2)$$

Consider first the case in which extraction of A in the second period would be economically feasible. In this case, $(1-k)V(A)-C_2(A) > 0$. That is, extraction of ore A in the second period would cover costs and taxes. Then the tax has resulted in a subtraction of $k \cdot V(A)(1 - 1/(1+r))$ from the left hand side of inequality (7-1), and of at least $k \cdot V(B)(1 - 1/(1+r))$ from the right side of (7-1). Since $V(A) < V(B)$, the inequality is preserved and indeed strengthened. So, if extraction would be economically feasible in the second period, it will never be economically justifiable to delay extraction of A. So, the sequence A, B will be unaltered after imposition of the tax. Consider now the second case, $V(A) - C_2(A) < 0$. That is, given known, constant prices ore A cannot be economically extracted if extraction is delayed. Therefore if an output tax causes ore B to be mined in period 1, then A is dropped from the extraction plan.

The above analysis obviously can be extended to consideration of a sequence of more than two ore grades. Therefore we have the following interesting result:

If an output tax causes a delay in the extraction of a low-grade ore from a mine of fixed capacity, then recovery of such ore will not be economically justified at a later date, given constant prices and technology.

So whether or not these bypassed ores are technically recoverable is irrelevant. Economically they are "lost forever" given the assumption of known, constant prices and technology, and further assuming that the new tax is permanent. However, their loss is offset by the earlier extraction of ores of higher value. Accordingly, in the case where the extraction sequence is altered by an ad valorem output tax, the social cost of the high-grading due to the tax is the aggregate of three components:

1. the net social value of the ore which is bypassed
2. the net social value of ores towards the end of the no-tax extraction sequence, which are left unexploited because of the tax.

From the sum of 1 and 2 must be subtracted:

3. the increase in the discounted social value of the ores which remain in the after-tax sequence, and which are extracted sooner because of the tax. This third component is nonzero only where the extraction sequence is altered by the tax.

The social cost of high-grading is, as emphasized above, much less than the current sale value of the foregone ore. In cases where investment is fixed, high-grading can be reduced by moving from an output tax to an income tax. Next we shall see how the two taxes measure up in the second polar "ideal type" of deposit, where no investment is currently in place.

Taxes and Development of New Prospects: Micro Considerations. The entire discussion up to now has assumed that mine capacity is fixed. In fact, this assumption is essential to the conclusion that the output tax causes high-grading relative to extraction under an income tax. If we turn to consideration of the case of previously unexploited prospects where all inputs are variable, we obtain a very different set of results. These results can be easily generalized to the long-run effects on existing mines.

In this case even the reference income tax affects the rational decisions of profit-maximizing producers, for the income tax is a tax on the return to equity capital and profits. In a small open economy like Bolivia, the marginal source of capital is from foreign portfolio or equity investment. The supply of such capital may be very elastic, the supply price being determined by the yield on alternative investments in other countries. If this is true, the required gross rate of return on new investment under an income tax will rise by about the rate of tax. If capital cannot obtain this higher return it will be invested in available alternatives—low-taxed domestic sectors or overseas. The incidence of the tax is fully shifted from capital owners.

Therefore by raising the cost of capital, the income tax reduces the growth of the mining sector. Investors might attempt to minimize the impact of an income tax in at least two ways. First, if debt payments are tax deductible, they might reduce tax payments by increasing the debt-equity ratio. This ratio is affected by other considerations, however, and in practice has appeared to be rather unresponsive to tax considerations, except in Britain. But in the case of a subsidiary of a multinational corporation, loans from the parent may be substituted for equity infusions, thus avoiding tax by raising the debt-equity ratio, or by taxing debt payments like return to equity.

Another response to an income tax is to decrease the capital-labor ratio chosen for the mine. If capital and labor are priced at their true opportunity cost, the tax can cause wasteful distortion in production by diverting factors from uses with highest social value. In developing economies with surplus labor, the social cost of labor is often claimed to be below the wage rate. Where this is true the income tax may improve the choice of technology by providing an offsetting distortion (i.e., by raising the cost of capital). However, in the case of many large-scale mining investments it may well be that the available technology allows little variation in factor input proportions, regardless of relative factor prices. If the debt-equity ratio and the capital-labor ratio are in fact fixed (and debt is charged its "arms-length" price), then the income tax is exactly equivalent to an equal-yield tax on output, for a given mineral prospect. If either ratio is a function of the cost of capital, then the income tax induces input distortions not associated with an output tax.

But the income tax still has one great advantage, in that it makes allowance for differences between different mineral prospects in the costs of noncapital items: labor, energy, materials and services. Since its revenue is more closely related to cost differences between deposits, the income tax will cause fewer new prospects to remain unexploited, relative to an equal-yield output tax. This advantage probably far outweighs any distortions in capital intensity which it may induce. It should be remembered, however, that the quantitative significance of these tax-induced distortions may well be small relative to profits and tax revenues on mineral production. It is quite possible that the difference between the welfare cost of an income tax and an output tax might be outweighed by the difference in costs of administration and evasion. In any case the income tax revenue is more closely related to differences between deposits and more likely to involve a larger amount of resource rents available for appropriation by the government. Though both an income tax and an output tax lead to underinvestment in development of new prospects, the effect of the output tax will probably be more severe.

Though this section has concentrated on new, unexploited prospects, the same results hold for long-run maintenance and expansion of capacity of existing mines.

Taxes under Non-Constant Prices. The discussion up to now has utilized the simplifying assumption of known, constant output prices and input costs. But in reality, both mineral prices and to a lesser extent costs of inputs are subject to sharp movements, as noted elsewhere in this study.[3] In this section we briefly consider the implications of nonconstant prices for mining taxation.

In recent years there has been a revival of interest in the economic theory of natural resource extraction, which was first outlined by Harold Hotelling nearly a half-century ago.[4] This theory shows that if prices are changing, it may not be optimal to encourage production of all mineral deposits which would yield positive rents (in the Hotelling usage of the term "rents"[e]), contrary to the prescription under constant prices and technology. From the point of view of a price-taking exporting country, the argument is that as long as the increase in the rental on a deposit in the ground exceeds the return on an equivalent sum in the best alternative investment, the deposit should remain unexploited.

The argument can be illustrated as follows. Assume a cost function $C = C(X)$, $C_X > 0$ where C is cost per unit of output, and X is cumulative output.

This cost function implies that future resources can be produced only at increasingly higher cost per unit of output. One example of this would occur in mineral production where remaining mineral deposits are known to have successively lower grades. An optimal plan would direct that the most valuable (those with highest grade and lowest cost) be developed first, then the next most valuable, and so on. When cumulative production reaches X^*, the next unit of output will cost $C(X^*)$. If the expected rate of price increase is constant at λ, i.e., $\frac{1}{P}\frac{dP}{dt} = \lambda$, where P is output price per unit, then the rate of increase of mine rent on this unit is:

$$\frac{1}{P - C(X^*)}\frac{dP}{dt}$$

which will decline over time. If the rate of increase of rents on this unit currently exceeds the rate of interest, r, it pays to delay extraction of this unit until:

$$\frac{1}{P - C(X^*)}\frac{dP}{dt} = \frac{P\lambda}{P - C(X^*)} = r.$$

(It is clear that extraction will never occur as long as $\lambda > r$.)

Now a unit tax T will alter the above condition to:

$$\frac{P\lambda}{P - T - C(X^*)} = r.$$

[e]"Hotelling rent" may be defined as the *net* price resource owners receive after payment of all costs of extraction.

Clearly the tax will cause a delay in exploitation of new deposits of a given grade. If the country is a price-taking exporter, this delay will have some social cost, as long as the social rate of time preference is positive. But if the country as a whole (not the individual mines) has some monopoly power, such delays may actually increase national welfare.[5] Further investigation of this question for any specific mineral would necessitate a careful study of the international market structure. It may be that a uniform international mineral tax would be a useful instrument for international mineral cartelization.

Though much more can be said about these intertemporal natural resource models, their practical importance for our purposes may be rather limited. In mining, the major output decision is the installation of fixed capacity which has a life of many years. In this context attempts to follow year to year fluctuations in the price of output (or interest rates) would be rather misguided. What would be more important are the long-run price and cost trends. But in metals the historical price trends are very hard to identify, and appear in any case to be rather gently sloped. For copper, Herfindahl found no significant price trend from the close of World War I to the mid-1950s.[6] Exhaustion was offset by exploration and technological developments. For tin, Smith and Schink identify a trend line which shows a less than 30 percent real price rise in the twenty years of 1956-1975.[7] Costs may well have had a steeper upward trend over this period, so rents may have been increasing even more slowly than prices over this period. World commodity price fluctuations in the period 1972-1977 should not be allowed to obscure the fact that the historical basis for expectations of long-term increases in rents for most minerals is not very secure.

If prices are not expected to rise in the long term, the resource theory outlined above would predict rapid exhaustion of mineral deposits, which would cause upward pressure on prices. Historical observations imply that the little-understood, time-dependent processes of technological change and exploration confound the predictions of simple resource models.

Finally, price uncertainty can affect the desirability of output royalties. If mineral prices, though uncertain, have no predictable trend, i.e., they follow a random walk, then the margin of exploitation should not include properties which would yield close to zero rent if currently exploited. These properties represent an option on future rents which might accrue if prices rise—therefore the social value of such properties is greater if they are not exploited currently. If for other reasons the state allows a corporation only limited tenure of a development site, the corporation may attach a lower option value to such prospects, since the relevant price-path ends at the close of the lease. In this institutional setting it may be possible to counter premature exploitation of such properties, by imposing a per-unit output royalty. Hence, even under an income tax a small per-unit output royalty, with no deductions, may be a sensible component of the lease contract.

Significance of High-Grading in Bolivia. Although the extent of tax-induced high-grading cannot be determined with any degree of precision in Bolivia, interviews with managers of mines suggest that some high-grading does in fact occur. The concept of a cutoff grade, below which ore will not be taken, is well established. At least some of the medium miners have a thorough grasp on the kinds of calculations that a profit-seeking manager would make. Moreover, they know that a change in the royalty would change the cutoff grade. A shift away from mining taxes geared to output is likely to lower the cutoff grade for the better managed medium mines. The effect on the small miners is likely to be less. Most do not use more than very rudimentary management tools to evaluate alternative actions, and their technology may limit amounts taken of low-grade ores more than does the tax system. However, it is likely that some will respond eventually, as they emulate the larger mines. The state minerals enterprise presents a more difficult case. Although its size would suggest that it would have rational management tools that would allow it to respond to changes such as those in taxes, available evidence suggests that the enterprise's control systems are such that the response is not likely to be quick or predictable. (See Chapter Nine.)

In fact, for the state enterprise, cutoff grades for all important mines are calculated at headquarters and relayed to the individual mine sites. However, managers at the mine sites also make separate cutoff calculations, which in two of the enterprise's largest mines were 40 percent lower than those prescribed by headquarters, due to the fact that the headquarters calculations included both direct and indirect mining costs, whereas the mine site managers (correctly) used only direct costs in the computations. In both cases, average grades taken were governed by cutoff grade decisions at the mine-site. Thus, the response to a change in the tax system may not occur quickly in the cutoff instructions, but it is likely to occur at the mine level, after a period of trial and error.

Present Taxes and Decisions Concerning Concentration. Many government officials and private miners in Bolivia and elsewhere are concerned with the issue of low-grade export concentrates. Particularly in Bolivia, some public officials argue that the government should adopt a policy of discouraging export of low-grade concentrates, primarily on grounds that national returns on such exports are inadequate. Four reasons for this inadequacy are most often cited:

1. Low-grade concentrates bear excessively high freight charges, since heavy transport costs are paid on what amounts to waste.
2. Treatment costs abroad are said to be excessively high, since there are few foreign smelters designed for treatment of low-grade concentrates and those that do exist may extract monopoly profits on such ore. Further, the penalties for impurities (arsenic in zinc, sulfides in tin, etc.) are said to be especially large for overseas smelting of low-grade concentrates.
3. The market for low-grade concentrates is said to be shrinking relative to overall markets (especially in the case of copper).
4. Bolivian exporters do not typically get paid for the complex minerals present in exports of low-grade concentrate.

However, the largely unintended thrust of present mineral tax policy in Bolivia is to encourage export of low-grade concentrates. The objective of this special treatment—providing tax relief for small-scale mining—is praiseworthy in broad social terms, but the effects are in conflict with other policy objectives and involve some significant social costs as well.

Four elements of the fiscal system affect the choice of technology and/or process intensity in concentration of ores. These are the *regalía*, the export tax, the excess production incentive and the subsidy to domestic consumption of petroleum. The prime effect is to induce producers to export low-grade concentrates with potentially greater economic losses than those of high-grading of ore bodies. In evaluating this issue, it is important to remember that the *regalía* and the export tax are not applied until after the concentration stage.

Consider first a very simple model of a concentration process: a series of shaking tables with the necessary milling stages.[f] After the ore is passed through the first step, a commercial concentrate of perhaps 20-40 percent is obtained. The remaining material can be divided into middlings and tailings. The manager must decide whether the middlings should be reprocessed to yield more metal.

[f]Note that the various steps in concentration may extend beyond the simple addition of more shaking tables, as in the illustration. The additional steps might consist of a volatilization plant, for example.

There is a cost, of course, to running the middling through perhaps another grinding process and then through another shaking table. The manager must decide whether the return is sufficient to justify the additional investment in concentrating equipment. In calculating the return, the *regalía* and export tax will appear as an additional cost, since they are a function of the concentrates actually produced, *not* of the ore taken from the mine. The higher the royalty, the fewer concentration steps will be justified and the more contained metal will be wasted.[g]

The above remarks apply to the major output-related taxes in general. In addition, certain specific features of the *regalía*, the export tax, the incentive for excess production and a petroleum subsidy affect concentration decisions. The *regalía* for copper is reduced substantially for intermediate and low-grade concentrates (see Chapter Six). Under the export tax, the applicable rate for tin and copper is less for low-grade concentrates for *all* producers. In the case of all major metals, the export tax is generally levied at lower rates for small miners (and even lower rates for cooperatives) than is the case for COMIBOL and the medium mines. Within this group, export taxes on tin and copper are much lower for low-grade concentrates. A major part of the rationale for lower export taxes on the smaller producers is that they tend to work lower-grade *deposits* than is the case for COMIBOL and the medium mines. This rationale has been called into question, and evidence is presented in Chapter Eight that indicates that this is not, by and large, a valid assumption. To be sure, many small miners and cooperatives export lower-grade concentrates because the concentrates arise from low-grade ore bodies. But it is also true that rich deposits of large mining firms can yield low-grade concentrates and that poor deposits of small firms can yield high-grade concentrates.[h] Much depends on the types of incentives offered for further concentration efforts. The above mentioned special features of the *regalía* and the export tax serve as significant incentives to produce low-grade concentrates, and there is some evidence that the private mining sector has responded strongly to these incentives, especially in the case of tin.

The differential tax upon tin of high metal content has apparently

[g]The manager of the concentrating facilities in one large tin mine indicated that a 13¢ increase in the price of tin (from \$3.07 per fine pound to \$3.20) would make further beneficiation efforts profitable. Reduction of royalty would have the same effect.

[h]For the most endangered small mines and cooperatives working the once-fabled Cerro Potosi, it is, however, clear that low-grade concentrates derive from very low-grade (*baja ley*) ore bodies.

helped to reduce the effort expended on concentration by private miners, as shown in Table 7-4. For the large private mines,[i] exports of grades below 35 percent (the break-point between "high" and "low" concentrate categories) increased markedly from 1973 (the first full year of existence for the export tax) while exports of all higher grades decreased. For small mines, there appears to have been a transfer from the 25-34.9 percent category into the 15-24.9 percent category, partly, if not wholly, in reaction to the very high differential of four percent of the value of output for low-grade tin produced by this group of enterprises.

It is interesting to note that the state enterprise appears to have adjusted in the wrong direction, increasing its exports of ore in the 35-39.9 percent group, and decreasing exports in the 25-34.9 percent group. This reaction implies either that tax incentives do not affect the enterprise's decisions or that other considerations overrode the tax incentives. If tax incentives did not in fact affect its performance, this indicates a pattern of response which has to be taken into consideration when predicting the response of the enterprise to other tax schemes.

The excess production incentive described in Chapter Six reduces the export tax and royalty by 50 percent for sales of concentrates in excess of the average of 1971 and 1972. Since this incentive is based

Table 7-4. Classification of Exports by Grade 1973 and 1974 fine tons of metal (TMF)

Grade of Concentrate %	*State Minerals Enterprise*			*Medium Mines*			*Small Mines*		
	1973	*1974*	*Change*	*1973*	*1974*	*Change*	*1973*	*1974*	*Change*
0-14.9	82	498	416	0	14	14	0	0	0
15-24.9	1288	2667	1795	351	910	559	221	971	750
25-34.9	1590	824	–766	932	1541	609	782	298	–484
35-44.9	1930	4034	2104	1221	102	–1119	499	158	–341
45-54.9	4973	7027	2054	714	698	–16	1219	1708	489
55-64.9	4708	174	–4534	815	228	–587	15	0	–15
65-74.9	26	0	–26	175	53	–122	0	0	0
75-84.9	0	0	0	0	0	0	0	0	0

Source: Ministerio de Minería y Metalurgia.

---

[i]Several private firms indicated that in order to take advantage of lower tax rates on lower grade concentrates, they merely mixed low-grade concentrates with high-grade concentrates, when original concentrate grade was not far above the point at which the rates were reduced. The chief technical defect of this provision, then, is the use of just two tax rates on one "break-point."

on metal content, it would appear to encourage production of low-grade concentrates; for as the grade of metal is increased by concentration, a certain percentage of metal is lost in the process. Therefore, the incentive tends to discourage high concentration. Accordingly, the figures in Table 7-4 may partially reflect the reaction to the excess production incentive.

Of course, the failure to concentrate output sufficiently may increase the cost to the refiner-smelter, if the smelter process is adjusted to the previous average grade. This is true in the case of ENAFBOL, the state smelter. (Also, there is a significant loss in the extra transport charges required to ship the same amount of metal because of the lower grade of concentrate.) Before introduction of the present export tax scale, average grade of input in ENAFBOL was, in 1971-72, about 42 percent.[8] After introduction of the new scale, ENAFBOL's average input grade fell to 39.2 percent. Part of this decline (which may or may not be statistically significant) may reflect the effects of the incentives in the tax scales.

The incentives to produce lower-grade concentrates are also enhanced by a heavy subsidy on domestic consumption of petroleum products[j] which makes transport cheaper, thereby giving an additional differential subsidy to low-grade concentrates. Therefore, the export tax regime, certain features of the *regalía* and the excess production incentive all increase the economic cost of the petroleum subsidy by increasing domestic consumption of crude, while the subsidy increases the costs due to the tax incentive.

Effects on Investment in New Capacity: Macro Considerations

There is a common and strongly held view in Bolivia among both government officials concerned with mining issues and managers of mining enterprises that investment in new capacity in the sector has been grossly inadequate for several years. (See Chapter Two.) A particularly frequent assertion is that investment in exploration and development has been and continues to be well below "optimal" levels, and tax measures should be taken to rectify the situation. In this context, two major questions arise: (a) what would be "optimal" levels of exploration expenditure (public and private) and development, exploitation and processing expenditures over time?; (b) what is the role that can be played by tax reform in helping to ensure

[j]The magnitude of the subsidy for most of 1975 was approximately $14.00 per barrel, but was reduced to about $7.45 per barrel in 1976. The effect was that Bolivian firms and consumers paid, in general, only about 57 percent of the opportunity costs of petroleum fuel utilized for nonautomotive purposes in 1976.

these levels? While it is difficult to know what the proper level of investment should be, we do know the direction of the effect of mining investment of the distortions created by the present tax and subsidy system. The amount of investment will move in the right direction to increase the economic welfare of the country if these distortions are decreased.

It was noted in Chapter One that there are no compelling reasons why investment in mining should not be generally subject to the same criteria that normally would govern investment in any other type of capital goods. Thus, allocations of investment should be made based on the criterion of "whether a prospective investment will pay"; that is, what the investment will be worth to society and what has to be given up (the value of other products) in order to undertake the investment.

It is often claimed that investment in the mining sector merits favored treatment, including tax policy, because of an inherent tendency for "underinvestment" in the sector. This tendency is generally attributed to the special risks that mining allegedly faces, and to the "public good" nature of a portion of expenditures on exploration.[9] That is, the information generated in exploration activity can be used by any number of firms without reducing the amount of utility of such information available to the exploring firm. The argument about special risks has merit only if expected returns from mining are indeed highly variable and if mine owners are repelled by risk to a greater extent than investors in other activities. Some experts believe that the riskiness of mineral exploration has been exaggerated, particularly when possibilities for pooling risks among firms as, for example, in exploration syndicates, are present.[10] Even if these risks have not been exaggerated, it is still difficult to argue that special tax concessions are necessary to compensate for alleged underinvestment in mining. It is probably true that expected returns from investment in mining are highly variable in Bolivia.[k] It is certainly true that the tax system contributes to this variability, along with many other aspects of past and present government policy. Insofar as "special risks" for investment in mining are traceable to unintended effects of *government policy*, they are largely correctable. But the presence of policy-induced risks does not argue for offsetting tax concessions to mining investment; rather, it argues for reform of the offending policies, whether tax or other measures.

[k]Precise statements on this point would require access to company books, or failing that, evidence on the incidence of bankruptcies in the mining sector relative to other sectors, returns from equity investment in the Bolivian mining industry, and interest rate differentials charged in "arms-length" loans to mining vis-à-vis other sectors. None of this data is readily available in Bolivia.

With the important qualifications that apply if social policy considerations dictate a gradual shift in resources away from the mining sector, the optimal volume of mining investment in the long run should in general be viewed on the same footing as investment in other sectors: in terms of prospective economic returns. To proceed otherwise is to reduce potential economic welfare below the economy's capabilities.

Exploration and Development. Over the long run, additions to new capacity in the mining sector require outlays on exploration and development of new ore bodies. Information cited in both Chapter Two and this chapter suggests that investment in both types of activity has been stagnant in Bolivia for several years. Public discussion of this issue in Bolivia has tended to focus primarily upon the lack of incentives for investment in exploration. But investment in developmental activities (preparation of a known ore body for extraction) is no less critical. In fact, data from economies with mature and viable mining sectors indicate that outlays for exploration are but a small fraction of development outlays. For example, in 1960 expenditures for exploration in the United States mining sector (exclusive of hydrocarbons) amounted to only 8.5 percent of combined expenditures for exploration and development (or 9.2 percent of total development outlays alone).[11] Data on which to base an estimate of the optimal mix between exploration and development in Bolivia are lacking, but tax reform geared only to revitalization of exploration efforts would clearly be wide of the mark. While tax reform intended to stimulate only exploration would be administratively feasible, it would likely have limited impact relative to a reform focused upon both exploration and development. Unfortunately, however, the distinction between development costs and general mining costs is not nearly so sharp as that between exploration and general mining costs. Hence any attempt to encourage investment outlays on mine development in a program of sectoral revitalization would, in the absence of rather arbitrary rules for determining eligible development costs,[1] be administratively difficult under present conditions.

The present tax regime provides disincentives for both exploration and development, relative to the situation under a proportional income tax yielding the same amount of real revenue, as the present

[1]Such arbitrary restrictions might include restricting eligibility of only development costs to *newly discovered* ore bodies, restriction of eligibility to capital items only (labor costs are a substantial portion of development costs), etc.

system is unable to take any account of differences in costs among various enterprises. Thus, those firms that spend money on exploration and development do not have a lower tax under the present system, although they would under an income tax.

The problems that the present tax structure generates if mine costs vary are obvious. Consider two medium mines for which *presumed* costs under the *regalía* are identical. If the real costs in the two mines are different, the effect of the present royalty and export tax system is to place a tax on the higher-cost mine that is a greater percentage of its profits than on the low-cost mine. Such a system can place such a burden on the high cost mine that it may either be forced to close, or fail to generate a sufficient cash flow to make needed investments in exploration and development, or have insufficient cash flow to provide an adequate base for borrowing to finance exploration and/or development.

Whether there is a large variance in costs among mines of the same category is an empirical question that can only be answered with field research: the evidence gained from our survey of the medium mining sector and COMIBOL suggests that variance in costs between firms is substantial (see Chapter Three).[m] In fact, possibly two of the largest potential sources of variance in costs among mines are expenditures for exploration and mine development. Under the existing system, the mine that spends large sums on these activities pays the same taxes as a mine that spends none. Whether mines will actually explore more if they can pay less taxes because of expenditures made on exploration is a moot question. However, private miners indicate great eagerness in securing exploration money from abroad. That money is supposedly more forthcoming because the foreign investor can deduct such expenditures in calculating his income tax at home.

The present incentive for excess production was originally intended as compensation for some of the depressing effects of the existing tax system upon investment in exploration and mine development. However, quite apart from the very substantial administrative problems involved in implementing and policing this system, the incentive is ill-suited for encouraging such investment. There are more effective and less expensive ways to provide incentives for exploration and development.

An income tax system gives an incentive only to the extent that exploration and development are actually undertaken. Under the

[m]The data utilized may suppress potential interfirm cost variance arising from differential effort in exploration and development, as firms may explore and develop less under the existing system than they would with income taxes.

current incentive system used in Bolivia, the effect is only to offer the equivalent of higher prices for minerals. The miner may simply exploit known reserves more rapidly rather than increasing the reserves, for he receives an equal tax break no matter which path he follows. Under the alternative income tax system the tax break is earned only for efforts to increase reserves. It is better to gear tax reductions to money spent on exploration and development rather than to volume of production. An income tax system would either allow costs of exploration and development to be expensed as they are incurred or would allow a predetermined portion to be amortized over a set period. Some theoretical and empirical work has indicated that a more effective incentive is provided if the amount of money at risk is reduced, rather than if the potential gain is increased. Writing off expenditures as current costs under an income tax system reduces the money at risk, as does, to a lesser extent, rapid amortization of such outlays, while the reduction of *regalía* and export taxes increases the possible gain.

Finally, the structure and level of the existing system of mining landrents serves to reduce the scope and dampen incentives for exploration. The landrent rates are so low as to make it profitable for a person to hold as unexploited reserves a large number of potentially rich mineral tracts. When combined with the very sizable sections of the country's mining belt held idle by the government in the fiscal reserves, this has meant that much of the most promising areas has been unavailable for exploration. Since annual land rent levels for a medium-sized exploration claim (10,000 hectares) amount to only U.S. $600 and since it is a simple task for one person to control as many as ten medium-sized claims, the holding of large tracts of idle claims is not an unduly expensive proposition. A higher rate for exploration land rents, particularly in a land rent structure requiring escalation of rates from year to year, would provide strong incentives for either active exploration by the claim-holder so as to generate income to pay the land rent, or release of the tract to those who will explore.

Exploitation, Beneficiation and Smelting-Refining. The principal problems of the present tax system for exploitation and processing activities are the system's effects on use of existing capacity, choice of technology and process mix, discussed earlier in this chapter. In addition, relative to a proportional income tax, the existing tax structure serves to discourage some types of investment in new capacity at the exploitation and beneficiation stages, while the administration of the major mining taxes slightly favors investment in smelting and refining.

Relative to a proportional income tax, the present tax structure discourages investment in exploitation in much the same way as it dampens incentives for exploration and development. Under the present system, if there are substantial cost variations (between mines of the same category) due to the fact that some spend substantial amounts on, say, investment in shaft ventilation, provision of emergency access tunnels, or other investment related to health and safety standards, while other enterprises spend little or nothing on such installations, the former will pay the same taxes as the latter. This is so even though the real costs of the safety-conscious mine are higher. If the level of presumed costs under the present *regalía* were high enough to cover all such outlays, then investment in such installations would not be discouraged in any *absolute* sense, but a relatively greater volume of investment resources would tend to continue to flow to mines where such investment is not undertaken, and there would be an implicit (and expensive) governmental subsidy for doing so. The subsidy would arise in the form of the tax revenues that would be foregone by attempting to set presumed costs at levels sufficiently high to accomodate safety-conscious firms.

In this sense, the present tax treatment of the mining sector contains an unmistakable message for mine owners: investment geared to improving health and safety conditions in underground mines carries a negative premium and is to be discouraged; firms that invest minimal amounts in ventilation, emergency access tunnels and worker safety are to be rewarded with maximum returns. This amounts to a decision rule that states that human resources are, relative to natural resources, to be viewed as cheap and expendable. It might be argued that this situation would rectify itself over the long run, as labor costs in the more "unsafe" mines would rise relative to those for "safe" mines, owing both to higher compensation for risk that might be received by labor in the former, and to the lower productivity of infirm or injured labor in such mines. Such a result would obtain in the long run, provided labor markets were perfect and in the absence of substantial surplus labor in the overall economy. But quite apart from the fact that the long run is likely to exceed the work span of most present miners, labor markets in Bolivia are far from perfect and surplus labor promises to be a continuing problem for the relevant future.

The present tax structure discriminates against investment in exploitation and beneficiation in yet another way, also related to the human resources question. Mining enterprises of any appreciable size are required by law, once exploitation begins, to provide housing, primary education facilities and health care for workers and their families at the mine site. To the extent that investment in such facilities is more or less uniform as between firms of the same category, and is on the average recognized and reflected in the level of presumed costs allowed by the government in calculation of *regalía* liability, such investment is not discouraged in any absolute sense. However, for various reasons (differences in labor bargaining power at different mines sites, differences in the extent to which mine owners attempt to fulfill governmentally mandated "social" requirements), the real costs for these purposes can differ widely between firms, with the same results for investment in exploitation and concentration as observed for exploration when large cost variance is present. As in the other cases, a proportional tax on net income (after expensing or amortization of such investment) would not penalize investment to anything like the present extent.

Finally, the structure of the present system serves to dampen investment incentives in *all phases* of mining activity, relative to an income tax with provision for loss carry-forward and carry-back. Under the present system, taxes are due on each shipment and are assessed without regard for profits or losses on previous shipments. Tax analysts are unanimous regarding the negative effects on risky investment implicit in *income* taxes without loss carry-forward provisions when *annual* accounting measures of income are used, rather than longer periods. The disincentive effects for risky investment are even stronger when output rather than income taxes are relied upon for substantial amounts of revenue and where tax is assessed on per *shipment* values.

Provision for loss carry-forward serves to encourage investment by reducing the long-run difference in tax rates between more and less risky enterprises. Lack of some means for allowing carry-forward of losses constitutes a formidable barrier to investment in activity with variable profit and loss experience over a period of years, as has been true for mining activity in Bolivia.[12]

The manner in which the present system is administered serves to provide some differential incentives to smelting and refining. Two points are particularly relevant: apparent widespread evasion (and/or "tax farming") by private smelters, and the income tax treatment of the state-owned tin smelter and bismuth refinery. Private smelters are not only subject to the generally applicable income tax, but must collect and withhold *regalía* and export taxes on sales to them by mining enterprises. The private smelters evidently pay no income tax and often fail to remit all *regalías* and taxes they collect on their purchases. Together, these subsidies should amount to a potent incentive to existing smelters for expansion of capacity. Aside from such administrative/compliance questions, a lower (20 percent) rate is applicable to the profits of the state smelter (ENAFBOL) than to business in general (30 percent). This constitutes a small tax incentive for investment in smelting by the enterprise.

Other Tax-Related Questions. An income tax offers two other advantages over royalties or presumptive income taxes: a sense of stability or security to the entrepreneur, and tax credits abroad for the foreign investor. Most studies of investment behavior strongly indicate that uncertainty is a major factor in discouraging investment. Taxes can provide a source of uncertainty. Uncertainty as seen by the enterprise is especially great if the form of tax is an export tax on gross value or a tax like the *regalía*. Under such a fiscal regime, the entrepreneur bears the risk of having to pay a tax even if his venture is not profitable. Or he may have to pay a tax that is greater than his profits, even if he has net earnings. By contrast, he need pay an income tax only if he makes profits. This is without doubt a major reason for the general shift from royalty-type taxation to income tax in mining agreements in the 1950's throughout the developing world. With the risk attached to royalties, the investor was ready to accept a higher expected tax payment under an income tax regime.

Note that it is the system, not any rigidity in rates, that leads to the above conclusion. No matter what changes are made in the income tax rates, the tax cannot exceed profits; and no tax will be paid if losses are incurred. In fact, the entrepreneur will probably expect an income tax rate of not greater than, say, 60 percent or 70 percent at the most, while he will fear an effective tax rate of 100 percent or more under a system where (as in Bolivia) exclusive revenue reliance is placed on gross output taxes. (See, for example, Figure 6-3 in Chapter Six.)

A high degree of variability in tax laws can in itself also be a source of enterprise uncertainty, quite apart from questions of tax structure. Tax changes serve to increase uncertainty on the part of enterprise owners regarding the future reliability of tax laws and this in turn may lead to a curtailment of investment in activities affected by such changes. Further, this effect obtains even if short-term tax collections do not increase as a result of the policy change.[13] The more frequent the changes, the greater the uncertainty. The Bolivian tax climate was particularly unstable from 1952 through the early seventies: tax bases or tax rates were altered with great frequency and with little or no prior warning.[14]

The issue of tax credits is important for foreign investors and, by extension, their possible joint venture partners in Bolivia. In general, capital-exporting countries allow their foreign investors to subtract from home taxes due on foreign source income those taxes already paid abroad on such income. These credits may be given unilaterally or by tax treaty. However, such credits are usually available only if the tax in the overseas country takes the form of an income tax. A few capital-exporting countries, e.g., France, exempt *all* profits earned abroad from corporate income tax. But when such profits are distributed to the French shareholder, they are subject to the French personal income tax.

Whether Bolivia's *regalía* or some improved version of a presumptive income tax would qualify for credits is dependent on the ruling of the tax authorities in the foreign investor's home country. Having to depend on a ruling adds another element of uncertainty for the potential foreign investor. A true income tax would be automatically creditable in, say, the United States. The *regalía* and similar taxes in lieu of income taxes are questionable. The creditability of such a tax is governed by various criteria established by the tax authorities in the foreign investor's home country. In the United States, such factors as the legislative history of taxation on the mining sector would be considered. Other criteria would be the presence or absence of a specific exemption from generally applicable income taxes, the form and substance of any tax in lieu of the generally applicable tax, and the reasons for which mining was exempted. It is known that many taxes similar to the present Bolivian *regalía* were ruled noncreditable by U.S. tax authorities in the early sixties, and that some capital-exporting countries are more liberal than others in allowing tax credits. In particular, the European countries and Canada would be more likely to allow credits for *presumptive* income taxes on mining than would the United States. (For further commentary on the tax credit issue, see the final chapter of this book.)

Effects on Government Revenues

Although taxes bearing upon the mineral sector have furnished a large share of government revenues in recent years, both directly and indirectly, revenue performance of the system leaves much to be desired. It has failed to appropriate for the state much of the large resource rents that have been generated in years of rising minerals prices (e.g., 1973 and 1974), and has become oppressive upon enterprises in years of poor market conditions (e.g., 1971 and 1975), taxing many on income they did not have. The system has not shown a high degree of responsiveness to inflation and, when there have been discretionary adjustments to compensate for inflation, they have occurred after very long lags. Thus, the effects have, at various times, tended to be traumatic for the Treasury (when sharp declines in expected revenues have occurred, as in 1975) and for the mining enterprises (when stiff new taxes have been suddenly imposed, as in 1972). However, not all of these features are inherent in the existing mineral tax structure; some are due to departures from the basic structure as initially established. To purge the system of structural defects, thorough going reform would be required, but the nonstructural defects could be dealt with through relatively minor adjustments. The following discussion attempts to separate the structural from the nonstructural defects insofar as revenue performance is concerned.

Structural Issues. The most glaring structural defect of the system is the incentive for excess production. While relative to total revenue the actual revenue loss from this incentive was not large in the year following its creation in 1973, future revenue losses promise to be substantial if the incentive is retained.[15] But more important, the 50 percent reduction in *regalía* and export taxes under the incentive negates the efforts to collect a larger share of windfall gains for the government when world market prices are high. High world prices of metals are likely to induce increased exports. Under the present incentive for excess production, an increase in exports caused by favorable prices will give rise to a reduction in *regalía* and export taxes at precisely the time those taxes should be serving to extract some of the high profits that result from high prices. A proportional income tax would automatically capture a substantial part of such windfall gains.

A second structural revenue problem is the heavy reliance upon mineral export taxes, which contribute nearly a third of total mineral tax revenues, and over 10 percent of total central government revenues. Although the export tax is ad valorem, it is levied at a constant rate regardless of market developments. Thus, under conditions of rising minerals prices, the ratio of the tax to income (presumed or actual) falls dramatically. For example, in the case of tin the effective rate of the tax on presumed income falls from 20 percent at a $2.40 price to 12 percent at a price of $4.00. The same pattern holds for antimony, wolfram, zinc, copper and lead. A proportional income tax would capture a higher share of windfall profits than the present export tax, as the effective tax rate would be constant at all income levels.

A third structural defect is the very favorable *regalía* and export tax treatment given to some minerals, including bismuth and several "minor" minerals, such as iron ore and almost all nonmetallic minerals.[16] For all these minerals, the *regalía* and export tax are levied on gross values at very low rates relative to, say, tin and antimony. Unless the ratio between gross value and net income is far lower for these minerals than for the more heavily taxed ones, a proportional income tax levied would generate substantially more revenues than the existing system, except at very low prices.

Fourth, whatever their merits on nonrevenue criteria, the various forms of tax relief provided (under both the *regalía* and the export tax) to exports of low-grade concentrates serve to weaken the revenue potential of the present mining tax system under both good and bad market conditions. While some exporters of low-grade concentrates may indeed have low income relative to gross value of their exports, it is clear that many others, particularly some of the larger medium mining firms, do not. Tax relief in this form erodes the revenue potential of the present system relative to a proportional income tax *and* relative to the potential of the present system.

Non-Structural Issues. Other revenue defects of the present tax regime of mining taxes are not inherent in the system; rather, they stem from either the way in which the system has been implemented and/or little-noticed limits placed on maximum effective rates of the *regalía*.

The most obvious problem is the limitation placed on effective rates of *regalía* for all the major minerals, but most particularly lead and zinc. For tin, antimony, wolfram and copper, the maximum effective rate of *regalía* on presumed income (*regalía* as a percent of presumed income) is 38 percent. For zinc the maximum is 14 percent, while for lead it is 16 percent. These provisions severely limited the government's ability to capture a larger share of the windfall income generated in the sector in 1973-74, when minerals prices exceeded those given in the *regalía* schedules by substantial amounts. In fact, for most minerals these provisions have meant that the marginal *regalía* rate at very high prices is lower than that at low prices. There is nothing inherently wrong with this, on revenue grounds, but this effect was clearly unintended. For example, in the case of tin, the marginal rate for the *regalía* (the rate applicable to increments in presumed income) climbs from 27 percent at a price of $1.60 to 51 percent when prices reach $2.90. But at higher prices and higher levels of presumed income, the marginal rate falls to 38 percent, the limit on effective *regalía* rates.

For zinc and lead, the limits placed on maximum effective rates

(14 percent and 16 percent) were apparently designed to compensate these producers for the inability of the government to adjust presumed costs to account for all "realization costs."[n] For these minerals, realization costs are not only much higher as a percent of gross value, but are partially geared to prices in world markets for lead and zinc. The establishment of such low maximum rates of tax for lead and zinc was clearly an inferior method of dealing with the problem. (See Chapter Eight.) The effect was that substantial revenues from these minerals were foregone in the 1973-74 minerals boom and even at early 1975 prices revenue losses from this limitation were significant.[o]

A second nonstructural problem has been the very long lags between discretionary adjustments in presumptive costs in the face of rising actual costs.[17] This has not served to reduce revenues, but to make them higher than they would have been in the absence of presumed cost adjustments to reflect actual cost developments. However, the long periods between adjustments in presumed costs (as long as nine years in the case of tin) have meant that *regalía* revenues per unit of export declined sharply in periods just after adjustments were made. Fortunately for the Treasury, and for national macroeconomic stability, two of the upward adjustments in presumed costs occurred in periods of rising mineral export prices (early 1970 for antimony, wolfram and copper; early 1974 for all minerals). However, the most recent adjustment announced in September 1975 (effective January 1, 1976) took place in a period of depressed minerals prices.

Administration and Compliance

Although the basic elements of the present mineral tax system undoubtedly involve fewer inherent administrative difficulties and complexities than would a proportional tax on net income, the system is not without such problems. For the *regalía* and the export tax, administrative difficulties are primarily due to past efforts to provide tax relief to certain types of minerals and to specific categories of the producers. As noted in Chapter Six, this has resulted in a bewildering pattern of export tax rates. For the mineral tax structure as a whole, such problems abound, largely due to the multiplicity of other taxes on mining.

---

[n] As discussed in Chapter Six, realization costs include costs of transport from the mine site, other handling costs, and smelting costs.

[o] A limitation of 14 percent on effective rates for zinc also results in lower marginal rates of tax at high prices (and levels of presumed income) than at low prices. The marginal rate on zinc rises from 20 percent at lowest prices to 26 percent at higher prices. At still higher prices, the marginal rate declines to 14 percent.

One of the major administrative problems of both the *regalía* and the export tax has been a lack of verification by customs of assay as reported by mineral exporters at the time of export. The customs agency has not checked the grade of export concentrates until "reliquidations" are made several months later, after final payment is made by foreign purchasers. Failure to verify assay at the time of export results not only in loss of interest on underdeclared funds. It also opens up great scope for collusion between exporters and foreign purchasers to understate values for reliquidation purposes, given the ease with which exporters and foreign buyers (even when unaffiliated) can arrange for "offshore" payments to evade export taxes. Such practices have been well documented in other countries.

The incentive for excess prodoction also gives rise to a number of very thorny administrative problems which require that careful checks are made to verify whether the entirety of a shipment represents an increase in the production of the exporting firm or merely purchases by the exporter from other firms that would otherwise be ineligible for the incentive. Finally, the multiplicity of earmarked taxes and "nuisance" taxes discussed in Chapter Six bring in infinitesimally small amounts of revenues. No fewer than 40 such taxes are assessed on mining exports, claims and property, but together they account for less than one-half of one percent of mining sector taxes.

NOTES TO CHAPTER SEVEN

1. For a review of some of the considerations involved in selecting an appropriate reference tax, see Malcolm Gillis, "Reference Tax Structures in the Appraisal of Development Strategies," (unpublished manuscript), Development Advisory Service, Harvard University, April 1971).

2. See, for example, IBRD, *The Mining and Metallurgical Sector: Bolivia*, vol. II, Annex One: *Exploration and Mining Survey*, Nov. 1971, p. 17; and Eugenio da Vila, "Consideraciones Sobre el Crisis Minero en Bolivia," article presented at National Seminar on Mining Policy, La Paz, June 24, 1975, p. 4.

3. See Chapter Two.

4. Harold Hotelling, "The Economics of Exhaustible Resources," *Journal of Political Economy*, vol. 32, April 1931, pp. 137-75.

5. See J.E. Stiglitz, "Monopoly and the Rate of Extraction of Exhaustible Resources," *American Economic Review*, vol. 66, no. 4, September 1976.

6. "There are sizable gaps in the record of normal years after World War I, but . . . the years that remain yield no suggestion of upward or downward drift over the period as a whole." See p. 28 of Orris C. Herfindahl, "Deterioration, Output and the Price of Copper," selection 10 in David Brooks (ed.), *Resource Economics: Selected Works of Orris C. Herfindahl* (Baltimore: Resources for the Future, 1974).

7. These authors used the deflated L.M.E. (London Metal Exchange) price (1963 U.K. £ per metric ton). See G.W. Smith and G.R. Schink, "The International Tin Agreement: A Reassessment," *Economic Journal*, vol. 86, December 1976, Figure 2, p. 724. They claim that the International Tin Council buffer stock made little difference to the price path. Unfortunately, national stockpiles are ignored entirely.

8. ENAFBOL, *Annual Report* (La Paz, 1974).

9. For a further discussion of this point, see Orris Herfindahl in Brooks, *op. cit.*, p. 239.

10. See Chapter Four. See also A.C. Harberger, *Taxation and Welfare* (Boston: Little Browk, 1972), Chapter 11, pp. 215-216.

11. The percentage of exploration to total outlays on exploration and development in metallic minerals ranged from a low of 0.2 percent in bauxite firms to 13.1 percent in copper firms. For phosphate rock, then in a period of substantial expansion, the percentage was 3.68 percent. Source: U.S. Congress, House Ways and Means Committee, *Hearings on the President's 1963 Tax Message*, Exhibit 10, "Depletion Survey, 1958-1960," Table 9c, p. 321.

12. For a brief but cogent analysis of the desirability of loss carry-forward provisions in another developing country in Latin America, see Richard Slitor, "Reform of the Business Tax Structure," in Richard Musgrave and Malcolm Gillis (eds.), *Fiscal Reform for Colombia* (Cambridge, Mass.: International Tax Program, Harvard University, 1971), pp. 501-502.

13. For a clear discussion on this range of issues, see Martin Feldstein, "On the Theory of Tax Reform," Harvard Institute for Economic Research Discussion Paper No. 408, April 1975.

14. A very good illustration of the unstable tax climate is the history of the incentive for excess production since it was first announced in October of 1973. See Chapter Six.

15. See Chapter Six for estimates of revenue loss from this incentive.

16. See Chapter Six for tables showing the light tax treatment given to these minerals.

17. See Chapter Six.

Chapter Eight

Alternatives for Mineral Tax Reform

Malcolm Gillis, Glenn Jenkins,
Louis T. Wells and Brian Wright

MINING TAX REFORM IN PERSPECTIVE

The entire Bolivian tax system is in need of thoroughgoing reform.[a] A series of ad hoc adjustments in the system over the years, primarily made in response to recurring short-run fiscal emergencies, has yielded a patchwork overall tax structure that is not only chronically unable to satisfy narrow revenue goals, but in many cases is inconsistent with, and often inimical to, achievement of commonly accepted nonrevenue objectives of tax policy. Such goals might include improved resource allocation, longer-run economic stability, greater equity in income distribution and assurance of adequate national returns on the exploitation of natural resource endowments (see Chapter Four). Although the precise weights attached to each objective are rarely specified, different societies will attach different weights to these and other legitimate goals of tax policy. Even so, it is important that tax reform not proceed in the abstract, but rather should be undertaken with explicit goals in mind and with a fairly clear notion of the extent to which tax reform may contribute to achievement of these goals. That this statement would apply to initiatives for overall tax reform

[a]In late 1975 the Bolivian government created a blue-ribbon commission of international experts, organized and directed by Professor Richard A. Musgrave, to examine virtually all aspects of Bolivian economic policy, and to present alternative proposals designed to enhance the effectiveness of such policies in achieving national goals. The fiscal system, and in particular the tax system, constitutes the prime focus of this commission, which is expected to complete its final report by June of 1977.

appears obvious, but it applies with special force when dealing with sectoral tax issues, where the outcome of tax reform measures focused on one sector (say mining) will be dependent in no small part upon the nature and impact of taxes imposed throughout the economy. For example, to what extent are the "tax problems" of a mining sector a result of badly designed mineral taxes and to what extent are they attributable to the ineffective taxation of other sectors, such as commercial agriculture? In the Bolivian case, the presence of a sizable tax-free sector not only heightens revenue constraints on the scope for a more flexible approach to mining taxation, but also serves to drain resources from the mining sector that might otherwise have been invested there.

It is essential to have a clear idea of what is expected from mining tax reform, whether in Bolivia or elsewhere. First, the tax problems of the sector must be placed in perspective with other difficulties affecting mining, including world market conditions, managerial and technical difficulties, infrastructure problems, other laws and regulations affecting the sector and other factors. It may be concluded from previous chapters that while the tax structure has undoubtedly played a role in stifling growth in productivity and income in Bolivian mining, there is little basis for assigning the greater share of the responsibility to inappropriately designed and administered taxes, rather than to other retarding factors.

Second, no tax reform package geared to overall public policy goals can be entirely free of internal conflict. Thus, the objective of increasing tax revenues from the mining sector may be somewhat at odds with the objective of reducing national vulnerability to instability in world markets, so that tax measures that are geared to both objectives cannot fully satisfy both. Fortunately, many goals of mining tax reform in Bolivia tend to be complementary, so that progress toward achieving one does not necessarily conflict with others. For example, the revenue goal is clearly consistent with that of increasing overall national returns from mineral exploitation and processing. Nonetheless, it is important to avoid overstating the possible benefits from mineral tax reform. For example, mineral tax reform per se cannot be expected to do much to directly improve income distribution. However, if as a result of mining tax reform the state can secure greater revenues by capturing a higher proportion of mineral resource rents in years of favorable market conditions, the scope for direct improvements in income distribution through the expenditure side of the budget may be vastly improved. In another vein, it does not seem useful to think of tax reform as a principal means of insuring the long-term survival of small-scale mining or the continued operation of ancient, highly marginal large-scale mines. If it is government policy that both aims should be pursued, there are more appropriate and more effective policy instruments than taxation at the government's disposal, particularly if the real target is that of employment opportunities rather than encouraging marginal mining activity having a very high social cost.

Third, it is necessary to underline the important distinction between optimal tax *reform* and optimal tax *design*. The reform alternatives offered in this chapter differ from the type of tax structure one might wish to design if it were possible to start anew, with a "clean sheet of paper." Whereas sensible tax reform has no choice but to take the existing tax system as its point of departure, the range of possible options that could be generated by extensive attention to principles of optimal tax *design* is almost limitless. However, the alternative reform measures offered here recognize that optimal taxation depends on the historical context and that tax reform is best viewed as a process of change, not as a once-and-for-all solution.[1]

ALTERNATIVE RECOMMENDATIONS FOR REFORM

It is evident from Chapter Seven that the present mineral tax system in Bolivia compares unfavorably with a proportional income tax of equal yield. Reform of mining taxes is advisable, not only for revenue reasons, but to encourage more efficient and rational use of existing capacity (the efficiency goal) and to create one of the necessary conditions for reversing stagnation of investment in the sector (the growth goal). A major reform effort should focus on all three considerations. But there are some important constraints which limit the range of feasible alternatives for tax reform in Bolivia.

The primary constraint is that imposed by administrative and compliance considerations, problems well worth repeated mention. The administrative machinery in the Ministry of Finance and Ministry of Mines does not have sufficient manpower or adequate structure to implement immediately a highly ambitious, once-and-for-all reform of mining taxes. Furthermore, the management and accounting infrastructure in the mining sector itself (and particularly within the state enterprise and small-producers subsectors) is not at present adequate to allow full compliance with the provisions of highly sophisticated mining tax legislation. Thus, while it is essential that immediate tax reform measures be undertaken with an ultimate "first-best" objective in mind, realism necessitates that progress toward this objective be achieved in discrete steps, over time.

A second constraint, at least for the short term, is imposed by the extraordinary degree of governmental dependence on mineral taxes as a source of revenue. This constraint is not entirely independent of the administrative-compliance constraint; with a more highly developed infrastructure of administration and compliance, the revenue risk involved in switching to a more desirable mineral tax structure would be considerably relaxed. It would also be lessened significantly by reform of taxes on other sectors of the economy, to the extent that such reform served to increase tax revenues from industrial, commercial and agricultural activities.

Given these constraints, there appear to be two basic alternatives for reform of taxes on the mining sector. The first is a true income or profits tax, which would clearly be superior in terms of improving efficiency, reducing waste and encouraging investment growth (both domestic and foreign investment). The second alternative is a bona fide presumptive income tax. Such a tax would be superior on revenue grounds, at least in the short-to-medium term, and could be implemented within a short period of time. Enactment of a bona fide presumptive income tax would also result in substantial progress toward removal or amelioration of many of the adverse effects of the present tax system on efficiency in use of existing capacity and would (although to a much lesser extent in the case of a true income tax) facilitate expanded investment in the sector.

A true income tax for firms in the mining sector, then, represents the ideal toward which mining taxation in Bolivia should evolve. If the long-term vitality of the mining sector is to be assured and if the sector is to continue to effectively perform its role as the major supplier of fiscal and foreign exchange resources for a diversified and balanced economic and social development program, an eventual shift to reliance upon income taxation appears essential. No other reform alternative is nearly so consistent with the ultimate and proximate policy objectives enunciated in Chapter One. However, successful implementation of income taxation would require a far more advanced tax administration and compliance infrastructure than is now present in Bolivia. Such an infrastructure cannot be developed overnight, but requires very careful advance preparations, not to mention substantial government investment in improving the quantity and quality of resources available to the tax administration. To be sure, an expensive crash program might succeed in creating, in rapid order, the kind of tax administration machinery that would be capable of meeting the heavy demands inherent in the operation of income taxation of mining activities. But the caliber of the tax administration is only one side of the coin of effective taxation of mining profits. The compliance capacities and the compliance propensities of the principal mining enterprises must be considered as well.

Chapter Nine demonstrates that compliance capacity in the state enterprise sector is not sufficiently strong to allow immediate implementation of profits taxation for that sector, which by itself accounts for over 60 percent of mineral export values. In the private mining concerns, perhaps four to six of the larger firms could conceivably develop sufficient internal capacity for adequate income tax compliance by 1978, but this would require a truly concerted effort on the part of most of these firms. Any attempt to implement a true income tax (even one limited to these larger enterprises and to the public sector) over the near term would, in the absence of careful groundwork to strengthen tax administration and improve conditions for compliance, involve considerable risks of near-term revenue loss. Furthermore, the magnitude of such a loss is virtually impossible to predict with any degree of precision. This risk of substantial revenue loss could be narrowed materially by carefully laid preparations for an ultimate shift to income taxation, following the enactment of a bona fide presumptive income tax as a first step toward minerals tax reform.

For these reasons, it would perhaps be more appropriate to label the income tax as the medium-to-long-term alternative, and the bona fide presumptive income tax (hereinafter labelled BPT) as the best available short-term alternative. Implementation of the BPT as proposed would in fact facilitate an eventual shift to a true income tax system, but it is important to note that judgments on the appropriate timing of a shift to a true income tax for firms in the mining sector cannot be made independently of income tax developments in sectors other than mining. Especially given the government's heavy dependence upon mineral tax revenues, it is essential that an income tax on mining be meshed into a well running general income tax system. In short, enactment of income taxation for mining enterprises should await evidence of effectiveness in applying income taxes in the rest of the economy, which has been nominally subject to such taxes for a number of years. A premature and hurried effort to install a true income tax on the mining sector may be worse than none at all. The country has already had one brief and unsatisfactory experience (1952-57) with the consequence of haste in implementing net income taxes on mining.

One point merits special mention. If an LDC such as Bolivia wishes to secure substantial inflows of foreign investment into mining, it would be necessary to frame its tax reform legislation so as to insure that its principal tax on the mining sector is creditable against the income taxes of the important capital exporting countries. Bolivian income taxes paid by foreign firms under a true income tax would clearly be creditable against income taxes in all countries where foreign tax credits are allowed. The creditability of the bona fide presumptive income tax is somewhat doubtful.[b]

Shift Toward Income Taxation: An Ideal System and A Shorter-Term Alternative

An Ideal System. While immediate implementation of a true income tax on mining appears inadvisable, there appears to be considerable merit in a brief presentation of the basic features of an ideal income tax for mining (ideal in the sense of its consistency with national goals, but not necessarily in the sense of its feasibility of implementation). This is done not only to provide better perspective for assessing the merits and limitations of the BPT, but to place in sharp focus the benefits to the development of the mining sector that, in the short term at least, must be forgone because of constraints imposed by the present state of tax administration and compliance.

The basic features of the ideal system are: (1) a true entity-level income tax, assessed at a flat rate on net income of all mining firms, based on actual costs and actual revenues during the taxpaying period (one year), plus (2) a separate tax, much like the BPT, to capture extraordinary gains from high minerals prices. When finally implemented the ideal system would *initially* be applied only on COMIBOL and the six to eight medium mining firms which keep adequate records and are subject to external audit by internationally recognized auditing firms. The transition for other medium mining firms would have to be undertaken in gradual fashion. The possibility of applying the income tax to small mining firms with wholly inadequate records cannot be realistically contemplated over the next decade. The details of the components of the ideal system are outlined below.

True Income Tax

Rates: The tax rate might be set at 30 percent to 35 percent of net revenue of mining firms. This level is slightly higher than that

[b]See Chapter Nine for further discussion of this point. The above does not imply that Bolivia or any other country should necessarily shape its major mining taxes to fit the mold preferred by the capital exporting countries. But in cases where governments count heavily upon foreign investment to help develop the sector, the creditability issue merits careful consideration.

applicable to economic sectors other than mining. Residual revenue needs would be met by a separate tax intended to facilitate capture of a portion of mineral rents.

Treatment of Amortizable Expenditures

There should be no provision for percentage depletion allowances such as those that have been contained in past U.S., Canadian or Philippine legislation.[2] Rather, eligible exploration, development, infrastructure and preproduction expenditures would be amortized over periods set by the Ministry of Mines. Given the difficulties involved in verifying past expenditures of this type, it might be necessary in the initial stages to restrict eligibility for amortization to exploration and development expenditures on newly discovered deposits or new projects in general, when these are located in sites away from existing operations.

Other Deductions

1. *Depreciation*: Depreciable assets in the mining sector should generally be subjected to the same depreciation rules as used for comparable assets in other sectors. The benefits to the mining sector of applying special depreciation rules would most likely be outweighed by the costs to the tax administration in devising and applying special rules for this sector.
2. *Other Necessary Expenses*: Other deductions from gross income to arrive at net income are commonly permitted for necessary expenses for carrying on an activity. Such necessary expenses include wages and salaries, repairs, interest, bad debts, certain taxes, and the like. Except as noted just below, all such expenses should be deductible from gross income under the true income tax.
3. *Limitations on Certain Deductions*: Certain deductions for necessary expenses may have to be limited, so as to reduce the scope for large-scale evasion and avoidance, or to achieve other policy goals. These include salaries for owner-managers, payments to foreigners, travel and representational expenses and certain taxes.
   a. *Limits on Deductibility of Salaries*: The majority of medium mining firms in Bolivia are closely held enterprises.[c] A device commonly used elsewhere by such enterprises to avoid income taxation is to pay high salaries to those who control the firms, thereby reducing reported profits. A simple way of handling

[c]Closely held firms are those in which one person or one family holds a sufficient number of shares to affect the firm's actions on any given issue.

this problem is to set a limit on salaries of owner-managers that can be deducted for tax purposes.[d] Appendix D sets forth criteria for applying such a rule, together with alternative methods for determining deductibility limits. Whatever the limit, it should be the same for the mining sector as for business activity in general.

b. *Payments to Foreigners*: In many countries that employ net income taxes on business firms, substantial revenue loss has occurred through loopholes in the taxation of payments made to foreigners. Such payments include dividends, interest, royalty payments and the like. For expenses of this type, no restrictions on deductibility need be applied. Rather, a withholding tax should be applied, similar in structure to the present 30 percent Bolivian withholding tax. In any case, the handling of this problem should follow the same rules as set for Bolivian income taxes in general.

For other types of payments made abroad, such as salaries and home office expenses, the general rule should be followed that such expenses are deductible only if they are for production of Bolivian income. In addition, deductibility of such expenses might be limited to some maximum (say 10 percent) of the taxpayer's taxable income, so as to close off a possible avenue for significant evasion.

c. *Taxes*: Specifically enumerated taxes should be deductible from gross income when these taxes are related to the production of income. Consideration could be given to allowing deductibility of the progressive tax for capturing rents which is outlined below, but if this tax is made deductible, rates of true income tax higher than the 30 to 35 percent range would have to be considered.

Loss Carry-Forward

Without much doubt, the Musgrave Commission proposals for reform of the present income tax will make some provision for carry-forward of losses in all types of business activity. Clearly, such provisions should apply to mining as well, once the shift to income taxation is accomplished for this sector.

A Progressive Tax for Capturing Rents

A proportional income tax by itself would perform very weakly as a means of capturing extraordinary rents generated for the mining

[d]Note that such a provision does not prevent owner-managers from receiving high salaries—it only limits the amount that can be deducted for tax purposes.

sector when prices rise precipitously. Other countries have experimented with various tax devices for capturing such rents, but these measures utilized elsewhere suffer from a number of defects.[e]

If and when a true income tax system is imposed on mining, a complementary levy that is almost identical to the BPT could be utilized for capturing extraordinary rents. The only difference would be that when used in conjunction with an income tax, rates for the complementary levy would be set at much lower levels than proposed for the BPT. This is reasonable, since prime revenue reliance under the ideal system would be placed on the flat-rate net income tax. Rates for the special progressive tax for capturing rents could then range from zero on the lowest brackets of "presumed" income to, say, 40 percent or 50 percent on the highest brackets. But in the short term, in which the BPT must be the prime source of revenue, rates will have to range from 23 percent on the lowest bracket (alternative rate pattern A) to 75 percent on the highest bracket (alternative rate pattern B).

Thus, overall tax rates under the ideal system (true income tax and progressive tax for rent capture) could vary from 30 percent at very low minerals prices to 55 or 60 percent at very high prices. But the firm would always be faced with a 30 percent marginal tax rate, so far as its evasion or avoidance behavior is concerned.[f]

Interim Reform Measures: Bona Fide Presumptive Income Tax

General Considerations. Given the serious constraints on moving to a more ideal mining tax system in the short term, consideration should be given to an interim reform program. The most attractive option, from the point of view of consistency with overall and sectoral policy objectives, is a set of measures involving two principal features: a bona fide presumptive income tax coupled with a low-rate production royalty on all minerals.

The new bona fide presumptive income (BPT) tax would replace the present *regalía*. The low-rate production royalty would replace the present ad valorem export taxes on gross mineral values. The new system would make no distinction between the state mineral enter-

[e]The Philippine measures of 1974 and the British Columbia device implemented in the same year are two good examples. (See Chapter Five.)

[f]In other words, the firm can take steps to evade the 30 percent net income tax, but it cannot take steps to evade the presumed income tax unless it resorts to smuggling of a major fraction of its output (this does not occur on a substantial scale now, under the existing *regalía*/export tax situation, except perhaps for gold and silver).

prise, COMIBOL, and the medium mining firms for tax purposes, but would have provision for tax relief for small mines.

Under such a system, different metals would, as at present, have different presumed costs. But in the new tax regime, differences in presumed costs would, to a much greater extent than is now the case, reflect differences in actual costs. And beyond that, the system would not provide for special tax treatment for any metal. Implementation of the new tax would require abandonment of the present concept of presumed costs used in the *regalía*. It would be replaced by a new cost concept, *standard total costs*, which would be the sum of two distinct cost items: *standard mining costs* and *standard realization costs*. Further, the level of standard mining costs would, unlike the presumed costs under the present *regalía*, be changed on a regular basis by application of a *mining cost adjustment factor*. As noted in subsequent sections, the level of standard realization costs would be directly determined for each metal from standard sales contracts and other external evidence.

Symbolically the new cost concepts may be expressed as follows:

Let S_i = standard total costs per unit of a given metal

X_i = standard mining costs per unit

Y_i = standard realization costs per unit

Z_i = production royalty per unit

Then:

$$S_i = X_i + Y_i + Z_i$$

Standard mining costs are defined as *base-period* standard mining costs (X_b), times one plus the mining cost adjustment factor (F). The items that enter into base-period standard mining costs for any given metal (i) are:

X_{i1} = materials costs

X_{i2} = labor costs

X_{i3} = energy costs

X_{i4} = transport costs to point of shipment

X_{i5} = insurance

X_{i6} = depreciation

Thus, $X_i = X_{bi}(1 + F)$

The components of standard realization costs are:

Y_{i1} = standard smelting costs (including smelter penalties or bonuses)

Y_{i2} = standard transport costs beyond collection points within Bolivia

Y_{i3} = standard other costs (insurance, port charges and fees)

Unlike the present *regalía*/export tax system, the *bona fide* presumptive income tax would be truly progressive in nature, with progressivity a function of per unit presumed income. Presumed income would be defined for any metal as the difference between the *official price per unit* (i.e., world market prices) of that metal and the standard total costs for that metal.

There is a strong reason for utilizing a standard total cost concept that distinguishes between "mining" costs and "realization" costs. Mining costs for any given metal cannot be determined with any precision in Bolivia, given the state of cost accounting in mining enterprises. Such costs can only be estimated for some base period, using data such as that compiled and processed in the survey of mining firms undertaken for purposes of this study, and described in Chapter Three. Although such surveys are very expensive and administratively burdensome, periodic reestimates of mining costs are necessary if it is desired to avoid growing divergences between standard and actual mining costs. The cost increases for most mining inputs in Bolivia (and elsewhere) have been substantial in recent years (see Appendix A). Thus, it is essential to apply a mining cost adjustment factor to base period standard mining costs on a regular basis until the next survey establishes new estimates. Application of a cost adjustment factor would not be necessary for standard realization costs, which can be determined directly for any six-month period, with some precision. Standard smelting costs can be derived from a sample of standard smelting contracts. The data required for computing a standard cost for ocean freight are available from a variety of external sources, as are those for port charges and insurance.

Any standard cost concept used for tax purposes is bound to have some degree of arbitrariness. The problem is to limit this arbitrariness to a minimum. The splitting of standard costs into standard mining costs, which cannot be estimated with precision at present, and standard realization costs, which can be determined with some confidence, helps to do that.

Details of the Interim Reform Measures

Presumptive Income Tax. A bona-fide presumptive income tax would replace the present *regalía*.

1. *Rates*

a. The tax would involve progressive rates on per unit presumptive income, with progressivity a function of per unit presumed income, not of price. The progressive feature is intended to serve as the principal means by which the state can appropriate a significant share of resource rents generated by high minerals prices.

b. *Brackets*: i. *General Comments*. If all minerals are to be taxed equally under the new system, it will be necessary to have different bracket sizes for minerals with different levels of presumed costs. Thus, if presumed costs per unit of the mineral (*X*) are half those of mineral (*Y*), then each bracket for presumed income will be twice as big for mineral (*Y*), in terms of dollars.[g] By the same token, if mineral (*M*) has presumed costs of $3.00 per unit and mineral (*N*) has presumed costs of $4.00 per unit, then bracket widths for mineral (*N*) should be 33 and 1/3 percent greater than for mineral (*M*).

For minerals for which the "commercial unit" is not the fine pound (such as antimony and wolfram), the brackets must first be constructed in terms of "fine pounds" and then converted to "long units" for assessment purposes.[h]

ii. *Benchmark for bracket widths and number of brackets*. There should be a benchmark bracket width for one mineral, against which bracket widths for other minerals can be determined. Since tin is by far the most important mineral export it would be advisable that the

[g]Example: Suppose metal (*X*) has a presumed cost of $1.00 per fine pound, and mineral (*Y*) has a presumed cost of $2.00 per fine pound. If the rate brackets in the case of mineral (*X*) are U.S. $0.25, then the brackets for mineral (*Y*) should be twice as large, or U.S. $0.50, if the progressive rate structure is to weigh equally on presumed income of both metals.

[h]For example, consider units of mineral (*J*) expressed in fine pounds. Mineral (*J*) has a cost of $2.00 per fine pound. Consider also mineral (*K*), which has a presumptive cost of $22.40 per long unit. To achieve equality of taxation of presumed income for both metals, one must first convert presumptive costs of mineral (*K*) to fine pounds. Since one "long unit" is equal to 22.4 fine pounds, the presumptive cost of one fine pound of mineral (*K*) is $1.00. Therefore, brackets for a *fine pound* of mineral (*K*) should be only half as wide as mineral (*J*). Thus, if bracket size for mineral (*J*) is $0.10, then bracket size for a *fine pound* of mineral (*K*) should be only $0.05. But when converted back to "long units," bracket size for mineral (*K*) would be 22.4 × $0.05, or $1.12.

benchmark be set for tin. But the setting of the benchmark width is a technical question, the answer to which must be geared to the desired degree of progressivity on presumed income. This, in turn, must be related to expectations of probable price swings in the world market for the metal. If the bracket width is too narrow, then progressivity will be too steep. If bracket width is set too wide, then the government will not succeed in capturing much of the windfall rents that accrue when prices rise rapidly. The practice in Bolivia under the current *regalía* is to have brackets for tin of U.S. $0.10 in width, and there are no strong reasons for changing this practice.[i]

iii. *Number of brackets*. Once required bracket width is determined for each metal, the decision on the number of brackets is largely set. The only question relates to the medium-term prospects for extremely large price increases. The greater is the expectation of substantial price increases, the more brackets will be required, given bracket widths. In the case of tin, provision should be made for no less than 20 brackets, to take care of the eventuality that prices might exceed $6.00 per pound by 1980 (compared to $3.00 in early 1976). Twenty brackets should be sufficient for all other metals as well.

iv. *Degree of Progressivity*. The desired degree of progressivity[j] is dependent upon (a) minimal revenue requirements, (b) the govern-

[i]An illustration, based on presumed cost levels announced for Jan. 1, 1976 may help to clarify the point. To keep the explanation simple, the example is confined to presumed costs for medium miners.

Mineral	*Presumed cost per commercial unit (1)*	*Presumed cost per fine pound (2)*	*Required bracket width per fine pound[ii] (3)*	*Required bracket width per commercial unit (rounded) (4)*
1) Tin	$1.80	$1.80	$0.100	$0.10
2) Antimony	$9.00[i]	$0.40	$0.022	$0.50[iii]
3) Wolfram	$36.00[i]	$1.61	$0.089	$2.00[iii]
4) Lead	$0.215	$0.215	$0.012	$0.01
5) Zinc	$0.17	$0.17	$0.009	$0.01
6) Copper	$0.70	$0.70	$0.038	$0.04

[i]The commercial unit for antimony and wolfram is the "long unit," or 22.4 pounds.

[ii]Required bracket width is determined in relation to the ratio of costs per fine pound of any given metal to cost of tin per fine pound.

[iii]Column (3) times 22.4

[j]Under a personal income tax, a high premium would need to be placed on a smooth progression of rates, so as to reduce the possibility of inducing unintended behavioral changes in taxpayers. However, smooth progressivity is not nearly as critical in a tax levied on per unit presumptive income.

ment's view of its appropriate share in resource rents, (c) the extent to which the government wishes to provide relief to the mining industry in periods of depressed minerals prices, and (d) the desired amount of investment in the sector over time. The latter, in turn, depends greatly on judgments on the extent to which mining profits will be reinvested in the sector.

But, in the end, the selection of the desired degree of progressivity is very much a political decision. No claims are made for the "correctness" of the alternative rate scales presented below. They represent two among a range of alternatives and are based partly on the twin assumptions that the government places some premium on avoiding substantial revenue loss from reform but also wishes to capture a higher proportion of resource rents when prices are higher than has been the case in the past.

At the present time, the *regalía* for tin is 25 percent of presumed income in the bracket just above the level of presumptive costs (i.e., at a price of $1.60 per pound). The *regalía* and the export tax together are, however, 145 percent of presumed income in this lowest bracket, and the combined effective rate of the *regalía* and the export tax declines to 49.3 percent at a price of $4.50. Chapter Six discussed the problems inherent in the effective rate pattern of the present *regalía*/export tax structure. Such a pattern should not be perpetuated.

Alternative rate pattern (A) for tin is given in Column 1 of Table 8-1. Rate pattern (A) is suitable for a lower revenue target and results in a lesser degree of progressivity. Rate pattern (B) is geared to higher revenues. In the case of rate pattern (A), the effective rate of the BPT would climb from 23 percent at the lowest level of presumed income to a maximum of 63 percent at the highest levels. Rate pattern (B) begins with a 25 percent effective rate and reaches a maximum of 74.5 percent at the highest levels of presumed income. This contrasts with the present *regalía* (see Column 3), for which the lowest effective rate on presumed income is 25 percent and the maximum effective rate is 38 percent.[k]

Table 8-1 does not present a complete picture, as it does not include the effects of the complementary low-rate production royalty designed to replace the ad valorem export tax, the details of which are discussed below. Table 8-2 illustrates the pattern of liabilities for

[k]It might be argued that the maximum rate for the per unit presumptive tax should be held to perhaps 50 percent or 55 percent, to avoid possible adverse effects on "business psychology" that might arise when the government takes much more than an even split in profits. This argument may be valid in the case of a true profits tax, but does not merit serious attention in the set of proposals under consideration, for reasons explained below.

Table 8-1. Alternative Patterns for BPT[a]

Brackets of presumed income per unit for tin (US $)	(1) Alternative Rate Pattern (A)		(2) Alternative Rate Pattern (B)		(3) Present Pattern (as of 1975)	
	Effective rate (%)[b]	Presumptive tax due per unit	Effective rate (%)[b]	Presumptive tax due per unit	Effective rate (%)[b]	Tax due per unit
$0.00-$0.099	23	$0.012	25.0	$0.013	25	0.125
0.10-0.199	23	0.025	27.0	0.027	25	0.025
0.20-0.299	27	0.054	29.5	0.059	26	0.052
0.30-0.399	29	0.087	32.0	0.096	27	0.081
0.40-0.499	31	0.124	34.5	0.138	28	0.112
0.50-0.599	33	0.165	37.0	0.185	29	0.145
0.60-0.699	35	0.210	39.5	0.237	30	0.180
0.70-0.799	37	0.259	42.0	0.294	31	0.217
0.80-0.899	39	0.312	44.5	0.356	32	0.256
0.90-0.999	41	0.369	47.0	0.423	33	0.297
1.00-1.099	43	0.430	49.5	0.495	34	0.340
1.10-1.199	45	0.495	52.0	0.572	35	0.385
1.20-1.299	47	0.564	54.5	0.654	36	0.432
1.30-1.399	49	0.637	57.0	0.741	37	0.481
1.40-1.499	51	0.714	59.5	0.833	38	0.532
1.50-1.599	53	0.795	62.0	0.930	38	0.570
1.60-1.699	55	0.880	64.5	1.032	38	0.668
1.70-1.799	57	0.969	67.0	1.139	38	0.646
1.80-1.899	59	1.062	69.5	1.251	38	0.684
1.90-1.999	61	1.159	72.5	1.378	38	0.722
2.00 and above	63	1.26 and up	74.5	1.49 and up	38	0.760

[a]On tin of 35 percent grade or more.

[b]Calculations for first bracket made on bracket midpoint. On all other brackets, effective rates calculated on lower limit of bracket.

the production royalty when levied on tin at a rate of 3.0 percent on standard export value (gross export value minus standard realization costs). This table also compares the effective rate of the production royalty on presumed income with that of the present export tax, using 1975 presumed costs. As may be noted from Table 8-2, the effective rate of the production royalty on presumed income falls steadily from 38 percent at very low prices to 3.7 percent at very high prices. But the effective rate of the existing export tax is 120 percent at very low prices, remains above 35 percent for slightly

Table 8-2. Comparison Between Effective Rates of Tax on Presumed Income: Proposed Production Royalty and Present Export Tax (illustration for tin concentrates of 35% grade or more)[a]

Tin Prices (US \$)	*Presumed Income (with 1975 presumed costs) per unit (US \$)*	*Proposed Production Royalty (at a rate of 3% on standard export value–gross export value minus standard realization costs)[a] (US \$)*			*Present Export Tax (7.5% of gross value) per unit (US \$)*	
(1)	*(2)*	*(3)*	*(4)*	*(5)*	*(6)*	*(7)*
		tax base per unit	*tax due per unit*	*tax as % presumed income*	*tax due*	*tax as % presumed income*
1.60	0.10	1.25	0.038	38.0	0.120	120.0
1.70	0.20	1.33	0.040	20.0	0.128	64.0
1.80	0.30	1.40	0.042	14.0	0.135	45.0
1.90	0.40	1.48	0.044	11.0	0.143	35.8
2.00	0.50	1.56	0.047	9.4	0.150	30.0
2.10	0.60	1.64	0.049	8.2	0.158	26.3
2.20	0.70	1.72	0.052	7.4	0.165	23.6
2.30	0.80	1.79	0.054	6.8	0.173	21.6
2.40	0.90	1.87	0.057	6.3	0.180	20.0
2.50	1.00	1.95	0.059	5.9	0.188	18.8
2.60	1.10	2.03	0.062	5.6	0.195	17.7
2.70	1.20	2.11	0.064	5.3	0.203	16.9
2.80	1.30	2.18	0.066	5.1	0.210	16.2
2.90	1.40	2.26	0.068	4.9	0.218	15.6
3.00	1.50	2.34	0.070	4.7	0.225	15.0
3.10	1.60	2.42	0.073	4.6	0.233	14.6
3.20	1.70	2.50	0.075	4.4	0.240	14.1
3.30	1.80	2.58	0.077	4.3	0.248	13.8
3.40	1.90	2.65	0.080	4.2	0.255	13.4
3.50	2.00	2.73	0.082	4.1	0.263	13.2
4.00	2.50	3.12	0.097	3.7	0.300	12.4

[a]Assumes standard realization costs of 22% of gross export value for tin of this grade. Realization costs for other metals such as lead and zinc would constitute a significantly higher proportion of gross export values.

higher prices, and finally declines to about 12 percent at very high prices. Although both the production royalty and the export tax are regressive with respect to presumed income, the production royalty is much less so.

Table 8-3 (Columns 3 and 4) illustrates the combined burden of the interim system (bona fide presumptive tax plus production royalty). This pattern may be easily contrasted with that of the

Table 8-3. Comparison of Effective Rate Pattern, Proposed System (Bona fide Presumptive Tax—Rate Pattern A—Plus Production Royalty) and Present System (*Regalía* plus Export Tax) (tin of 35% or more grade)

		Proposed System		*Existing System*	
Tin Price per unit (US $) (1)	*Presumed income per unit (US $) (2)*	*Combined Effective Rates of Tax & Royalty on Presumed Income (allowing for deductibility of production royalty (3)*	*(4)*	*Combined Effective Rates of Regalia & Export Tax on Presumed Income (5)*	*(6)*
		Tax & royalty due per unit (US $)[a]	*Combined effective rate of tax & royalty*[b]	*Regalia & export tax due per unit (US $)*	*Combined effective rate of regalia & export tax*
1.60	0.10	0.054	54.0	0.145	145.0
1.70	0.20	0.083	41.5	0.180	90.0
1.80	0.30	0.117	39.3	0.216	72.0
1.90	0.40	0.157	39.3	0.255	63.8
2.00	0.50	0.197	39.4	0.295	59.0
2.10	0.60	0.242	40.3	0.338	56.3
2.20	0.70	0.292	41.7	0.382	54.6
2.30	0.80	0.345	43.2	0.429	53.6
2.40	0.90	0.403	44.8	0.477	53.0
2.50	1.00	0.464	46.4	0.528	52.8
2.60	1.10	0.529	48.1	0.580	52.7
2.70	1.20	0.598	49.8	0.635	52.9
2.80	1.30	0.671	51.6	0.691	53.2
2.90	1.40	0.747	53.4	0.750	53.6
3.00	1.50	0.828	55.2	0.795	53.0
3.10	1.60	0.913	57.1	0.841	52.6
3.20	1.70	1.000	58.8	0.886	52.1
3.30	1.80	1.100	61.1	0.932	51.8
3.40	1.90	1.190	62.6	0.977	51.4
3.50	2.00	1.309	65.5	1.023	51.2
4.00	2.50	1.610	64.4	1.250	50.0

[a]This column is *not* the *sum* of presumptive taxes due per unit in Column 1 of Table 8-1 plus production royalties indicated in Column 4 of Table 8-2. Rather, taxes due per unit in Column 3 of this table were adjusted to reflect the fact that the production royalty is a component of standard total cost under the proposed new system, and may be viewed as a deduction from presumed income under the BPT.

[b]Alternative Rate Pattern (A)

present system, shown in columns 5 and 6 of Table 8-3. Whereas under the present system, the combined effective rate of *regalía* plus export tax *falls* from 145 percent of income at a tin price of $1.60, to 50.0 percent at a tin price of $4.00 per fine pound, the combined effective rates of the proposed system *rises* from 54 percent at the lowest indicated price to 64.4 percent at a price of $4.00 per fine pound. The pattern of effective rates under the proposed system is, therefore, superior to the existing system on two counts:

(1) The proposed system avoids the oppressive fiscal burdens inherent in the present system when minerals prices are very low or decline rapidly.

(2) The proposed system captures for the state a higher share of the gains that now accrue to producers when prices are high or rise markedly.

Both of these features are readily evident from an examination of Table 8-4. This table compares tax liabilities, effective rates of tax and per unit profit after tax under the present system and the proposed system. Columns 1-3 are provided for reference purposes. They portray the situation for tin under the present tax system, with presumptive costs as in effect in 1975, at various possible prices. Columns 4-6 indicate total tax liabilities (*regalía* plus export tax), effective rates of tax and presumed per unit after-tax profits of enterprises, at various possible prices in 1976, using present tax rates and a presumptive cost of U.S. $1.80, which took effect on Jan. 1, 1976. Columns 7-9 show possible situations under the proposed tax system, assuming the same level of presumptive costs as in columns 4-6.

If tin prices were to remain near the U.S. $2.80 level reached in early December of 1975, then presumed after-tax profits of mining enterprises would, under the present tax system (and at the U.S. $1.80 level of presumptive costs) be U.S. $0.450 per unit. Under the new system (and at the U.S. $1.80 level of presumptive costs) after-tax profits would be U.S. $0.522 per unit.

But if tin prices were to reach an average level of $3.60 in any given year, then the present system would result in after-tax profits per unit of U.S. $0.846. Thus, an increase in average prices of only 28.6 percent would result in an increase in per unit after-tax profits of 88 percent. Under the proposed system, whenever tin prices reach an average level of $3.60, per unit after-tax profits would be U.S. $0.718. Thus, an increase in price of 28.6 percent would give rise to an increase in after-tax profits per unit of 30 percent, an outcome that is more consistent with the objectives of mineral tax policy as outlined in Chapter Four.

Under the proposed system, mineral producers would continue to receive growing net revenues per unit of output in periods of rising prices, in spite of the fact that rates of the BPT are more progressive than under the present system. However, under the proposed system, producer gains from higher prices are not as great as under the

Table 8-4. Comparisons of Tax Liabilities, Effective Rates of Tax (on Presumed Income) and Profit After Tax: Present System and Proposed System, Using Different Cost Assumptions (tin concentrates of 35% grade or more)

Possible	*Situation under Tax System Prevailing in* July 1975 *& Presumed Costs of $1.50 per unit*			*Situation under Tax System* Announced *for Jan. 1 1976 & Presumed Costs of $1.80 per unit*			*Situation under* Proposed *Tax System with Standard Total Costs of $1.80 per unit*		
	(1)	*(2)*	*(3)*	*(4)*	*(5)*	*(6)*	*(7)*	*(8)*	*(9)*
Tin prices per unit (average for year) (US $)	*Total taxes per unit*[a]	*Effective rate on presumed income (%)*	*Presumed enterprise after-tax profit per unit*	*Total taxes per unit*[a]	*Effective rate (%)*	*Presumed enterprise after-tax profit per unit*	*Total taxes per unit*[b]	*Effective rate (%)*	*Presumed enterprise after-tax profit per unit*
2.80	$0.691	53.2	$0.609	$0.550	55.0	$0.450	$0.448	44.8	$0.552
3.00	0.795	53.0	0.705	0.657	54.8	0.543	0.578	48.2	0.622
3.20	0.886	52.1	0.814	0.772	55.1	0.628	0.724	51.7	0.676
3.40	0.977	51.4	0.923	0.863	53.9	0.737	0.886	55.3	0.714
3.60	1.068	50.9	1.032	0.954	53.0	0.846	1.082	61.1	0.718
3.80	1.159	50.4	1.141	1.045	52.3	0.955	1.273	64.6	0.727
4.00	1.250	50.0	1.250	1.136	51.6	1.064	1.423	64.7	0.777

[a]Includes *regalía* and export tax.

[b]Includes bonafide presumptive income tax and production royalty, with presumptive income tax rate pattern (A) and a 3.0% production royalty on standard export values.

existing system, as the proposed system yields a higher share for the government when prices are high. But in periods of depressed prices, per unit gains to the producer would be significantly greater under the proposed system than under the present system. This would serve to substantially reduce fiscal pressure on the firms when metal markets are depressed and also would reduce the chances that underestimated presumptive costs (standard total costs) will result in large after-tax losses. The effect should be to diminish overall uncertainty facing mining operations. While the proposed system will reduce the scope for very large windfall gains for producers when markets are flourishing (as occurred in 1974), it will also diminish the possibility of economic ruin of producers when prices are badly depressed (as in late 1975).

2. *The Problem of the Taxpaying Period*

Under the present *regalía*, the taxpaying period is defined in terms of each shipment, and rates of *regalía* are applied on each shipment. The proposed system also contemplates collection of tax on a per shipment basis, as a means of limiting the scope for mineral tax evasion. However, use of this expediency creates problems if mineral prices vary significantly within a year (as in 1975). Under a progressive tax levied on a per shipment basis, a firm exporting at different months during the year would pay vastly different taxes per unit of mineral exports. If the aim is that of taxing the firm on its total presumed income for the year, application of a progressive tax on a per shipment basis will not allow this goal to be achieved, unless prices are stable throughout the year.

Definition of the taxpaying period as one year, rather than a longer period, is an arbitrary measure. But setting the period to coincide with every shipment is even more arbitrary, particularly given use of progressive rates. In fact, this is one of the chief arguments against the use of progressive rates under either a presumed or a true income tax, especially in the absence of a system providing for loss carry-forward.

The BPT can incorporate a feature to deal with this problem in an effective manner. Under this arrangement, mineral exporters would still pay presumptive income tax on each shipment, at rates similar to those in patterns (A) or (B). But at the end of the year, the firm should be allowed to compute average presumed income on all units exported during the year. The prevailing rate schedule (such as (A) or (B)) would be applied to average presumed income, then compared to actual presumptive taxes paid during the year.[l] If after this comparison is made by the tax administration, it is found that firms have overpaid taxes, they would receive a credit to be applied to BPT tax liability for the following year. If the comparison shows that they have underpaid, they would have to pay the differences to the Treasury.

This year-end reassessment would involve more burdens for the tax administration, particularly at year-end, than is now the case under the *regalía*, but would not be nearly so burdensome as a shift to a true income tax. It also means that enterprises should be less concerned over the possibility of being subject to presumed tax rates in excess of 50 percent. The higher rates would apply only during periods of sustained mineral price boom, lasting longer than a year. The measure would also represent a very effective means of laying the groundwork for an ultimate shift to a true income tax, as very careful and detailed year-end assessments would also be required under the income tax.

3. *A Partial Tax Credit for Exploration and Development Outlays*

Partially to compensate for the inability of a presumptive income tax to recognize outlays for exploration in computing tax liability, a tax credit for specified investment expenditures in exploration and development activities at new mine sites might be considered. The credit could be set at a specified fraction of qualifying investment in exploration and development of new mine sites. As with all such devices, the credit would be subtracted from the firm's tax liability, as computed at the time of year-end reassessment.

For administrative reasons, eligibility for the credit would have to be confined to exploration and development outlays on new mine sites.[m] It is difficult enough in new mine sites to distinguish

---

[l]The procedure for finding average presumed income during the year would be as follows: At the end of the year, the tax administration would, for each exporter and *each mineral* exported by him:

(1) Find average price per unit of mineral exported
(2) Find average presumed cost per unit
(3) Subtract (2) from (1)
(4) Multiply the remainder by total export volume for each mineral.

[m]To limit the scope for abuse, the credit should be denied to firms which install mining stopes during exploration and development. However, as a firm explores a mine and finds ore, it cannot avoid getting the ore in the exploration headings. Some firms may, of course, find it advantageous to do active stoping

accurately between exploration and development activities on the one hand, and general mining expenditures on the other. In the Bolivian context, the problem would become intractable with already-existing mine sites. Eligibility for the credit should be based upon the number of meters of advance in new mine sites. Measurement and verification of meters of advance (of shafts, drifts or tunnels) reported by the firms should present no insurmountable administrative difficulties to the tax administration, if the incentive is confined to new mine sites. A standard cost per meter of advance (probably between U.S. $40.00 and $45.00 in 1975) could be established after a careful review of this cost item in COMIBOL and the medium mines. If the purpose of the incentive is to stimulate exploration and development activity beyond what would normally be expected, eligibility for the credit could be limited to the number of meters of advance per year, in excess of some minimum norm that would be established by the Ministry of Mines.

Once the norms are established, it would then become necessary to decide upon a rate for the credit (i.e., the fraction of eligible exploration and development outlays that may be used as a tax credit). Choice of the rate depends on (a) the extent to which it is desired to stimulate exploration and development and (b) the amount of revenue loss from the incentive that is deemed acceptable by the government. The higher the rate of credit, the greater the incentive effects and the corresponding revenue loss. A rate of perhaps 20 percent might be considered for the first year of operation if it is desired to reduce the tax revenue risk associated with the incentive. However, a higher rate, say 40 percent, is more likely to stimulate exploration and development.

A simple example may be helpful in illuminating the basic features of such a credit: consider a medium-size firm with gross sales of U.S. $2 million per year and with presumptive income tax liabilities (after the year-end reassessment) of U.S. $400,000. Further, suppose that during the year the firm has completed 1,200 meters of advance in exploration and development activity of a new mine site, at a cost of $45 per meter. For illustrative purposes, further assume that the government has set a value of $40 as the standard cost per meter of advance and has adopted 500 meters as the "norm" for meters of

and withdraw and sell minerals in excess of the amounts removed from exploration headings. In such cases, the credit should not be available, or should be reduced peso for peso by the additional revenue derived from active stoping of an exploration claim. A further means of limiting the scope for abuse of the credit might be to initially confine eligibility to outlays on exploration and development made by contract to *third parties*. The contractor would then, of course, have to declare such income under the generally applicable income tax.

advance in exploration and development of new mine sites. Also, assume that the tax credit rate set by the government is to be 30 percent.

The firm in this example would be eligible for a tax credit of U.S. $8,400 on the 700 meters of advance over and above the norm set by the Ministry of Mines.[n] In this particular example, the credit would serve to offset 26.7 percent of the cost of any actual exploration and development in excess of the norm ($8,400 in credits divided by actual costs of total meters of advance in excess of 500 meters: $31,500).

4. *Definition of Presumptive Revenues*

Under the present *regalía*, presumptive revenues at the time of export are determined by means of "official prices," which are geared to metals prices on major world minerals markets. There appear to be no compelling reasons for any basic changes in this system.

5. *Definition and Determination of Presumptive Costs*

In the past, presumptive costs of major minerals have been adjusted only after very long delays, and failure to make timely adjustments in presumptive costs has resulted in serious problems for both the mining sector and the Treasury. Adjustments in presumptive costs should be placed on an explicit, regular and predictable basis. These should reflect actual developments in markets for major mining inputs and in realization costs, and should be easily understandable by tax administrators and mining enterprises alike.

The following subsections and accompanying appendices present a step-by-step description of the way in which these cost adjustments can be made. As noted, the base of the proposed per unit tax is the difference between the official price of the metal and *standard total costs*. Standard total costs are the sum of what we have called standard mining cost, standard realization cost and the production royalty. As the preliminary steps in a full description of the computation of the tax liability, the following two subsections show how standard mining cost and standard realization cost can be calculated for a given metal. The computation of tax liability per unit is illustrated in the third subsection below.

a. *Calculation of Standard Mining Cost*. The tax administration requires explicit methods for adjusting standard mining costs in order

[n]The basis for computing the credit would be U.S. $28,000 (700 meters of advance times U.S. $40.). With a tax credit rate of 30 percent of the basis, the credit would be U.S. $8,400.

to closely reflect actual levels of these costs as they change through time. This adjustment may be achieved by means of a *mining cost adjustment factor.*

The first task in the process is that of selecting a base period for calculation of standard mining cost per unit of metal. Then, base-period mining costs must be estimated. As noted earlier, mining costs are defined as all costs, including beneficiation and purchased transport, up to collection points within Bolivia. Once these have been obtained, we have *base-period* standard mining costs per unit of each metal. To determine standard mining costs (X_i) in any given period after the base period, the mining cost adjustment factor (F) is applied to the base-period standard mining cost (X_{bi}). The level of the mining cost adjustment factor at any given time depends on the level of prices of the components that make up total costs and upon the shares of each component in total costs. The mining costs are broken down into six categories (materials, labor, energy, transportation to points of shipment within Bolivia, insurance and depreciation).

For standard mining cost a base-period estimate must be made for each metal, and once made should be an adequate basis for adjustment for a number of years. While accuracy is desirable, it is shown below that an error of, for example, 10 percent in the estimation of base-period costs will not materially affect the burden of the tax, even in a period of great cost increase, provided that the rate structure for taxation is similar to those suggested in Table 8-1.

Every six months, the standard mining cost for a given metal would be calculated as:

Base-Period Standard Mining Cost = (1 + (Mining Cost Adjustment Factor))

$$\text{or } X_i = X_{bi}(1 + F)$$

Appendix E provides a detailed methodology for calculating the mining cost adjustment factor.

b. *Calculation of Standard Realization Costs.* The Standard Realization Cost for a particular metal is the sum of Standard Smelting Costs (including typical bonuses or discounts), Standard Transportation Costs, and Standard Other Costs (insurance, port charges and fees). Of these three items the most important component to be measured for each mineral is the Standard Smelting Costs.

Standard smelting costs per unit of the refined mineral must be determined for each six-month period for each mineral. If there are important differences in smelting costs for different grades of concentrates, then an average must be taken of the smelting costs of high- and low-concentrate ores. Transportation and other costs would be determined as an average for each unit of refined metal, with a distinction made between the costs of shipping via Chile and Peru. A detailed methodology for determination of all elements of Standard Realization Costs is provided in Appendix F.

c. *Computation of the Bona Fide Presumptive Income Tax*. The procedures outlined in the previous subsection (and described in detail in Appendixes E and F) are preliminary steps in the accurate calculation of the base of the bona fide presumptive income tax per unit of a particular metal. The calculation of the tax base can be illustrated in the following example for conditions prevailing in January, 1974. The tax rates are based on those of Option A of Table 8-1. The cost figures are consistent with those presented in the exercises in Appendix E.[o]

1. Official Price	$3.00	
2. Less Standard Realization Cost (including standard bonuses or penalties)	0.95	
3. Standard Export Value (see the subsequent section)		$2.05
4. Less Production Royalty (0.03 × line 3)	0.06	
5. Less Standard Mining Cost (base period standard mining costs times a cost adjustment factor of 1.438)	1.22	
6. Base of bona fide presumptive income tax		$0.77

Then calculation of the tax liability per fine pound of mineral proceeds as follows:

7. Base of bona fide presumptive income tax (line 6)	$0.77	
8. Tax rate (see Table 8-1 showing rates as a function of the base)	0.370	
9. Bona fide presumptive income tax liability (line 7 × line 8)		$0.285
10. Production royalty liability (line 4)		0.06
11. Total tax liability per pound (line 9 + line 10)		$0.345

[o]The Standard Realization Cost in line 2 is an additional arbitrary figure constructed for the purposes of illustration.

The advantages of the proposed tax system over the present one do not depend crucially upon precision in the estimation of base-period mining costs.[p]

6. *Other Principles and Features of the Bona Fide Presumptive Tax*

a. *Elimination of Preferential Tax Treatment of Certain Minerals.* The present system of mineral taxes is replete with provisions for favorable tax treatment of lead, zinc, and bismuth. In the cases of lead and zinc, the preferences are apparently intended to compensate producers for systematic underestimation of presumed costs. But the preferences go further than can be justified on economic policy grounds. In the cases of bismuth and other metals, there appear to be no compelling arguments for continuation of any of these arrangements, either on strictly technical grounds or any other grounds. To the extent that costs of extraction, transportation and treatment of some minerals are higher, on a per unit basis or as a percent of gross value, such differentials should be reflected in the level of presumptive costs. That is, presumptive costs should reflect actual costs as closely as possible. Then, no special tax treatment would be warranted for any mineral. Removal of these largely unintended tax subsidies for certain minerals would not only increase revenues, but would remove the bias that can lead to misallocation of investment.

b. *Elimination of Preferential Tax Treatment for Low-Grade Minerals Concentrates.* Low-grade concentrates of tin and copper receive preferential tax treatment under the *regalía* and the export tax. This aspect of minerals tax policy is not only in conflict with other government policy objectives, but involves some significant social costs as well.[3] The rate schedule of any tax, such as the

[p]If, for example, the estimate of base-period standard mining costs for January, 1974, should have been higher by 10 percent (i.e., U.S. \$0.935 rather than U.S. \$0.85 per lb.), then the estimate of mid-1975 standard mining costs is too low by U.S. \$0.122 [(0.935 − 0.85) × 1.438] since the mining cost adjustment factor is 1.438 as defined in Appendix E). Therefore the tax base in lines 5 and 6 above should be approximately U.S. \$0.65 rather than U.S. \$0.77, and the total tax liability in line 11 should be U.S. \$0.286 rather than \$0.345. In this example, then, a 10 percent error in standard mining costs in the base period combined with very great subsequent cost increases results in an increase of tax payments of U.S. \$0.059 per lb. By comparison, if no attempt had been made to adjust for mining cost changes since January, 1974, the tax base in lines 5 and 6 would have been U.S. \$1.50, and tax payments would have been \$0.828. Under the *current* tax system of mid-1975, total tax payments (*regalía* plus export tax) would have been US. \$0.795, more than double the liability in line 10 of the example above. Therefore, even if standard mining costs can only be estimated within a reasonable margin of error, the greater part of the existing tax difficulty caused by very large changes in mining costs is largely overcome in the proposed tax system.

proposed presumptive income tax, should not distinguish between high-grade and low-grade concentrates for tax purposes.[q] It is often claimed that small miners tend to export concentrates of lower grade than the average for the medium mines and that, since higher transport costs and penalties on low-grade concentrates mean that returns per fine pound are lower in the small mines, concentrate exports by small miners should be subject to lower tax rates.

The first part of the above assertion is incorrect, according to published data of the Ministry of Mines, at least as far as tin is concerned. The average grade of tin concentrates exported by the medium mines in the first six months of 1975 was 31.57 percent. The average grade of concentrate exported by all small mining firms as a group in the same period was 35.45 percent.[4] Thus, the medium miners, not the small miners, are the prime beneficiaries of the lower *regalía* and export tax rates for low-grade tin concentrates.

c. *Treatment of Small Mining Enterprises and Cooperatives.* The small mining sector in Bolivia, like small-scale mining enterprises all over the world, faces serious problems.[5] Because this group of firms accounts for between 30,000 and 40,000 jobs in the mining sector (over half the total), these problems cannot be taken lightly. Very small firms in this subsector, particularly in Potosi Province, do tend to mine low-grade ore bodies (and tailings) and do tend to export very low-grade concentrates. For these reasons, and because of their primitive technology, their costs per ton of metallic content are appreciably higher than for the medium mines or COMIBOL; therefore presumptive costs for the remaining "small" mining firms should, for each mineral, continue to be set at levels somewhat higher than those established for COMIBOL and medium miners. (In 1974 and 1975, presumptive costs for small miners under the *regalía* were about 6 percent higher than for the medium mines.)

Low-Rate Production Royalty. The prime component of the interim tax reform package is the bona fide presumptive income tax (BPT). The other principal feature of the proposal is a *production royalty*, which would be levied at a low uniform rate on mineral production, including both that portion exported and that used in the domestic economy. The production royalty is intended as a replacement for the present export tax and is intended primarily to provide a floor on revenues when world mineral prices fall to levels near standard total costs. The proposed production royalty would

[q] Most of the differences in returns to an enterprise in exporting a ton of low-grade vs. high-grade concentrates will, of course, be reflected in the revenues received by the exporter, since the tax will be assessed on the metallic content of concentrates, as is now the case.

not produce as much revenue as the present export tax, but this should *not* be viewed as a drawback of the proposed measure. To the extent that revenue goals are important, it is the revenue performance of the entire package, not any of its components, that is important.

The base of the present export tax is gross export value, but the base of the proposed production royalty should be standard export value, or gross export value minus standard realization costs. For each mineral, standard realization costs would in any case be computed for purposes of calculating standard total costs under the presumptive income tax. There would thus be a ready-made measure of *standard export value* in each period that would enable the production royalty to be easily assessed on a net, rather than gross, value. This is an important consideration, since the ratio of net to gross value differs greatly among minerals, owing to great variations in realization costs. For example, realization costs are typically less than 25 percent of gross export value for high-grade tin and can be as little as 15 percent for antimony. But for low-value, high-bulk mineral concentrates such as lead and zinc, realization costs are often as high as 60 percent of gross value. Therefore, assessment of a uniform-rate production tax on gross values of all mineral concentrates would result in much heavier burdens upon minerals with high realization costs (most of which are incurred abroad) than those with low realization costs.[r] However, it is not administratively feasible to calculate actual net values for each shipment of minerals. An attempt was made to do this in the first year after the present export tax was adopted (1972). This led to intractable problems of administration and compliance, and the tax was converted to a gross basis in late 1973. The proposed production royalty would not require calculation of net values for each shipment.

The royalty would be levied at a uniform rate on the standard export value of all minerals produced by all firms. The rate could be between 2.5 and 3.5 percent, depending on revenue needs. The rate should not exceed 3.5 percent, as much higher rates would result in oppressively high combined rates of presumptive tax and royalty at low levels of presumed income (a prime weakness of the present export tax). In the event of fiscal crisis, the rates of the bona fide presumptive tax could be raised, but once set, the rate of production royalty should not be adjusted upward on revenue grounds alone.

[r]The present export tax attempts to deal with this problem by use of a highly complex pattern of rate differentiation for the present gross-value export tax (see Chapter Six).

Major Merits and Limitations of the BPT Relative to a True Income Tax

The next two subsections discuss first the merits and then the limitations of the BPT interim measures, relative to a true income tax.

Major Merits

1. The BPT is administratively more feasible than the income tax. The former does not involve as significant a departure from present practice, and could be implemented in short order.
2. The BPT involves less revenue risk for the government (particularly in the initial stages). The greater short-term revenue risk of the income tax is primarily attributable to administrative and compliance uncertainties, both of which would be substantial.
3. The BPT, if implemented properly, would do a better job of capturing resource rents for the state than would a proportional income tax alone.
4. The BPT would not require a separate tax regime for small miners and cooperatives, whereas this would be necessary under the income tax.

Major Limitations of the BPT Relative to a True Income Tax

1. "High-grading" of ore bodies and, to a much lesser extent, "creaming" of concentrates, inherent in the present *regalía*/export tax system would still be a problem under the BPT, to a greater extent than for an income tax. This is because the BPT package relies much more heavily on per unit taxes that may be viewed as output taxes by mining firms. The problem of high-grading of ore would not be quite as serious under the BPT as under the existing tax system, because effective rates of per unit taxes throughout most of the relevant price ranges for most minerals would be lower. Only at very high minerals prices (and very high levels of presumed income) would high-grading of ore bodies become a matter of serious concern.

Incentives for "creaming" of concentrates would be much less under the BPT than under the present system, largely because of the elimination of preferential tax treatment for low-grade concentrates. However, such "creaming" of concentrates under the BPT would still occur to a greater extent than under an income tax, because the former is a per unit tax.

2. The BPT and associated measures would not solve the various problems inherent in the inability of a presumptive income tax to recognize differential cost structures among firms producing the

same mineral, since no matter how sensitive is the presumed cost basis of the BPT, it still must represent an average. Cost differentials between firms are due not only to differences in relative efficiency in extraction and beneficiation, but to differences in exploration and development activity, metal content of deposits, rock formations, hardness of rock, altitude and available public infrastructure. Thus, the bona fide presumptive tax will be less favorable for expansion in investment for exploration and development, outlays for enhancing mine health and safety conditions, and other key investment items.

3. The BPT would probably not be creditable against income taxes of home countries of foreign investors, whereas a true income tax would be clearly creditable.

4. Implementation of provisions for loss carry-forward would be much more difficult under the BPT than under a true income tax.

5. To the extent that the BPT is less favorable to expansion of investment in the sector, the greater short-run stability of revenues inherent in the package could turn out to mean stability at a lower long-term revenue trend than under an income tax.

6. Finally, the BPT would be less conducive to long-term investment growth in the sector, because primary reliance upon taxes geared to output involves more uncertainty for mining firms than do taxes geared to net income.

TAX INCENTIVES

The major tax incentives affecting the mining sector are those provided by the incentive for excess production and those incentives available under the *investment law*.[6] No detailed comments on incentives under the investment law are offered here, as experience with provisions under the law is too limited to provide reliable indications of how the measures might work in practice. However, the incentive for excess production is so defective as to merit abolition as soon as possible. The incentive is ill-suited for achieving its intended purpose of encouraging exploration. The incentive benefits are geared, not to exploration and development actually undertaken, but to the volume of production. The effect is only to offer the equivalent of higher prices for minerals. As a result, the miner may simply exploit known reserves more rapidly, rather than increasing his reserves. The rationale for benefits from the incentive appears particularly dubious in situations where tin is under quota restrictions, as was the case for the last nine months of 1975. Second, the incentive serves to reduce government revenues most when minerals prices are at their highest. But this is precisely the time when taxes should be serving to extract some of the extraordinary gains from high minerals prices. Third, the incentive serves to heighten incentives for "creaming" of concentrates and is administratively unworkable (see Chapter Seven). There are more effective and less expensive ways to provide incentives for exploration and development.

If the BPT should be adopted (or for that matter, the true income tax on mining), then it would be advisable to phase out this defective incentive,[s] on grounds that the new system would provide significantly more stimulus for investment in exploration and development than is now present under the existing system. Phasing out of the incentive might be accomplished on a three-year basis.

LAND RENTS

The drawbacks of the present system of land rents have been discussed in detail in Chapters Six and Seven, and need not be reiterated here. In general, reform of the system of land rents should be focused upon efforts to discourage the immobilization of claims in the hands of speculators. This would require not only an initial increase in land-rent rates, but provision for their escalation over time.

Specifically, for *exploration* claims, consideration might be given to a tripling or quintupling of land-rent rates, from their present levels of only U.S. $0.15 per hectare per year (for the very largest claims) to perhaps U.S. $0.50 per year for the first year. Then, the general level of land-rent rates could be allowed to escalate by, say, 33 percent per year, from the first to the fourth year of the claim. Land rents of this level would still not be out of line with rates prevailing in other important mining nations. Exploration claims held beyond four years might be auctioned, to insure that other potential buyers who wished to engage in systematic exploration would have the opportunity to do so.

For *exploitation* claims in metallic minerals, the general rate of land rents might be established at U.S. $1.00 to $1.50[t] per hectare per year. Since deposits for most nonmetallic minerals and placer deposits of metallic minerals are typically spread over a much larger

[s]It would be inadvisable to abolish the incentive program outright. First, at least some firms have predicated their production plans over the next year or so to availability of the incentive. Second, mineral tax legislation, and this provision in particular, has already been too much distinguished by a stop-go character that has not been conducive to investment in the sector in the past.

[t]The present maximum rate on the largest claims in U.S. $0.30 per hectare per year.

exploitation area than for metallic nonplacer minerals, consideration should be given to application of land-rent rates half as high as for nonplacer metallic minerals. Finally, to prevent the erosion of real values of land rents during inflation, the general level of land-rent obligations per hectare could be allowed to rise at the same rate as the presumptive cost adjustment factor that would be applied under the proposed BPT.

EARMARKED TAXES ON MINING

The only earmarked tax on mining that is at all significant in revenue terms is the 18 percent share of *regalía* revenues that goes to subnational governments. In the event of adoption of the BPT, subnational governments could continue to share in revenues from the tax on the same basis as at present. In the event of adoption of a true income tax, allocation to subnational governments on the basis of mineral production in each geographical jurisdiction would be much more difficult, and a new formula for revenue sharing (perhaps based on past proceeds from earmarked *regalías*) would have to be devised. Other earmarked taxes on the mining sector have been developed without regard to their effects on the mining sector, although many serve very laudable purposes. There appears to be no logical basis for the remaining forty-odd earmarked taxes on mining, and consideration should clearly be given to a drastic reduction in the number of such taxes. If the functions financed by these earmarked taxes are deemed meritorious, it would be better to finance such activities through normal budgetary channels. The excessive number of earmarked taxes at present only adds to complexity in administration of mineral taxes, and many of them bring in such negligible amounts of revenue that it is entirely possible that costs of collection of most of these taxes exceeds the revenues gained from them.

TAX ADMINISTRATION MEASURES

One important avenue of evasion of mineral taxes lies in the failure to verify assays reported by enterprises at the time of export. This leads to loss in tax revenues and facilitates the buildup of illegal offshore balances of Bolivian mining firms. Certain steps could be taken to reduce the scope for this abuse. First, the customs agency could be provided with the kind and quantity of laboratory and other equipment necessary for accurate verification of assay. Second, new procedures need to be adopted with regard to verification of waybills, contract terms and assay of export shipment of contracts with foreign smelters or other foreign buyers. At present, the foreign smelters and other buyers can legitimize contracts, waybills and assay of shipments from Bolivia merely by obtaining the stamp of approval of the nearest Bolivian Consul. Clearly this has not halted substantial understatement of assay, especially since it does not provide the Bolivian Ministry of Finance with any legal recourse against the foreign smelter or buyer in the event it later discovers falsification of reported assay and/or contract terms. This defect could perhaps be remedied by requiring the foreign smelters (and other buyers) to have such contracts, waybills and other documents notarized in the home country of the buyer, under home country laws. Bolivia could then seek redress in the courts of that country, in the event falsification was discovered.

Another major problem lies on the import side: very long lags were common in 1975 and 1976 in clearance of imported mining machinery and equipment. Such delays in clearance serve to disrupt production schedules and benefit no one. Again, the solution to this problem lies in provision of additional administrative resources to the customs agency, which claims (with some reason) that it is both understaffed and undersupplied with critical equipment.

Either of the reform packages suggested in the report will require major improvement of the quality and quantity of administrative resources in the Internal Revenue Service. For these and other reasons, it is particularly important that plans for mineral tax reform be closely coordinated with plans for overall tax reform under consideration by the Musgrave Commission in 1976 (see Chapter Seven).

Finally, the most important taxpayer in the mining sector is, of course, COMIBOL. Little can be expected from any kind of mineral tax reform until a thoroughgoing reform of COMIBOL's accounting and control system has been carried out. Chapter Nine discusses these problems in some detail.

REVENUE IMPLICATIONS

Projected Revenues in the Absence of Reform

In assessing the likely revenue implications of any set of alternatives for tax reform, it is important to have as accurate an idea as possible of the potential revenue yield of the system in the absence of reform. That is, revenues from the existing tax structure, with an unchanged tax base and at constant tax rates, must first be projected.

Such projections are summarized in Table 8-5. The basis of the

Table 8-5. Summary of Revenue Projections for Identifiable Taxes Paid by the Mining Sector: Present Tax System in the Absence of Reform, 1976-1980 (millions of pesos, in nominal terms)

	1974 actuals	*1975 estimates*	*1976*	*1977*	*1978*	*1979*	*1980*
I. *Central Govt.*	1,515.5	863.6	1,286.5	1,551.3	1,675.6	1,800.4	1,909.9
A. *Regalías*[a]	881.8	449.8	740.7	888.9	959.6	1,014.3	1,069.0
B. Export taxes	481.4	290.4	435.6	508.3	550.0	608.5	560.9
C. Land rents	1.1	1.2	1.3	1.4	1.5	1.6	1.7
D. All other central govt. taxes on mining[b]	151.2	122.2	138.0	152.7	164.5	176.0	188.3
II. *Noncentral govt. earmarked taxes*	193.4	124.9	235.2	247.9	269.3	288.2	299.0
Total	1,708.9	988.5	1,521.7	1,800.2	1,944.9	2,088.6	2,208.9

Source: Appendix G.

[a]Central government share of *regalías* only.

[b]Including the tax collected by the Internal Revenue Service for the benefit of ENAFBOL. See Table 6-1.

projections, including assumptions for mineral prices and output, are set forth in considerable detail in Appendix G. The forecast for the existing system assumes no major changes in tax rates or in the tax base, with one exception: it is realistically assumed that presumptive costs under the existing *regalía* will not be held constant, but will be permitted to rise (with a lag) to reflect inflation in input costs.

Revenue Implications of Reform of Mining Taxes

As adoption of a true income tax cannot be seriously considered for another five years, at the earliest, revenue projections for that alternative have not been undertaken. Adoption of the BPT and its companion measures is, however, clearly feasible for the short term, and estimates of the revenue impact of that set of proposals were made for the period 1976-1980. Even with the lower-rate Schedule A and with only a 3.0 percent production royalty, the reform package would generate more revenues over the entire period (1976-1980) than the present mineral tax system if left unchanged. This statement holds both for central government revenues and for subnational government revenues. For 1976, the package would, under conservative assumptions, bring in slightly less than the present system. By

1977, the new system would begin to outperform the present system and by 1979 would be generating about 14 percent more revenues than the existing system (see Tables 8-6 and 8-8). This should not be surprising, since mineral prices are expected to continue to be somewhat depressed throughout most of 1976, but are expected to recover thereafter, and the reform package is explicitly intended to provide some tax relief when prices are low and to appropriate more resource rents for the state when prices are high.

By itself, the proposed bona fide presumptive income tax would generate substantially more revenues than the existing *regalía* system over the five-year period. This is because of several factors: (a) the higher ceiling on effective rates of taxes in the proposed system compared to the present *regalía*, (b) the elimination of preferential treatment for such metals as bismuth and lead, and (c) the abolition of relief for export of low-grade concentrates.

It is to be noted that the projections for the present *regalía* system (see Table 8-5) and for the reformed bona fide presumptive tax system (see Table 8-6) are based on the same assumptions regarding average yearly presumptive costs and average yearly world market prices. The proposed system is simply much more responsive to

Table 8-6. Revenue Projections for a Reformed Mining Tax Structure, Central Government Only (millions of Bolivian pesos)

Tax	*1976*	*1977*	*1978*	*1979*	*1980*
A. Presumptive income tax [rate schedule (A)]	942.2	1,207.5	1,358.6	1,505.2	1,651.5
B. Production royalty at 2% of gross export value	147.0	177.9	195.8	231.0	352.0
C. Land rents	2.5	3.0	3.8	4.9	6.6
D. Other central government taxes	142.8	152.7	164.5	176.0	188.3
E. Recoupment of losses from abolition of incentive for excess production	8.1	18.7	53.4	61.4	70.6
F. Revenue loss from new tax credit for exploration and development	–5.0	–10.0	–30.0	–50.7	–65.9
Totals (central government)	1,237.6	1,549.8	1,746.1	1,927.8	2,103.1

Source: Appendix H, which applies the proposed lower BPT rate structure (A) and the 3% production royalty to the economic magnitudes summarized in detail in Appendix G.

Table 8-7. Revenue Projections for a Reformed Mining Tax Structure (Subnational Government Units and Others), millions of pesos

Revenue Source	*1976*	*1977*	*1978*	*1979*	*1980*
A. Subnational government share of royalties[a]	206.8	265.4	298.2	330.4	362.5
B. Educational institutions[b]	8.8	9.4	10.0	10.6	11.3
C. Earmarked taxes for Mining and Metallurgical Institute[b]	10.2	11.0	11.8	13.0	14.0
D. Others[b]	2.5	2.7	2.9	3.1	3.3
Totals	228.3	288.5	322.9	357.1	391.1

[a]From Appendix H.
[b]From Appendix G.

Table 8-8. Mining Sector Taxes, Comparisons: Projections for Existing System vs. Projections for Reformed System, Conservative Assumptions (millions of pesos)

	1976	*1977*	*1978*	*1979*	*1980*
A. *Central government taxes*					
1. Interim reform package[a]	1,237.6	1,549.8	1,746.1	1,927.8	2,103.1
2. Present system[b]	1,286.5	1,551.3	1,675.6	1,764.4	1,909.9
3. Difference	−48.9	−1.5	+70.5	+163.4	+193.2
B. *Subnational and earmarked taxes*					
1. Interim reform package[a]	228.3	288.5	322.9	357.1	391.1
2. Present system[b]	225.2	247.9	268.3	288.5	299.0
3. Difference	+3.1	+40.6	+54.6	+68.6	+92.1
C. *Total taxes (A + B)*					
1. Interim reform package	1,465.9	1,838.3	2,069.0	2,284.9	2,494.2
2. Present system	1,511.7	1,799.2	1,943.9	2,047.7	2,218.9
3. Difference	−45.8	+39.1	+125.1	+237.2	+275.3

[a]From Table 8-6.
[b]From Table 8-5.

changes in presumed income than is the present system and performs much better in garnering tax revenues when mineral prices are high than does the present system. The proposed system would do no worse in revenue terms when minerals prices are low. The proposed production royalty would yield substantially lower revenues than the present ad valorem export tax, as intended. However, the higher additional revenues generated by the proposed presumptive income tax (over and above revenues expected from the *regalía*) more than offset the difference for the five-year period as a whole, and are just about sufficient for this purpose even in 1976.

The reform proposal also involves slightly higher revenues from land rents, though these would still be an insignificant proportion of total revenues, and there is a small increase in revenues arising from the abolition of the incentive for excess production (see Table 8-6).

Subnational levels of government would also fare much better under the reform package, if they continue to be entitled to 18 percent of presumptive income tax revenues, as they are now entitled to 18 percent of *regalía* revenues (see Table 8-7).

In sum, reform of the type proposed here would perform well on revenue grounds, particularly in the medium to long term. Sensible reform would involve a continued heavy revenue role for the mining sector, relative to other productive sectors in the economy. But it would seem highly inadvisable, over the near future, to look to the mining sector for further tax revenues much beyond those that are projected in Table 8-6. To do so could further endanger the long-term vitality of the sector, to the point where its long-term revenue potential could shrink to intolerably low levels. If after mineral tax reform there continue to be significant residual revenue needs, it would be much more worthwhile to look to those sectors and activities that up until now have borne a very light tax burden. These include commercial agriculture, the professional class and retail and wholesale trade in general.

NOTES TO CHAPTER EIGHT

1. This distinction between optimal tax reform and optimal tax design, and its implications for the types of tax measures that may be proposed, is ably discussed in Martin Feldstein, "On the Theory of Tax Reform," (Harvard Institute for Economic Research, Discussion Paper #408, April 1975).

2. For a thorough discussion of the arguments against percentage depletion, see *Blueprints for Basic Tax Reform* (Washington, Department of the Treasury, Jan. 17, 1977).

3. See Chapter Seven for a discussion of this point.

4. Ministerio de Minería y Metalurgia, Departmento de Estudios Económicos, *Boletín Estadístico*, Año II, June 1975, p. 1.

5. See Chapter Two for a detailed discussion of this point.

6. Incentive provisions affecting the mining sector are discussed in Chapter Six.

Chapter Nine

Issues and Policies Related to Mining Tax Reform

Malcolm Gillis,
Ulrich Petersen,
and Louis T. Wells

This chapter briefly covers a number of other, primarily nontax domestic and international issues and problems which have a strong bearing upon the design, implementation and prospects of success of tax reform in the mining sector. Topics discussed include such domestic issues as the tax compliance infrastructure in state and private enterprises in the mining sector; the effects of certain features of the design and implementation of mining codes upon mining sector development; and international issues such as policy toward foreign investment and commodity agreements, as well as problems related to regional economic integration.

THE TAX COMPLIANCE INFRASTRUCTURE

Successful operation of any reformed system of mineral taxation requires more than improvements in the government's tax administration machinery. Improvement in the compliance capacities of the affected firms must be sought as well. In turn, the compliance capacities of enterprises depend greatly upon the state of their internal accounting and control systems. The government, as the only stockholder in the major state enterprises in mining, can take direct initiatives to strengthen the compliance capacities in them. The government can promote improvements in the compliance infrastructure in private sector enterprise through indirect means, such as denying tax privileges from those who do not meet specified accounting or auditing standards, and by installation and enforcement of an administrable system of fines and penalties for firms that provide inadequate documentation of essential income and cost information. It is important to emphasize the word "administrable," as fines and penalties set at levels perceived as unrealistically high by taxpayers tend everywhere to be viewed similarly by tax administration officials, who in the end do not enforce them.

The problems associated with the tax compliance capacities of the major enterprises in mineral exploitation, both private and public, are discussed in the next two sections.

The Private Sector

A tax system will generate the desired response from management only if attention is paid to the kinds of accounting and control system used by management, as these systems determine, to a great extent, the tax compliance capacities of enterprises. In Bolivia, capacity for compliance with tax measures appears very uneven within mining concerns in the private sector, ranging from reasonably adequate in the largest of the thirty-odd medium mining firms to minimal among most of the small miners. In contrast to capacity, willingness to meet fiscal obligations is, in any case, extremely difficult to gauge. However, as elsewhere in the world, we would ordinarily expect somewhat higher standards of taxpaying responsibility on the part of the larger firms, if for no other reason than that they have more to lose if discovered flaunting the tax laws.

Among the medium mining firms, perhaps only a half-dozen of the larger firms possess an accounting and management infrastructure comparable to their medium-sized counterparts in the developed countries. Where it exists, this infrastructure has usually developed in response to the internal needs of the firms themselves. But in some cases the pace of improvement in bookkeeping, control and management methods has been accelerated in recent years by a recognition that needed foreign equity and technical assistance inputs would not be forthcoming in the absence of a modern accounting and managerial infrastructure within the Bolivian firms. For the handful of firms already involved in joint ventures with foreign partners, the foreign participants themselves have tended to insist upon high standards of audit and control for the joint ventures so as to assure their full participation in any gains that might accrue.

Perhaps six medium mining firms subject themselves to external audit by auditing firms capable of meeting international standards. But even within these enterprises, there is concern regarding their management information systems. These concerns center on the treatment of depreciation, inventory control, blurred distinctions between deferrable costs and current costs, and overall lack of cost accounting within the firms. Nevertheless, the five or six largest firms could, within three to four years, be capable of meeting minimal *internal* accounting standards required for application of profits taxes if they were to make a concerted effort to do so. Beyond this, there are perhaps another ten medium mining firms and three to five small and "other" mining firms that could conceivably satisfy these standards over the next three to four years, but to do so would require a drastic overhaul of their accounting systems and practices. Thus, the number of private sector firms having any possibility of complying with the bookkeeping demands and audit requirements of a true income tax would amount to no more than twenty firms by 1980 or 1981. However, these twenty firms now account for between two-thirds and three-quarters of the value of mining production in the private sector. If COMIBOL were also subject to income taxation by that time, then an income tax net would cover between 80 and 85 percent of the total value of nonfuel mineral production in Bolivia.

For all but the largest firms in the small mining sector, and for all of the cooperatives, income taxation cannot be considered in the next decade. For such firms, records (where they exist at all) are rudimentary and seriously incomplete. Many firms are unable to provide even reasonably reliable records of the total physical volume of their production during any given year, much less an accounting of gross income, or costs of earning that income. The only feasible means of including these enterprises in the overall tax net is by continued reliance upon BAMIN (the state mining bank) as a tax collection agent for the government. In the event that BAMIN's virtual monopoly on marketing of small mine production is terminated, considerable attention will have to be given to more effective tax surveillance of the private trading companies involved in the buying and selling of minerals, if the value of the mineral production of the small mines is to be effectively taxed, as there are a number of indications that many of the private trading concerns do not take their tax withholding responsibilities as seriously as does BAMIN.

The State Enterprise Sector

The technical capabilities of management in COMIBOL are well known in Bolivia. Most managers have a strong background in engineering, and both anecdotal evidence and observation indicate that technical competence has been high from almost the birth of the enterprise, especially relative to other public and most private sector firms in Bolivia. Further improvements in the enterprise's control, accounting and organizational structure would not only allow it to take maximum advantage of the technical talents of its personnel, but would serve to enhance its capacity for tax compliance as well. These issues are discussed below.

An accounting system should be designed, in the first instance, to yield information for the shareholders (in this case, the government) and the tax authorities (again, the government). But the accounting system should do more: it should provide the basis for the evaluation of management at various levels in the firm, for motivation of managers, and for the evaluation of certain decisions. The system in COMIBOL does not yet fully serve this broader "control" function in spite of improvements made since 1965. The system as it now stands does not provide a basis for an information system that would enable the enterprise managers to identify and respond effectively to ordinary and extraordinary problems arising from within the enterprise (e.g., waste in explosives and materials, machinery breakdown) or outside (e.g., rapidly rising costs of imported materials, adjustments in transport costs).

Some shortcomings in the control system are not inherent in the systems used. Lags in reporting by operating units to the central office are a major problem, as is noncomparability of much of the data generated in distinct parts of the enterprise.

But improvement in accounting per se is not by itself a solution to the control problem. The accounting system is not used as a control system. For example, mine managers do not appear to be evaluated on the returns and costs over which they have control. In fact, budgets and performance figures contain different mixes of controllable and uncontrollable costs such as head office expenses. With such a control system, it can hardly be anticipated that the mine manager will respond, in the manner desired, to changes in the tax structure.

Inventory control is still another weak link in the management system. Although the inventory control mechanism has been handled with electronic data processing for several years, and appears to perform reasonably well as a tool for logistical decisions (in terms of ensuring adequate stocks of key materials at each mine site), the system has not yet been adequately integrated into the accounting system. Little of the very detailed data generated by the inventory section of the central office is utilized in account preparation in the Accounting Department, and some of what is used is not relevant for accounting purposes.

Further, the enterprise's accounts do not consistently separate production costs from investment expenditures. Thus, it appears that the costs of extracting a ton of ore go up from time to time when the figures actually reflect expenditures for development. This appears to be the case particularly when mineral prices are high. The confusion of operating and investment expenditures makes management evaluation particularly difficult. For example, consider a manager whose budget includes some provision for mine development. At the end of the year it is not possible to ascertain from the accounts how much he actually spent on development, as the figures for labor costs, for example, may all reflect expenditures on production. Alternatively, a manager who spends more on development (in many cases a labor-intensive activity) may be severely penalized for having labor costs above his budget. The assumption is likely to be that the money was spent on production. In fact, in the internal reporting forms used, it appears as a part of the cost of ore sold. In addition, there are no accurate investment figures to tell how well a manager is doing, and it is not possible to calculate any kind of rate of return.

The structure of COMIBOL also makes control difficult. Superficially, the firm appears to be decentralized into three geographic divisions, but with some weight given to type of mineral, in terms of the classification. In practice, the structure appears to be quite different. The group managers, far from being the equivalent of division managers, are in effect managers of regional service centers. Although the service centers are no doubt very important, the result is that all mines, in effect, report to the Office of the Technical Manager in La Paz. With over twenty mining-related enterprises (some further subdivided into sections) reporting to him, the Technical Manager has a span of control that almost inevitably means little supervision. Any change in this situation would require that the regional managers be "general managers" with an adequate staff and with clearly delegated authority over certain decisions.

The authority of managers at various levels is very unclear. There appear to be no clear specifications of either type or size of expenditures that can be authorized at the mine level, the regional level, or at La Paz headquarters. Consequently, the current structure appears to result in many minor decisions being referred to La Paz. On the other hand, an aggressive mine manager who is willing to ignore the bureaucracy can make a lot of decisions himself.

The foregoing comments on problems of administrative and accounting infrastructure in the state minerals enterprise do not constitute, by themselves, a design for change. But they do point up most of the major problems that need to be considered in efforts toward effective restructuring of COMIBOL's administrative tools and mechanisms. Such a restructuring would not only allow the enterprise to generate higher surpluses from its ongoing and projected operations, but would make it possible to consider a future shift to a true income tax on mining, in which COMIBOL would be the single largest taxpayer.

THE MINING CODE

The Bolivian Mining Code (*Código de Minería*) that is now in effect dates from May of 1965 (Decree Law No. 07148). Since that time, there have been some amendments, largely in articles affecting taxation. Aside from these, none of the basic provisions have been altered to any significant extent.

Substantial efforts were underway within the government in 1975 to alter some key provisions of the Code dealing with the nation's fiscal reserves, including a five-kilometer belt around COMIBOL properties. Up until now, both types of territory have not been available for systematic exploration by private concerns, and (except for some exploration activities by GEOBOL on the fiscal reserves proper) have not been the object of state exploration activity either. These areas, which have been immobilized for the past decade, are potentially rich sources of future mineral income, and steps to make them more accessible for exploration could furnish a basis for strong resurgence in Bolivian mineral development, as occurred in Peru in the early fifties.

Revision of the 1950 Mining Code in Peru clearly played an important role in that country's subsequent mining boom. And there are enough similarities in the geology, history and economy of the two nations that a survey of the Peruvian experience with reform in the mining code could be useful in Bolivia's efforts to revise her code. It is difficult to say which features of the Peruvian mining code reform contributed most to the rapid development of that country's mining sector in the subsequent decade. The answer to this question is complex, because several factors played important roles, and at different times.

It is possible to argue first that the general steep rise in metal prices after World War II must have stimulated mining, even in the absence of new legislation. This may well have been a factor, but in order to be convincing one has to produce plausible reasons why this surge occurred in Peru and not in all other important mining countries, particularly Bolivia and Chile, or other Latin American countries having some mining tradition, such as Colombia and Argentina. But the latter two countries were prey to considerable turmoil (i.e., the *Violencia* in Colombia and the struggle between Peronistas and anti-Peronistas in Argentina). And among the traditional mining countries, Bolivia executed major nationalizations in

1951, and in Chile there were rigid and highly discriminatory exchange rates for the *grán mineria*, so that foreign capital, which provided most of the vitality for Peru's boom, concentrated mainly on this latter country, which was, relatively speaking, politically quiet under the military government of General Odría. Another factor that may have favored Peru over other countries was the spectacular (237 percent) devaluation of the sol (from 6.50 to 15.43 soles/dollar) at the beginning of the Odría government.

Nevertheless, the initial factors mentioned in the preceding paragraph were sustained over the longer term by several features of the Peruvian Mining Code per se. There is general agreement in Peru that the following considerations contributed to rapid development of that country's mining sector in the fifties:

1. The fact that the "rules of the game" were essentially guaranteed to remain relatively stable for the twenty-year period evisioned by the Code;
2. A shift from output-based taxes to income taxes;
3. Special provisions which allowed for the possibility of negotiating special contracts for low-grade ore deposits—a significant factor leading to the Toquepala contract in copper[1] and the subsequent development of this mine and of the important Cuajone mine;
4. Provision for depletion allowances;
5. The opening up of the "mining reserves" (comparable to the "fiscal reserves" in Bolivia) to exploration and exploitation;
6. The preferential status conferred upon mining over agriculture/farming concerning the use of surface lands, as well as the right to prospect anywhere except on private property;
7. The institution of practical, low-cost exploration claims and the ability to hold mining claims even when there was no evidence of mineralization—a critical provision if modern geophysical and geochemical prospecting methods are to be used;
8. The option of capitalizing or charging as current costs the construction of major roads or railroads.

The Peruvian experience of the fifties may or may not contain some lessons that are relevant for rapid development of the mining sector in Bolivia or elsewhere, if in fact rapid development is desired. But in any case, the following observations may be offered:

1. Private investment is unlikely to be forthcoming in substantial volumes if investors cannot perceive that the "rules of the game," particularly concerning taxation, are likely to remain relatively stable over a fifteen- to twenty-five-year period.

2. Primary reliance upon income taxes, rather than output taxes, reduces substantially the risk of financial ruin in an industry (such as mining) subject to drastic price swings. For reasons noted in Chapter Seven, an income tax is widely perceived by investors as far superior to taxation through royalties involving the same amount of revenue.
3. The possibility for negotiating special contracts is, in principle, a desirable feature. But, in the case of the large Toquepala copper mine, the special tax concessions obtained under the belief that this deposit was too low-grade to otherwise provide a reasonable rate of return, coupled with unusually high metal prices and unsuspected rich ore at the beginning of operations, led to profits which were later widely viewed as "excess profits." This suggests that some "safety clauses" to increase state participation in unexpectedly large profits should have been contemplated.
4. Although the depletion allowance was widely credited with stimulating mining development in Peru, it was also widely accused of leading to "excess profits" and is thus commonly singled out as one of the factors leading to the reactions embodied in the mining policies of the recent Revolutionary Government of Peru. However, it might be noted that the latter simply substituted for the depletion allowance an even more liberal "reinvestment allowance," thus, in effect, making only a nominal change in the treatment of a major component of mining investment.
5. There is little question that the abolition of the old mining reserves (fiscal reserves) was a very strong promotional step and that the reinstitution of such reserves by the government in recent years has had a strong negative effect on exploration.
6. The right to prospect everywhere, except on private lands, made it possible to use modern large-scale geophysical, geochemical and airborne prospecting methods.

FOREIGN INVESTMENT POLICY

Bolivia's experience with foreign investment in her mining sector has been unusual. Unlike most countries with mineral exports, Bolivia's mining activities have never been dominated by the major international mining enterprises. Rather, foreign ownership, although large at times in the past, has been in the hands of relatively small foreign firms and of families of Bolivian origin with headquarters abroad.

The unusual pattern has not been without cause. The kind of minerals exploited in Bolivia and the size of deposits have no doubt been important factors in determining the extent and character of foreign investment. The foreign markets for Bolivia's major export minerals have been relatively open, since the industries have not been dominated by vertically integrated, multinational firms. Thus, tin could be sold on the world market without close affiliation with a firm owning a smelter. Similarly, bismuth, antimony, silver and gold were generally bought by users from independent sources. In those minerals such as copper, for which vertically integrated firms were important, Bolivia's output was sufficiently small that she could dispose of her production without close ties to foreign buyers. In addition, the small scale of Bolivian mines has meant that major international firms had little interest. The cost in terms of management effort was prohibitive. A firm on the order of Kennecott would prefer to invest its resources in the large mines of Chile or the huge porphyry deposits of New Guinea, rather than the comparatively smaller deposits of Bolivia.

As a result of these factors, Bolivia has been able to bring most of her mining sector under national control with little outcry from abroad. She has been able to pay reasonable compensation for nationalized properties, without ruining her credibility among foreign investors.

If and when the government wishes to encourage some limited foreign participation in mining, it will be essential to have a clear view of the goals that might be achieved from foreign involvement in the mining sector. The goals that may be secured from foreign participation can be three: to provide access to markets abroad, to provide capital and to provide technology. To some extent, Bolivia may desire to draw on all three of these contributions in the future.

For most Bolivian minerals, access to the market is not a difficult problem except in the case of iron ore in the as yet undeveloped Mutun deposit. Owing to high transport costs and other marketing factors, the product would have to be sold to Argentine steel producers, which have a number of alternative sources. In such a situation, investment from those potential customers may be a *sine quo non* for the development of Mutun; the ownership tie may be the only way of assuring a market for Mutun's output.

Although Bolivia already obtains some foreign capital, this occurs primarily through borrowing by mining firms. However, foreign capital may also be sought by the private domestic mining firms for exploration. In this case, some kind of equity arrangement may be worked out. One goal of American investors will be to retain the tax write-offs that U.S. mining firms receive for exploration expenses, although U.S. laws are changing on this matter. Bolivia may also feel compelled to accept foreign equity on occasion to obtain the technology it can bring. The technology can be in the exploration, mining, or beneficiation stage.

Most likely, foreign investment in Bolivia will continue to be undertaken by relatively small foreign mining firms, for reasons noted elsewhere in this volume. These firms are likely to be willing to enter into joint ventures with Bolivian firms. First, smaller firms favor joint ventures more than do their larger competitors. Second, since the minerals concerned are not likely to be characterized by vertical integration, the foreign firm is less likely to feel the need for absolute control and hence complete ownership.

Assuming that some foreign investment will be accepted, several issues arise as to how that investment might be treated. Three are dealt with in this section: (1) special contracts vs. general legislation, (2) creditability of taxes paid in Bolivia, and (3) significance of Andean Group (Grupo Andino) decisions.

Special Contracts vs. General Law

Currently, Bolivia taxes mining under two radically different systems. The majority of mines are subject to the taxes specified in the general laws that govern mining. However, some operations are the subject of special concession agreements that provide terms specific to the particular mine. A special uranium contract, a ten-year-old gold placer agreement, and a few lapsed contracts for mining on COMIBOL land illustrate the genre.

Practice in other countries varies. In Liberia, Papua New Guinea, and Indonesia, the vast majority of mining operations are governed by rather extensive special contracts that specify tax provisions and, indeed, many other regulations that are to control the mine. In Canada, to take the other extreme, taxation at the federal level is handled entirely by general legislation. The provinces usually set their tax levies in general legislation as well. There are, of course, countries that take approaches between the two extremes. Australia, for example, sets federal taxes by general legislation, but the states typically negotiate individual contracts that levy royalty rates and other provisions that are tailored to the particular project. In fact, there is probably a trend toward some types of intermediate approaches, a point to which we will return later.

There are reasons why special contracts ("ad hoc") have been so popular, especially in the developing countries. Peculiar to certain of these countries has been the fact that the laws of general application were poorly developed. Income tax laws did not deal with the problems of mining and, especially, the complex problems associated with taxing foreign firms, which were usually the initial investors. As a result, the required laws were spelled out for each investor. For example, typical Liberian iron ore agreements specify not only the tax rate but the definition of gross income and the specific cost deductions that will be allowed. They also spell out safety and labor rules and rights of access to roads and railroads, along with many other details.

There is another important reason for the frequency of ad hoc contracts for mineral extraction in developing countries. An individual mining operation has, in many cases, been so important to the economy that the host government was unwilling to leave the arrangements to the general laws which were based on "averages." The possible gains from tailoring the contract to the particular circumstances appeared to be very great. Thus Papua New Guinea, whose laws (virtually copies from Australia's) were much more developed than those of Liberia, nonetheless initially negotiated and, after renegotiation, has retained an ad hoc agreement for the Bougainville mine, which accounts for some 20 percent of GNP.[a]

The pressures for specific contracts have come not only from governments but also from mining companies. Two concerns have tended to underlie their quest for ad hoc arrangements. In many cases the firms have sought special exemptions from the general laws. Arguing that their ventures face risks not provided for in the general laws and that they must undertake unusual investments in roads, ports, schools, hospitals, and the like, they have convinced governments that they should receive more favorable treatment than that provided in the general laws for investors in more conventional activities. In many, but not all cases, the argument was a sensible one. The second drive behind companies' desires for ad hoc arrangements was their search for predictability and stability in their arrangements. If they made their investment under the general laws, they ran the risk that those general laws might be very different in a few years' time. The foreign mining companies in particular were concerned about how the new laws might affect them. Indeed, at home they generally took the risk; but at home they tended to feel more confident that the political process would assure that any changes would be reasonable and nondiscriminatory. In many cases, they were confident of their access to the government to influence any changes. In the developing countries they were less certain of the

[a]Note that Bolivia is in a rather different position in this regard from Papua New Guinea, or even from neighboring countries such as Chile and Peru. In Bolivia, no single private sector mine accounts for a very large percentage of GNP or of exports.

political process and of their influence on it. Thus, many sought agreements that would promise stability for 20, 40, or even 99 years.

Many of the conditions that led to the dominance of ad hoc mining agreements are changing. In many developing countries, the general laws have been developed sufficiently that they address most of the special problems of mining and foreign investment. Today, governments often provide at least a portion of the infrastructure that the mines provided in the past.

Equally important, companies are now less enamored of the stability promised by ad hoc agreements. They have learned from experience. The stability of several decades promised in the agreements has in practice usually held only for a few years. As bargaining powers shifted, the political process in the host country has usually insured that the outdated agreements would be renegotiated. In fact, the special contracts have become foci of political unrest as governments have been accused by their opposition of corruption and ignorance in their negotiations. The accusations have, of course, not always been entirely without ground. Where the terms were too favorable to the company, redress was sought. In fact, the efforts to renegotiate the special agreements have sometimes gotten out of hand. Criticisms of the old agreements have frequently become so bitter that nationalization became an almost inevitable outcome of the political process, although both the investor and the host country might well have benefited from a continuing relationship.

Some mining projects still tend to present special cases which cannot be foreseen by the general laws. And particularly large ones present especially attractive targets for ad hoc negotiations. However, new mechanisms have been developed to allow for the negotiation of special terms needed in those cases without some of the problems associated with the old ad hoc concession agreements. The general theme of these arrangements (and they are not just for mining, but often apply to such fields as forestry) is that the companies will be subject to the general laws, but that very limited special provisions may be negotiated. These are usually a bonus payment by the firm at the outset (as illustrated by the U.S. in its offshore oil concessions), the royalty rate, or the extent of government ownership (or its share in the output of the project). Under such arrangements, income taxes and other fiscal charges are not negotiable. The company simply pays what other companies would pay. Among other advantages, such devices place some limit on the potential for corruption. The negotiations do not give special breaks for the company, but rather they determine how much *more* it pays than the general taxpayer. Since mining can generally be taxed more heavily than other activities,[2] this allows for the special cases of poor ores and remote locations.

Negotiations on ownership (or production sharing) are particularly common in developing countries. They offer considerable political appeal in that foreign ownership of property rights in minerals, a touchy issue, is reduced. Any step that reduces foreign ownership offers more hope of political advantage. Under participation or ownership arrangements, a certain percentage of ownership may be given to the government without charge, or the government may make some payment for the shares it acquires. If payment is made, it is usually made after the project has been proved. Thus, the government does not share in the high risk at the outset.

There are certain precautions to be taken in ownership negotiations. First, governments should not fail to recognize that an ownership share of, say, 50 percent, is *not* the equivalent of a 50 percent income tax. The dividends accruing to the government will almost always be less than the flows from a nominally equivalent income tax. Dividends come after repayment of principal on debt and after provision for reinvested earnings. There are possibilities, of course, of requiring payments to the government that would approximate an income tax. Second, the contract should clearly specify what is to happen if the project is to be expanded. Does the government keep its percentage ownership only if it puts up a share of the capital for expansion? The contract can specify that the government share remains constant without new funds from the government (this will, of course, lead the firm to prefer reinvested earnings to new injections of capital).

Bolivia can choose among the alternative ways of handling foreign investment. At the moment, most mines are subject to the general laws governing the mining sector as a whole. This seems appropriate to the usual size of mines in Bolivia and the state of the legal system. Exceptions exist for the fiscal reserves and for certain minerals. Bolivia might also want to make an exception for projects of over some particular size: say, an invested capital of more than U.S. $125 to $150 million.

In these cases, clearer rules on what is negotiable should be established. In particular, one suggestion would be that the usual fiscal provisions are not negotiable. The miner would be subject to the general tax laws. However, government ownership would be negotiable (and, in many cases, free to the government). Alternatively, the government's share in the output of the mine could be negotiated. The share of output could be over and above the income tax, or it could be viewed as a payment of income tax, with the minimum share being equivalent to what the government would receive under the income tax regime in force at the time of payment.

The nature of the likely foreign investment in Bolivia means that it will probably be unnecessary to draw up detailed "concessions contracts." Rather, foreign investment can usually be accepted under the general legislation. There may be times the government wants to insist on a portion of ownership, but the entry terms can otherwise be left to the general legislation. Special contracts, including those for such strategic material as uranium, would be the exception.

Creditability of Taxes

In most cases, the home country of a foreign investor will provide that investor with a reduction in taxes on foreign-earned income to account for the taxes paid by the investor to the foreign country on that income. The most common practice is to reduce the taxes due at home by approximately the amount of taxes already paid on that income abroad (the tax-credit method). This is roughly the rule applied unilaterally by the United States. A number of other countries use a similar approach, but most arrange for the tax reduction ("tax credit") through a double taxation treaty with the country serving as host. In a few home countries, there is simply no tax on foreign-earned income.

Most countries do not allow "tax credits" for just any kind of tax paid in the host country. The creditable taxes are limited to something that can be considered as a tax on income. The name of the tax plays some role in the matter of creditability, but the form is generally more important than the name. Thus, one cannot simply call a tax on output an "income tax" and expect all foreign investors to receive tax credits in home countries for payment of output taxes. The tax must actually be imposed on net income.

The matter of creditability is important. If a tax paid by an investor in Bolivia is fully creditable against taxes paid by that investor at home, the tax in Bolivia is not likely to hinder foreign investors. True, there are some details that must be taken into account. For example, at least through 1976, the United States has taxed foreign-earned income only after a remission of that income has been made. Thus, a tax rate in a host country that is lower than the rate in the United States allowed a U.S. firm to have a higher after-tax income for a period, as long as the income was retained where it was earned. But, at the moment a dividend was declared, the advantages the firm might have gained through a lower tax in the host country have generally disappeared. In a world of foreign tax credits, as long as the host country has the option of applying a tax that is creditable and one that is not creditable, it should choose the creditable tax, everything else being equal, because to an extent this would amount to a tax on the investor's home government rather than the investor.

Significance of Andean Group Decisions

Another set of potential problems for Bolivia in the attraction of, and taxation of, foreign investment revolves around the rules of the Andean Common Market, to which Bolivia belongs.[b] These rules place strict requirements on most foreign investors: foreign investments must be approved and registered; majority ownership must revert to local investors according to a specified schedule; dividend remissions, royalty payments to parents, interest on loans from parents, and so on, are limited; and acquisition of stock by foreigners in ongoing enterprises is restricted.

Mining projects (i.e., "primary" mining activities), however, can be exempted from a few of the rules of the Andean Group. Presumably, mining investments must be approved and registered like other foreign investments. Also, mining investments are subject to the same rules as other investments with regard to rules on reinvestment, capital remittances abroad, use of credit and transfer of technology. But the rules for mining do differ regarding divestment and profit remittances.

Article 40 of Decision 24 of the Andean Group provides the basic guideline. Under that Article, the government can exempt mining projects from the requirement that ownership be transferred to local investors. Article 39 allows mining projects to be exempted from the restrictions on dividend remittances. However, Article 40 imposes a major restriction: the term of a mining concession cannot exceed twenty years. Moreover, the provisions of the Article apply only for the first ten years that the Decision is in effect. (Nonetheless, Article 43 allows member countries to depart from these guidelines.) The Decision is silent on what rules apply to mining at the end of the period. One possibility is that foreign mining projects will be acceptable, but subject to the rules on ownership and dividends that apply to manufacturing projects. Another interpretation is that mining projects will, like other projects such as public utilities

[b]The Andean Common Market (Grupo Andino) was formed as a result of the Cartagena Agreement, signed by Bolivia, Colombia, Chile, Ecuador and Peru in 1969, and by Venezuela in 1972. The Agreement provides for freeing of trade for some, but not all products, implementation of an industrial programming scheme, an eventual common external tariff, special treatment for the two least developed members (Bolivia and Ecuador) and a variety of other measures to promote economic integration within the Andean Group.

included in the same Article, simply be closed to foreign investment. If this interpretation should prevail, it could serve as a serious constraint on the development of Bolivia's mineral resources.

The rules prohibiting acquisition of stock in ongoing enterprises appear at first glance to threaten the kinds of joint ventures being proposed in Bolivia. In these proposals, a foreign firm would join forces with a local mining enterprise, through the purchase of stock, to expand exploration and development. However, an exemption (Article 4) allows the acquisition of such stock if the result is an increase in stock and the resultant firm is still classed as a national enterprise or a joint venture.

Beyond the rules on foreign investment, the commitment to the Andean Group will have broader effects on mining in Bolivia, both foreign and domestic. The development of a common external tariff designed to protect manufacturers in the Group may have important consequences for Bolivian mining. Efforts to exempt machinery and equipment imports from duties may run afoul of the Andean rules. And miners may find higher prices for some of their inputs—for imports as external tariffs rise and for goods they must acquire within the Andean Group.

Overall, Bolivia has a comparatively unrestrictive policy toward foreign investment, but the uncertainties created by ambiguities in Andean Group rules might well dampen the effects of the otherwise liberal Bolivian legislation.

NEW PRODUCERS' ASSOCIATIONS AND COMMODITY AGREEMENTS

Prospects for future tax revenues from mining may, in some instances, be materially affected by the success or failure of a number of producers' associations or commodty agreements to which Bolivia now belongs or is contemplating joining. Bolivia has been an active participant in the International Tin Council, which involves not only the major producing nations, but all important consuming countries including, since 1976, the United States. Although the tin agreement has had its problems,[3] there can be little doubt that it has been helpful to Bolivian interests. It would appear to be in Bolivia's interest to continue to participate actively in that agreement, particularly as there are special perils inherent in continued dependence upon tin for exchange earnings and fiscal resources, since world consumption has been growing at a rate slower than for most major metals, and near-term prospects for an increase in consumption growth are not encouraging.

Efforts are being made to establish effective commodity agreements for other minerals that are of concern to Bolivia. The efforts that Bolivia, or any other country, can spend on developing new arrangements have an opportunity cost and should normally be allocated to the most likely prospects, political considerations aside. Important factors in evaluating the likely success of a producers' agreement for a mineral include the following:

1. the number of producers supplying the product, and concentration of sales among these producers;
2. the degree to which other products can be substituted for the product in question;
3. the resistance of the consuming firms, which will depend at least partly on the importance of the particular mineral in the cost of the final product;
4. the interest of the consuming countries, both economic and political;
5. the degree of common interest among the producers.

Since the most important case now facing Bolivia is that of a possible association of producing and consuming nations in tungsten, more space will be devoted to a discussion of the prospects for success of this arrangement than to the others.

Tungsten

A tungsten agreement would presumably have as its objective the creation of a less volatile price than has governed this industry in the past.[4] Moreover, it might have the additional goal of creating higher average prices than would otherwise prevail. An effective agreement would presumably use either a buffer stock or export quotas, or perhaps both, to enforce the price objectives.

The prospects for an agreement that will be effective over a long time period are not especially bright unless major consuming countries can be induced to provide strong support to the goals of the agreement. An effective agreement is likely to require a demand that is relatively price inelastic, a gradeable product, and either a small number of important producers or cooperative consumers. Tungsten can probably pass the first two tests, but is not likely to pass the final one.

Various UNCTAD studies argue rather convincingly that tungsten faces a price inelastic market in the short run.[5] The primary evidence comes from the wide price swings that occur when the market contracts slightly. A dramatic fall in prices does not much affect quantity demanded. Over the longer term the price elasticity of demand is more uncertain. Molybdenum appears to have made some inroads into the tungsten market. Its short-run substitutability appears to be rather low, and the market for molybdenum is characterized by more moderate swings. The higher relative price of molybdenum in recession periods does not seem to have led to an increased market for tungsten. Further, over the longer term, molybdenum has continued to displace tungsten in some uses.[6] It is uncertain whether the substitution is primarily a result of price differences or a result of the successful marketing efforts of the leading molybdenum producer. But even if marketing is the key, higher relative prices for tungsten would probably accelerate the process of substitution. On the whole, compared to many other minerals, the price elasticity of demand for tungsten does not seem to be such that efforts to control price swings would be hampered, but the substitution of molybdenum may put some limits on any longer-term efforts to raise the average price.

A major problem is raised by the large number of countries that supply tungsten to the world market.[7] The three most important developing country exporters in the non-Communist world are Thailand, the Republic of Korea, and Bolivia. These nations account for only 28 percent of world exports. The addition of the Socialist countries raises the figure to 40 percent. The nine principal developing countries account for about 40 percent. The proportions are considerably smaller, if the base is production rather than world trade, since several consuming countries are major producers. Unlike the case for oil and tin, there is no small group of developing countries with the opportunity of having a major influence on world prices if they act in concert. For tungsten, there is no Saudi Arabia or Iran as in oil, and there is no Bolivia, Malaysia, Thailand or Indonesia, as in the case of tin. In addition, there is no clear community of political interests among the major exporters or producers. The principal exporters include a Southeast Asian country, an East Asian country, a Latin American country, Communist China, Australia, and Canada. There is nothing like the Arab community to exercise an influence in holding together a large group of countries.

Moreover, there are likely to be strongly different economic interests among the producing countries. Some are important consumers, as well as producers. Some produce tungsten as a coproduct with other minerals. And for most, tungsten is a relatively small portion of their foreign exchange earnings. Unlike the coffee exporters, most are not dependent on a single product. In addition, tungsten supply does appear to expand over the medium term when prices are high. Thus, a successful agreement would mean that there is a temptation for an outsider to expand his output in response to the higher prices, thus bringing them down.

In some cases, large numbers of countries with somewhat disparate interests can probably be held together in an agreement if the consumers cooperate in enforcing the terms. This kind of cooperation was important in holding the African and the Latin American producers together in the coffee agreement. But that cooperation is unlikely to be forthcoming for tungsten.

It is true that tungsten does not represent a significant portion of the final cost of most products.[8] Thus, the political and the market reaction to higher prices is unlikely to be great, as it is for coffee. However, there are other problems. Most important, tungsten is viewed as an essential product for the defense industry. Thus, the United States is unlikely to encourage a producers' association that could impose an embargo. In fact, the United States keeps a large stockpile that could be used to disturb any agreement.[9]

In addition, the political rewards for the producing countries in supporting a tungsten agreement are not great. For coffee, the U.S. State Department fought hard to convince Congress that the coffee agreement would be a significant aid for potential Latin American friends. It would be much more difficult to argue as strongly in the same vein for a tungsten agreement. Only a few "friends" would be helped significantly, and some countries less friendly to the major consumer would be helped as well. For oil, the cooperation of the consuming countries was lacking, but the vertically integrated oil companies helped keep prices high and stable. For tungsten, such a market structure is not present, as the industry is not vertically integrated.

Most of the major producing countries, especially the developing ones, will most likely sign an accord for tungsten. However, the prospects would improve only if the industrialized countries further moderate their usual stance on commodity agreements and decide to actively support them. This is a possibility, given the interest in the "new international economic order," but it is a matter over which a single developing country has little control. If an agreement is signed, expectations for long-run success should be tempered by recognition of the market and political realities mentioned above. It is particularly doubtful that, even in the context of a tungsten agreement, Bolivia will be able to "export high taxes" on tungsten. True, the members of the International Bauxite Association (IBA) succeeded in exporting higher taxes on bauxite, at least in the early years of the

Association, but in that case most market conditions were favorable for substantial tax exporting.[10] In the case of tungsten, perhaps the best that can be hoped is creation of a somewhat greater degree of price stability than has been the case in the past for this mineral.

Other Minerals

Bolivia is already a member, with Peru and Belgium, of the Bismuth Producer Group, headquartered in La Paz. This organization concentrates its efforts on promoting uniform pricing policies among members, and on resolution of marketing problems in bismuth.

The prospects for a strong producers' association in this mineral are poor. Although there are few producers, their interests are very different. Bolivia is the only country that mines bismuth ore for its own sake. Peru, Mexico, the United States, Japan, Australia and other nations produce bismuth as a byproduct in the production of other metals. Thus, these countries have little ability to cut back production to stabilize prices, much less to export higher taxes on bismuth. The only possibility is for them to store bismuth, but bismuth is probably not sufficiently important to either the firms or the countries to induce them to take strong action.[11]

In the case of antimony, the number of important producers is relatively small, the principal suppliers being Bolivia, South Africa and China.[12] Other supplying nations include Mexico, Russia and Yugoslavia. For most buyers, antimony is a rather small portion of the final cost of the product. The principal problem in forming an effective association is likely to be the different political orientations of the principal producers. Thus, the outlook for a successful producers' association is fair to good for antimony, and Bolivia would be an important factor in the prospects for success of such an association.

Prospects for a successful association of iron ore producers are very poor. Previous efforts to establish a producers' agreement for iron ore have been unsuccessful. There are many producing countries, among them several industrialized nations, and the differing interests are not likely to be reconciled soon. Notwithstanding the lack of success of earlier attempts, a new producers' association came into being on September 15, 1975. This group, called the Association of Iron Ore Exporting Countries, included as members Algeria, Venezuela, Australia, India, Chile, Peru and Mauritania. In 1976 efforts were made to persuade Brazil, Canada and Sweden to join, but Canada has recently refused to do so and Sweden will probably not join.[13] Australia, Venezuela, Brazil and India account for the bulk of world exports of iron ore, but for a much smaller share of total world production. The Association would likely not enjoy any measure of success even if Brazil, Canada and Sweden were to decide to affiliate.[14]

A producers' association for copper already exists: the Council of Copper Exporting Countries (CIPEC). This group includes the four largest copper-exporting nations: Chile, Peru, Zambia and Zaire. The association has tried to impose quotas on its members, but such measures proved unsuccessful in halting the slide of copper prices in 1975. The major difficulty has been the entry of new producers which are expanding supply and are not willing to join the association if it limits their growing sales. Major new producers to enter in recent years or soon to enter the market include Indonesia (Freeport in West Irian), Malaysia (ore from Sabah), Iran and Panama. Bolivia is responsible for an infinitesimally small share of world copper production, and thus would not be an important factor in CIPEC even if it joined.

The prospects for a successful producers' association for zinc are also quite poor. Of total world zinc production, about 43 percent came from North America. An additional 23 percent came from Europe. In fact, only some 13 percent came from the developing countries. In any case, Bolivia produces a negligible share of the world's zinc (less than one-tenth of one percent in 1973) and would have no influence over major decisions made by any producers' association for the metal, even if a successful one could be created. However, Bolivia does already derive some benefit, at no cost, from market stabilization efforts of the major zinc producers.

In sum, optimism regarding the prospects for strong international agreements to protect Bolivia's interests in the export of most of her nonfuel minerals should be tempered with an awareness of the sometimes formidable obstacles in the way of successful conclusion and operation of such efforts. But even where the outlook for a strong association able to stabilize (or raise) mineral prices is not bright, there may well be much to be gained from broadening and strengthening contacts between producing countries on the one hand, and producing and consuming countries on the other.

NOTES TO CHAPTER NINE

1. For detailed information on the initial Toquepala contract (and subsequent revisions), see Raymond F. Mikesell, *Foreign Investment in Copper Mining* (Baltimore: Johns Hopkins Press, 1975), Chapters 2 and 3.

2. For a discussion of minerals as a distinct tax base, see Chapter Four.

3. For a critique of the results of the International Tin Agreement since

1956, see Gordon W. Smith and George R. Schink, "The International Tin Agreement: A Reassessment," *Economic Journal*, vol. 86, December 1976, pp. 715-728.

4. See United Nations Conference on Trade and Development, *Tungsten: Characteristics of the Market and the Feasibility of Price Stabilization Measures* (Geneva, Document No. TD18/C.1/Tungsten Com/WG/w, June 25, 1975), pp. 1-4.

5. For example, see UNCTAD, *Ibid.*, pp. 19-23.

6. *Ibid.*, p. 25.

7. United Nations Conference on Trade and Development, *Study of the Feasibility of Alternative Measures to Stabilize Tungsten Prices* (Geneva, Document No. TD/B1C.1/Tungsten/4, June 18, 1975), p. 7.

8. For a discussion of the importance of being unimportant in prospects for success in mineral producers' associations, see Malcolm Gillis and Charles E. McLure Jr., "The Incidence of World Taxes on Natural Resources with Special Reference to Bauxite," *American Economic Review*, vol. 65, No. 2, May 1975, p. 391.

9. See UNCTAD, *Study of the Feasibility of Alternative Measures to Stabilize Tungsten Prices*, p. 3.

10. Gillis and McLure, *op. cit.*, pp. 394-395.

11. For a further discussion of some of the strategic considerations involved in establishing a more effective bismuth producers' association, see Arthur D. Little, Inc., *Commercialization of Bolivian Non-Ferrous Metals and Minerals*, vol. II: Strategic Considerations (Cambridge, Mass., 1972), pp. 32-34.

12. See Chapter Two.

13. *Financial Times*, September 16, 1975, p. 31.

14. For further information on the iron ore association, see Kenneth W. Clarfield, Stuart Jackson, Jeff Keefe, Michaele Ann Noble and A. Patrick Ryan, *Eight Minerals Cartels* (New York: McGraw Hill, 1975), pp. 141-153.

Glossary

Alluvial (adj.) Deposit. Deposited by a stream. Especially relevant for tin and gold in Bolivia.

Assay To determine the amount of metal contained in an ore. Note: Difference between assay and analysis: In an analysis all of the chemical constituents are determined; in an assay only certain constituents, usually those of commercial interest.

Cassiterite SnO_2. Basic tin-bearing ore in Bolivia.

Caving A method of mining in which the ore, the support of a great block being removed, is allowed to cave or fall, and in falling is broken sufficiently to be handled; the overlying strata subside as the ore is withdrawn. *Block caving*: Similar to top slicing except that the slice which is allowed to cave is of much greater thickness and may, in fact, constitute the full thickness of the orebody.

Crosscut A level driven across the course of a vein or in general across the direction of the main workings or across the "grain of the coal."

Eluvial deposit Loose material resulting from decomposition of rock. Eluvial material may have slumped or washed downhill for a short distance but has *not* been transported by a stream (cf. *Alluvial*).

Exploration The work involved in looking for ore. May be done on the surface or may consist of underground work or drilling. Distinguished from exploitation and in some cases from development work.

Face In any adit, tunnel, or stope, the end at which work is in progress or was last done.

Main Source: Hugh Exton McKinstry, *Mining Geology* (Englewood Cliffs, N.J.: Prentice Hall, Inc., 1948).

Flotation A method of concentrating ore by inducing the particles of ore to float to the surface of water or other solution (usually buoyed up by air bubbles) while the gangue particles sink to the bottom.

Mica Name for a group of minerals, all complex aluminum silicates with K and OH, often also Mg, Fe''', and, in some varieties, Na, Li, and Fe'''.

Ore A metalliferous mineral, or an aggregate of metalliferous minerals, more or less mixed with gangue, which from the standpoint of the metallurgist can be treated at a profit.

Outcrop The coming out of a stratum (or vein) to the surface of the earth. That part of a stratum (or vein) which appears at the surface.

Plenemensurate orebodies Those capable of being fully measured and sampled at an early stage in the operations. Other types are, according to categories of risk involved: *Partimensurate orebodies*—those in which prospects for ore in addition to proved reserves remain a substantial element until the later stages of the life of the mines based on them; *Extramensurate orebodies*—those difficult to explore and measure much in advance of mining, in which the value of prospects for ore based on geologic evidence exceeds the value of proved reserves throughout most of the life of mines supported by them.

Placer Deposit of gold-bearing alluvial gravel. Also applied to similar deposits containing other metals such as tin, platinum, or tungsten.

Porphyry copper Copper deposit in which the copper-bearing minerals occur in disseminated grains (or in veinlets) throughout a great volume of rock. The rock need not be porphyry. Characteristic: large tonnage and relatively low grade.

Scheelite $CaWO_4$. A tungsten-bearing ore.

Shaft A well-like downward opening in a mine. Usually a shaft starts from surface; if started from underground workings it is called a winze, though if equipped with a hoist it may be called an underground shaft. A shaft may be vertical or inclined. In some districts a flatly inclined opening is called an incline, "shaft" being limited to vertical openings.

Shoot, Oreshoot That part of a deposit in which the valuable minerals are so concentrated that their utilization becomes possible. Body of ore as contrasted with enclosing subcommercial vein-stuff. (The alternative spelling, chute, is now usually reserved for the mining term—a shaft or trough through which broken ore falls.

Shrinkage stoping A modification of *overhand stoping* in which ore is mined beginning on a level and working upward, withdrawing only enough broken ore to leave a space between the broken ore and the back for the men to work. After the stope reaches the next level above (pillars being left to support the floor of the level), the broken ore is drawn from the stope.

Stannite A tin-bearing ore. Cu_2FeSnS_4.

Stope An underground opening from which ore has been or is being extracted. Usually applied to vertical or highly inclined veins. According to Raymond the term is properly restricted to a series of advancing horizontal faces in step-like arrangement, but this restriction is not in accordance with present-day usage. An *overhand stope* is made by working upward from a level into the ore above. An *underhand stope* is made by working downward beneath a level.

Wolframite (Fe, Mn)WO_4. A tungsten-bearing ore. An isomorphous series whose end-members are ferberite, $FeWO_4$ and heubnerite, $MnWO_4$. Monoclinic, G. 7.0-7.5. H. 5-5½. Black (ferberite) to brown (huebnerite). Lustre submetallic to resinous. Perfect cleavage.

Appendix A

Size Distribution of Firms in the Medium Mining Sector, by Size of Sales, Profits, Labor Force, Payrolls and Total Fixed Assets

Table A-1. Size Distribution of Medium Mining Firms by Total Sales, 1973 and 1974 (all firms sampled)

Total Sales in Bolivian Pesos (millions)	*Number of Firms*	
	1973	*1974*
Less than 10	5	6
10-39.9	9	9
40-79.9	2	5
80-119.9	2	2
120-199.9	1	0
200-299.9	1	1
300 and above	0	1
Total numbers of firms	20	24
Total sales (millions)	1,006.8	1,259.2
Average sales (millions)	50.3	52.5

Table A-2. Size Distribution of Medium Mining Firms by Total Sales, 1973 and 1974 (*included* firms only)

Total Sales in Bolivian Pesos (millions)	*Number of Firms*	
	1973	*1974*
Less than 10	5	4
10-39.9	9	7
40-79.9	2	3
80-119.9	2	2
120-199.9	1	0
200-299.9	1	0
300 and above	0	1
Total number of firms	20	17
Total sales (millions)	1,006.8	886.7
Average sales (millions)	50.3	52.2

Table A-3. Size Distribution of Medium Mining Firms by Total Profits, 1973 and 1974 (all firms sampled)

Reported Profits (millions of pesos)	*Number of Firms*	
	1973	*1974*
Less than 0	2	1
0-0.49	2	2
0.5-0.99	4	1
1.0-4.9	8	11
5.0-9.9	1	2
10.0-19.9	1	2
20-39.9	1	2
40-79.9	1	2
80 and above	0	0
Total firms	20	23
Total profits (millions)	120.7	244.5
Average profits (millions)	6.04	10.63

Table A-4. Size Distribution of Medium Mining Firms by Total Profits, 1973 and 1974 (*included* firms only)

Reported Profits (millions of pesos)	*Number of Firms*	
	1973	*1974*
Less than 0	2	0
0-0.49	2	2
0.5-0.99	4	1
1.0-4.9	8	8
5.0-9.9	1	2
10.0-19.9	1	1
20.0-39.9	1	2
40.0-79.9	1	1
80 and above	0	0
Total firms	20	17
Total profits (millions)	120.7	147.4
Average profits (millions)	6.04	8.67

Table A-5. Size Distribution of Medium Mining Firms by No. of Employees, 1973 and 1974 (all firms sampled)

No. of Employees	*Number of Firms*	
	1973	*1974*
0-99	1	2
100-299	14	16
300-499	0	2
500-799	4	1
800-1000	1	3
Total number of firms	20	24
Total employees	5,995	7,019
Average number of employees per firm	300	293

Table A-6. Size Distribution of Medium Mining Firms by No. of Employees, 1973 and 1974 (included firms only)

No. of Employees	*Number of Firms*	
	1973	*1974*
0-99	1	1
100-299	14	11
300-499	0	2
500-799	4	1
800-1000	1	2
Total number of firms	20	17
Total employees	5,995	5,230
Average number of employees per firm	300	308

Table A-7. Size Distribution of Medium Mining Firms by Total Payrolls 1973 and 1974 (all firms sampled)

Size of Payroll Exclusive of social security) (millions of Bolivian pesos)	*Number of Firms*	
	1973	*1974*
0-0.99	1	1
1.0-4.9	13	10
5.0-9.9	2	8
10.0-19.9	2	2
20.0-39.9	·1	2
More than 40	1	1
	20	24

Table A-8. Size Distribution of Medium Mining Firms by Total Payrolls 1973 and 1974 (*included* firms only)

Size of Payroll (exclusive of social security) (millions of pesos)	*Number of Firms*	
	1973	*1974*
0-0.9	1	0
1.0-4.9	13	8
5.0-9.9	2	5
10.0-19.9	2	2
20.0-39.9	1	2
More than 40	1	0
	20	17

Table A-9. Size Distribution of Medium Mining Firms by Total Fixed Assets 1973 and 1974, Year End (all firms sampled)

Amount of Fixed Assets (in millions of pesos)	*Number of Firms*	
	1973	*1974*
0-0.9	1	2
1-4.9	6	5
5-19.9	8	9
20-39.9	2	5
40-99.9	0	0
100-149.9	1	2
150-199.9	2	1
Over 200	0	0
Total number of firms	20	24
Total fixed assets, all firms (millions of pesos)	708.2	634.7
Average fixed assets per firm (millions of pesos)	35.4	26.5

Table A-10. Size Distribution of Medium Mining Firms by Total Fixed Assets 1973 and 1974, Year End (*included* firms only)

Amount of Fixed Assets (in millions of pesos)	*Number of Firms*	
	1973	*1974*
0-0.9	1	2
1-4.9	6	3
5-19.9	8	6
20-39.9	2	4
40-99.9	0	0
100-149.9	1	1
150-199.9	2	1
Over 200	0	0
Total number of firms	20	17
Total fixed assets, all firms (millions of pesos)	708.2	440.3
Average fixed assets per firm (millions of pesos)	35.4	25.9

Appendix B

Share of Three Major Materials Inputs in Total Material Costs, Medium Mining Sector, 1973 and 1974

PART 1—SHARE OF THREE MAJOR MATERIALS INPUTS IN TOTAL MATERIAL COSTS, MEDIUM MINING SECTOR 1973 AND 1974

Background

As indicated in Chapter Two, three particular imported materials inputs account for a very significant share of *materials* costs in the medium mining sector. These are

1. Ammonium Nitrate,
2. Sacks for metal concentrates,
3. *Barrenos* (for perforators).

The data derived from the Mining Sector Survey for both 1973 and 1974 (see Chapter Three) are of reasonably high quality, all things considered. Summaries of these data have been used to prepare tables showing:

1. The proportion of the costs of the three inputs above in total material costs, for 1973 and 1974.
2. A comparison of per unit cost changes for 1973 and 1974, for those three major inputs.

For 1973, 20 medium mining firms were included in the sample. For 1974, 26 firms were included. Cross-checks utilizing Ministry of

Finance import clearance documents were performed. The cross-check results confirm that the costs reported in the Survey were not overstated in 1973 or 1974.

Table B-1. Share of Three Materials Inputs in Total Materials Cost, 1973 and 1974 (costs at the mine site)

Materials input	*Quantity (thousands of kilos or pieces)*		*Value (B. $) (thousands of pesos)*		*Unit Cost (pesos)*		*Share of each input in total material (costs (%)*	
	1973	*1974*	*1973*	*1974*	*1973*	*1974*	*1973*	*1974*
1) Ammonium nitrate (kilos)	1,081.7	1,334.8	3,987.1	7,082.0	3.69	5.31	3.3	6.1
2) Sacks (per piece)	2,078.8	847.6	14,577.6	9,455.0	7.01	11.16	11.9	8.2
3) *Barrenos* (per piece)	15.8	19.6	4,565.2	6,630.5	288.30	337.36	3.7	5.7
Totals			23,129.9	23,167.4			18.9	20.0

Table B-2. Comparison of Per Unit Costs, 1973 and 1974 (costs at the mine site)

Material	*Average unit price (Bolivian pesos)* 1973	1974	*Difference Absolute (B. $)*	*% Increase*
1) Ammonium nitrate	3.69	5.31	+1.62	43.9
2) Sacks	7.01	11.16	+4.15	59.2
3) *Barrenos*	288.30	337.36	+48.53	16.8

PART 2—INPUT STUDY: AMMONIUM NITRATE

Background

Explosive ammonium nitrate has increasingly replaced dynamite and other explosives in Bolivian mining (especially in COMIBOL and the medium mining sector); however, a half-stick of dynamite is needed to detonate the charge. Also required are fuses, mercury caps and fuel oil (6 percent of weight of charge).

Ammonium nitrate accounts for a not inconsiderable share of total materials costs in such mines. In absolute terms, usage is quite high, and the larger medium mines usually order in 500-700 metric ton lots two or three times per year.

In Beza section of Siglo XX mine at Catavi, usage of ammonium nitrate in the first six months of 1975 averaged about one kilo per ton of ore cracked. (Note: 70 percent of ore won in this mine is obtained through block caving.) In one of the larger medium mines (tin-tungsten) where veins are worked in hard rock, usage in 1975 averaged 1.6 kilos per ton of ore cracked. Depending on hardness of rock and type of formation, usage can vary as much as 30 percent within a mine. In what follows, we will assume that 100 percent of ore cracked is removed from the interior of the mine.

Prices of explosive ammonium nitrate increased markedly from 1972 through 1974 (see below), but have more or less stabilized since January 1975. The product is, of course, based on natural gas feed, and is not notably different from nonexplosive ammonium nitrate used for fertilizer. Thus, prices of explosive ammonium nitrate tend to move in lockstep fashion with fertilizer prices.

Prices Since 1972

The following are typical C.I.F. charges, internal transport costs in Bolivia and costs at the mine site, for ammonium nitrate ordered directly in 500 to 600 metric ton lots.

Table B-3.

Date	*C.I.F. costs per metric ton (U.S.$)*	*Internal transport and handling costs per metric ton U.S.$)*	*Costs at mine site per metric ton (U.S.$)*	*Costs per kilo at mine site (U.S.$)*
Nov. 1972	122.00	28.00	150.00	0.15
April 1974	179.54	40.46	220.00	0.22
Dec. 1974	360.00	80.00	440.00	0.44
April 1975	362.75	82.25	445.00	0.445

In blasting operations for vein work, 46 kilos are used for one explosion. The ammonium nitrate is mixed with fuel oil and other substances before blowing the charge; thus, the cost of the charge will be somewhat in excess of that indicated above.

In Catavi in August 1975, the average grade of ore cracked and removed was 0.45 percent. Thus, the value of *contained tin* at the mine mouth per ton of ore (even before recuperation losses in concentration) was only 0.45 kilos. As a kilo of tin on August 22, 1975, was valued at $6.66, then the gross value of the tin won per ton of ore was only $3.00, even before recuperation losses.

Thus, explosives cost *alone* of $0.445 accounted for at least 14.8 percent of the value of contained tin in ore at the mine mouth.

Since recuperation losses in sink-and-float and flotation at Catavi's *ingenio* now runs about 50 percent, the value of tin actually won from one ton of ore was only $1.50 (50 percent × 3.00), in August of 1975. Thus, in Catavi, the largest mining installation, explosives costs alone, as a percentage of tin value per ton of ore cracked after concentration, are 29.7 percent ($0.445/1.50).

One may note that the cost of this input (at the mine site) increased by 200 percent from November 1972 to August 1975.

PART 3—INPUT STUDY: *BARRENOS*

Background

Barrenos are bit-like devices used in drilling. They come in three lengths of from about 80 cm. to about 240 cm. *Barrenos* are a highly important component of total materials costs in mining, at least for medium mining firms. For example, in one firm from 1972 to July 1975, the number of tons of ore broken per *barreno* ranged from a low of about 21 (in July 1974) to a high of 78 (in March of 1975). The average number of tons broken per *barreno* from 1972 to July 1975 appears to have been about thirty-six.

Prices

We secured a reliable price source for Swedish-manufactured *barrenos* in three different sizes. The prices in the following tables are given in Swedish Kroner and U.S. dollars (F.O.B.).

Table B-4. Prices for Swedish-Manufactured *Barrenos* in Three Sizes, 1973-1975

Barreno Size			*F.O.B. Price in Swedish Kroner*			
Bit size	*Model no.*	*Length in cm.*	*1973*	*1974*	*Jan. 1975*	*Aug. 1975*
08	714	40	74.75	90.85	104.65	115.0
16	714	39	82.80	101.20	115.00	132.25
24	714	38	100.05	120.75	138.00	155.25
			F.O.B. Price in U.S. Dollars (using recorded exchange rates)			
08	714	40	16.61	20.19	26.16	26.38
16	714	39	18.40	22.49	28.75	30.33
24	714	38	22.23	26.83	34.50	35.60

Table B-5. Percentage Increase in Costs of *Barrenos* for a Large Mining Firm from 1973 to August 1975[a] (*with* quantity discounts)

Barreno Size				
Bit size	*Model no.*	*Length in cm.*	*% Increase in Dollar Price F.O.B.*	*% Increase in Kroner Price (C.I.F.) F.O.B.*
08	714	40	58.8	53.9
16	714	39	64.8	59.7
24	714	38	59.6	55.24

[a]These percentages understate the actual cost increase for *barrenos* delivered to the mine site, since (as F.O.B. prices) they do *not* include higher costs due to increases in ocean freight costs or increases in domestic freight costs. Moreover, the firm in this example orders in *large lots* (10,000-20,000 pieces), receiving quantity discounts of 5 to 10 percent; thus, costs for smaller mining firms likely rose even more rapidly.

PART 4—SMALL MINING SECTOR: CHANGES IN INPUT COSTS, 1973 to 1974

Introduction

Other notes on input costs have largely dealt with the medium mining sector. This note, based on data supplied to the Ministry of Mines by BAMIN, provides comparisons of unit prices of some important imported material inputs supplied by BAMIN to small miners. The per unit costs here refer to *sales prices from BAMIN warehouses* (total value of sales of input (*j*) divided by quantity of (*j*) sold. Thus, these cost figures are not strictly comparable to earlier figures for medium mining.

Table B-6. Cost Comparisons, Small Mining Sector 1973-1974

Material	*Average per unit price U.S. $ 1973*	*Average per unit price U.S. $ 1974*	*% Change 1973 to 1974*
Dynamite (50 lb. bag)	17.20	27.41	+59.4
Mercury caps (each)	0.02	0.02	0
Fuses (waterproof)	0.04	0.03	−25.0
Nitrate of ammonium	?	12.24	?
Sacks—calcite	0.51	0.41	−19.61
Sacks—polypropyline	0.43	0.45	+4.70
Sacks—polyethylene	0.38	0.38	0
Steel bars (rockface work)	2.12	3.15	+48.58

It is known that the world price of, for example, fuses did not decline from 1973 to 1974. It is also very unlikely that the price of polypropylene (a derivative of crude oil) increased by only 4.7 percent from 1973 to 1974. What, then, can we make of some of these figures? We could conclude that BAMIN bought fuses and polypropylene at high end-of-1973 prices. Or, we could conclude that for such products BAMIN's pricing policy for sale of inputs to miners responded only very sluggishly to changes in C.I.F. import values. Alternatively, there may have been some recording errors.

Appendix C

Revenue and Expenditure Relationships for the Bolivian Mining Sector

Table C-1. Medium Mining Enterprises (Bolivian pesos)

	Item	*1973*	*1974*
1.	*Sales*		
2.	Domestic sales		329,345,219
3.	Foreign sales		702,277,609
4.	TOTAL sales	946,234,100	1,031,622,828
5.	*Labor costs*		
6.	Wages and salaries	123,291,490	123,096,476
7.	Nonwage benefits		
8.	Perdida pulperias	–	–
9.	Family subsidies	1,793,490	1,678,142
10.	Social expenses	10,703,732	11,760,426
11.	Indemnifications	11,609,757	15,027,716
12.	TOTAL nonwage benefits	24,106,979	28,466,284
13.	TOTAL labor costs	147,398,470	151,562,760
14.	*Nonmine costs*		
15.	Transportation costs	12,380,422	12,770,260
16.	Realization expense	153,788,850	247,953,132
17.	Insurance expense	4,278,391	1,602,281
18.	TOTAL nonmine costs	170,447,663	262,325,673
19.	Energy costs	14,970,809	16,207,982
20.	Other expenses	23,502,450	124,358,638
21.	*Material costs*		
22.	Domestic material purchases	18,827,547	9,692,130

Table C-1 (cont.)

	Item	*1973*	*1974*
23.	Foreign material purchases	90,877,023	62,698,089
24.	TOTAL material costs	109,704,570	72,390,219
25.	Depreciation expenses	37,032,100	32,171,533
26.	Mine rentals	3,683,027	837,641
27.	Total Costs [13+18+19+20+24+25]	503,056,605	659,854,446
28.	*Taxes*		
29.	Royalties	149,657,860	186,883,869
30.	Export taxes	N.A.	61,514,918
31.	Patentes and other taxes	N.A.	3,484,972
32.	TOTAL TAXES	206,043,180	251,883,759
33.	*Mine revenue* [4-18]	775,786,440	769,297,155
34.	Net cash flow [4-13-18-19-20-24] Gross of depreciation, mine rentals, taxes	480,210,140	404,777,556
	IMPORTANT RATIOS		
35.	*Composition of total costs*		
36.	Labor cost/Total cost [13/27]	.294	.230
37.	Realization expenses/Total cost [16/27]	.306	.377
38.	Transportation and ins/Total cost [15+17/27]	.025	.022
39.	Energy costs/Total cost [19/27]	.030	.025
40.	Other expenses/Total cost [20/27]	.053	.187
41.	Domestic material/Total cost [22/27]	.037	.015
42.	Foreign material/Total cost [23/27]	.181	.095
43.	Depreciation expense/Total cost [25/27]	.074	.047
44.	TOTAL, rows 36 to 43	1.000	1.000
45.	*Composition of mine revenue*		
46.	Labor cost/Mine revenue [13/33]	.200	.200
47.	Energy cost/Mine revenue [19/33]	.019	.021
48.	Other expenses/Mine revenue [20/33]	.047	.162
49.	Total material cost/Mine revenue [24/33]	.141	.094
50.	Depreciation expense/Mine revenue [25/33]	.048	.042
51.	Total taxes/Mine revenue [32/33]	.266	.328
52.	Total profits, interest, mine rental/ Mine revenue	.291	.153
53.	TOTAL, rows 46 to 52	1.000	1.000
54.	*Composition of net economic cash flow* gross of depreciation, taxes, mine rentals		

Table C-1 (cont.)

Item	1973	1974
55. Total taxes/Net cash flow [32/34]	.429	.622
56. Depreciation expense/Net cash flow [25/34]	.077	.079
57. Mine rental/Net cash flow [25/34]	.008	.002
58. Profits + interest/Net Cash flow	.486	.297
59. TOTAL, rows 55 to 58	1.000	1.000
60. *Other Ratios*		
61. Wage costs/Total labor costs [6/13]	.836	.812
62. Foreign material/Total material [23/24]	.828	.866
63. Royalties/Total taxes [29/32]	.726	.742
64. Export tax/Total taxes [30/32]	N.A.	.244
65. Patentes and other taxes/Total tax [31/32]	N.A.	.014

Table C-2. COMIBOL (Mining Operations Only: La Paz Head Office, Electric Plant, Smelter Excluded) (U.S. $)

Item	1973	1974
1. *Sales*		
2. Domestic sales		
3. Foreign sales		
4. TOTAL sales	190,450,441	221,132,640
5. Labor costs		
6. Wages and salaries	19,593,755	28,831,850
7. Nonwage benefits		
8. Perdida pulperias	4,819,085	12,378,655
9. Family subsidies	2,687,892	3,259,570
10. Social expenses	2,108,805	2,470,277
11. Indemnization	2,229,601	4,104,389
12. TOTAL nonwage benefits	11,925,378	22,212,891
13. TOTAL labor costs	31,519,133	51,044,742
14. *Nonmine costs*		
15. Transportation costs	793,549	918,642
16. Realization expenses	53,046,158	58,580,842
17. Insurance expense	82,114	107,714
18. TOTAL nonmine costs	53,921,821	59,607,198
19. Energy costs	3,240,682	3,720,506
20A Other expenses	3,765,025	4,913,889
20B Central office admin.	2,274,396	3,794,910

Table C-2 (cont.)

	Item	*1973*	*1974*
21.	*Material costs*		
22.	Domestic material purchases	N.A.	N.A.
23.	Foreign material purchases	N.A.	N.A.
24.	TOTAL material costs	15,270,382	18,605,058
25.	Depreciation expenses	3,077,613	3,621,765
26.	Rentals	137,029	171,654
27.	Total costs [13+18+19+20A+20B+24+25]	113,069,050	145,189,070
28.	Taxes		
29.	Royalties	20,948,604	38,835,210
30.	Export taxes	13,512,341	15,849,234
31.	Patentes and other taxes	6,479	37,913
32.	Total Taxes	34,467,425	54,722,358
33.	*Mine revenue* [4-18]	136,528,620	161,525,450
34.	Net cash flow [4-13-18-19-20-24] Gross of depreciation, mine rentals, taxes	80,459,020	79,565,340
	IMPORTANT RATIOS		
35.	*Composition of total costs*		
36.	Labor cost/Total cost [13/27]	.279	.352
37.	Realization expenses/Total cost [16/27]	.469	.403
38.	Transportation and ins./Total cost [15+14/27]	.008	.007
39.	Energy costs/Total cost [19/27]	.029	.026
40A	Other expenses/Total cost [20A/27]	.033	.033
40B	Central office admin./Total cost [20B/27]	.020	.026
41.			
42.	Total material cost/Total cost [25/27]	.135	.128
43.	Depreciation expense/Total cost [25/27]	.027	.025
44.	TOTAL, rows 35 to 43	1.000	1.000
45.	*Composition of mine revenue*		
46.	Labor cost/Mine revenue [13/33]	.231	.316
47.	Energy cost/Mine revenue [19/33]	.024	.023
48A	Other expenses/Mine revenue [20A/33]	.028	.030
48B	Central admin./Mine revenue [20B/33]	.017	.023
49.	Total material cost/Mine revenue [24/33]	.111	.115
50.	Depreciation expense/Mine revenue [25/33]	.023	.022
51.	Total taxes/Mine revenue [32/33]	.252	.339

Table C-2 (cont.)

Item	1973	1974
52. Total profits, interest, rentals/Mine revenue	.314	.132
53. TOTAL, rows 46 to 52	1.000	1.000
54. Composition of Net Economic Cash Flow Gross of depreciation, taxes, mine rentals	.428	.688
55. Total taxes/Net cash flow [32/34]	.038	.046
57. Rentals/Net cash flow [26/34]	.002	.002
58. Profits + interest + overhead/Net cash flow	.532	.264
59. TOTAL, rows 55 to 58	1.000	1.000
60. *Other ratios*		
61. Wage costs/Total labor costs [6/13]	.622	.565
62. Foreign material/Total material [23/24]		
63. Royalties/Total taxes [29/32]	.607	.709
64. Export tax/Total taxes [30/32]	.392	.290
65. Patentes and other taxes/Total taxes [31/32]	.001	.001

Appendix D

Limits on Deductions for Salaries for Owner-Managers Under an Income Tax

A commonly used method for corporations to avoid income taxation is for owner-managers to pay high salaries to themselves. Thus, they reduce the profits recorded by the firm.

A simple way of handling this problem, used in some other developing countries (e.g., Colombia) is to set a limit on the salaries of owner-managers that can be deducted for tax purposes. Note that this does *not* limit the amount of salary the enterprise can actually pay, only the amount that is deductible for tax purposes.

Two tasks exist: (1) to define an owner-manager, and (2) to set a reasonable upper limit on the salary that can be deducted as an expense in calculating taxable income.

An owner-manager is usually defined in terms of the amount of stock he and his immediate family hold either directly in the enterprise or through other enterprises. Fifteen percent is a reasonable stock/holding to qualify him as an owner-manager.

The maximum rate can be based on the size of the operation, under the assumption that managers of larger operations can be reasonably entitled to larger salaries. A simple approach is to decide what is an appropriate salary for managers of the smallest medium mine, what the maximum allowable salary should be and at what volume of production that salary should be allowed.

An example follows:

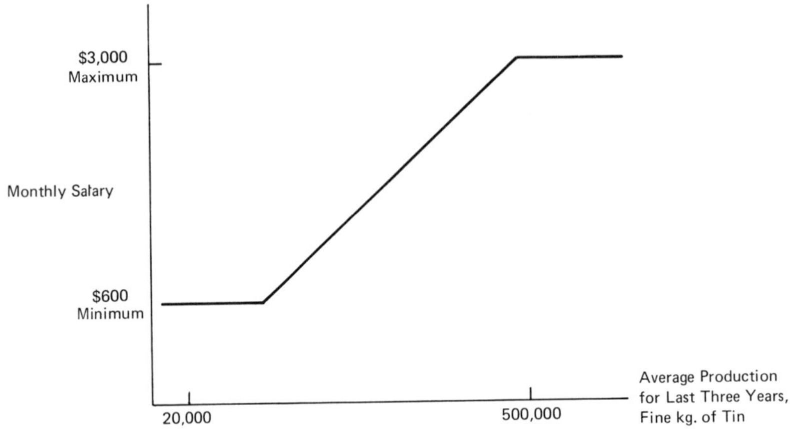

The equation that describes the maximum is the following: Upper limit on monthly salary = $\$600\ (1 + \frac{A}{120})$ provided A exceeds 20 thousand kilos. If A is less than 20 thousand, then the limit is \$600; where A is the excess of the average production of the firm for the past three years, in thousands of fine kilograms of contained metal, over 20 thousand (but A may not exceed 500). The \$600 base can be tied to a price index that reflects world inflation (such as the OECD member country index).

Note that the formula gives an upper limit to the amount of salaries that can be deducted. There will be reasons for allowing less than the upper limit. For example, if the manager spends little or only a small portion of his time in the enterprise, the deduction will be lowered accordingly. Or, if his responsibilities are not commensurate with the salary, the limit would be lower.

Appendix E

Methodology for Determining the Cost Adjustment Factor for Standard Mining Costs

The mining cost adjustment factor would be developed as follows. First, it is necessary to calculate the relative shares of each category of costs included in mining costs. These are shown in Table E-1.[a] The magnitudes of the relative cost shares are derived from information from the survey of medium mines and COMIBOL as summarized in Chapter Three.

The materials component is itself an aggregate of the seven major mineral inputs: ammonium nitrate, dynamite, drill steel and bits, chemicals, sacks for minerals, timber, and cement. The relative shares of these items in total materials costs are shown in Table E-2.[b] The major material inputs for which prices are readily available are recorded in Table E-3. The aggregate materials price increase is calculated in the following steps. First, a reliable measure of the rate of price increase for each component of the materials input adjustment factor must be identified. The rates for each component are then multiplied by the weights given in Table E-3, column 2, to obtain the aggregate materials price increase. This allows a determination of the rate of increase of the price of materials. Next, the price increases of the other components of mining costs listed in Table E-1 must be identified. Then, each rate of price increase is multiplied by the weights shown in Table E-1 to get the mining cost adjustment factor.

[a]These figures must be adjusted at regular intervals to reflect changes in the general structure of input costs.

[b]These figures are also derived from the survey of medium mines described in Chapter Three.

Table E-1. Input Shares[a] for Mining Costs for Production Cost Adjustment Factor

	Weight
Materials	.241
Labor	.616
Energy	.049
Transportation (purchased by mine) and insurance	.028
Depreciation	.066
	1.000

[a]Shares are derived from Appendix C, by excluding all realization costs, including transportation from the point of collection to the smelter, smelter costs and charges, insurance, and port charges, from total costs.

The mining cost adjustment performed by this method will be much more comprehensive and accurate than that which would be possible using the index prepared by the United States Bureau of Mines for the U.S. mining industry, which uses only three items: labor, materials and "other."[c]

The weights shown in Table E-1 and Table E-3 can be used for the next two to three years in calculating the mining cost adjustment factor as described above. It will be necessary to collect price information on all the major inputs listed in the tables every six months. Before making an initial operative calculation of production cost, this information must be collected. However, as an exercise to demonstrate the method of calculation, a heuristic calculation of the cost adjustment factor for late 1975 is provided.

The estimates of cost movements of major material inputs from January 1974 through mid-1975 are shown in Table E-4. Using the weights of Table E-3 discussed above, it may be determined that materials costs increased about 85.9 percent from January 1974 to late 1975. The cost movements of major materials as a group and other production inputs are shown in Table E-5.

Using the procedure described above, it can be determined that the mining cost adjustment factor for the period January 1974 to mid-1975 is 1.438. The use of the mining cost adjustment factor can be illustrated in the following exercise. Let us assume that a fair standard mining cost for tin, exclusive of all output taxes, was U.S. $1.50 per lb. in January 1974, of which U.S. $0.85 was mining costs

[c]See Edward Johnson, *Price Indexes for the Mining Industries* (U.S. Bureau of Mines Information Circular, 1965), pp. 42-43.

Table E-2. Share of Items in Materials Inputs—Medium Mines

Item	*1973*		*1974*		*Average Share 1973 and 1974*
	$b	*%*	*$b*	*%*	*%*
Ammonium nitrate	3,987,112	3.3	7,081,970	6.1	4.7
Calcium carbide	1,106,533	0.9	1,772,649	1.5	1.2
Sacks for minerals	14,577,616	11.9	9,454,973	8.2	10.1
Drill bits and drill steel	4,565,175	3.7	6,630,497	5.7	4.7
Carbide lamps	270,635	0.2	450,455	0.4	0.3
Rubber boots	1,229,898	1.0	1,609,535	1.4	1.2
Xanthate and other chemicals	5,760,630	4.7	7,267,526	6.3	5.5
Explosives (excluding ammonium nitrate)	5,777,215	4.7	8,514,282	7.3	6.0
Helmets, coats and safety equipment	1,854,729	1.5	3,017,775	2.6	2.0
Timber	5,289,808	4.3	9,980,242	8.6	6.4
Cement, mortar and plaster	9,108,558	7.4	3,829,627	3.3	5.4
General supplies	62,824,880	51.3	44,654,165	38.5	45.0
Other	5,966,128	4.9	11,674,310	10.1	7.5
TOTALS	122,399,910	100.0		100.0	100.0

Source: Survey of Medium Mines, 1973 and 1974.

Table E-3. Weights for Materials Input Adjustment Factor

	(1) Average percent of total materials 1973 and 1974	(2) Relative weight
Ammonium nitrate	4.7	0.11
Dynamite	6.0	0.14
Drill bits	4.7	0.11
Chemical products	5.5	0.13
Sacks for minerals	10.1	0.23
Timber	6.4	0.15
Cement	5.4	0.13
	42.8	1.00

Source: Calculated from data in Appendix C.

Table E-4. Estimated Cost Movements of Material Inputs, 1974-1975

		Jan. 1974	Apr. 1975	Increase %
Ammonium nitrate	ONIA	179.54[c,d]	362.75[c,d]	102.0
Dynamite	Tronador, 7/8" × 8", 60%	10.48[c]	27.44	161.8
Drill bits	16-714-39 cm.	20.44[a]	30.33[b]	48.4
Xanthate	Z-12	0.33[c]	0.56[c]	69.7
Sacks for minerals		7.01[d]	11.16[c,d]	59.2

[a]Average of 1973, 1974 dollar prices at plant, Swedish manufacturer. The price used for calculations of the mining cost adjustment factor *should* properly be inclusive of tariffs and taxes.

[b]August 1975 price at Swedish manufacturer.

[c]Provided by industry sources.

[d]Confirmed from data from import invoices.

as defined above and U.S. $0.65 was realization costs. The U.S. $0.85 figure would, in this example, represent base period standard mining costs. Then, using the mining cost adjustment factor derived above (1.438), we calculate that standard mining cost for tin in mid-1975 would be U.S. $0.95 × 1.438, which equals U.S. $1.22.

Table E-5. Cost Movements of Mining Inputs, 1974-1975

Labor	28.4%[a]
Transport and insurance (except as included under realization costs)	82.0%[b]
Energy	7.8%[c]
Materials	85.9%[d]

[a]This figure is an estimate of the increase in COMIBOL labor costs from January 1974 to April 1975, as reported in COMIBOL, *Informe de Labores, Septiembre de 1971-Mayo de 1975.*

[b]Rail freight.

[c]Price increase of electrical energy purchased by COMIBOL which occurred between September 1974 and March 1975.

[d]Calculated from materials cost estimates as reported in Table E-4, using weights shown in Table E-3.

Appendix F

Methodology for the Determination of Standard Realization Costs

INTRODUCTION

The Standard Realization Cost for a particular metal is the sum of Standard Smelting Costs (including typical bonuses or discounts), Standard Transportation Costs, and Standard Other Costs (insurance, port charges and fees). Of these three items the most important component to be measured for each mineral is the Standard Smelting Costs.

Standard smelting costs per unit of the refined mineral must be determined for each six-month period for each mineral. If there are important differences in smelting costs for differences in concentrates, then an average must be taken of the smelting costs of high and low concentrate ores. Transportation and other costs would be determined as an average for each unit of refined metal with a distinction made between the costs of shipping via Chile and Peru.

SMELTING COSTS

For mining firms that attempt to maximize net income, the decision as to the particular grade to sell is partially determined by the tradeoff between the increase in mining costs required to raise the grade of concentrate and the decrease in transportation and smelting costs that the higher grade of concentrate allows him to realize. Because the lower smelting costs for high concentrate ores are somewhat offset by higher marginal benefication costs if an average cost is used for standard mining costs, then an average smelting cost should also be used for all grades of concentrates for each individual metal.

To develop a measure of standard smelting costs for a metal that will be consistent over several periods of time, a specific set of parameters that are representative of the typical shipment must be defined and maintained through time. It is the smelting costs of this representative shipment that are used as the standard smelting costs component of presumptive costs.

For tin three representative shipments should be defined: one for low grades, one for medium grades and one for high grades. These could be defined as concentrates with 20 percent, 30 percent and 40 percent tin content, respectively. In addition, these representative shipments must have a specified level of impurities. The smelting contracts of COMIBOL should then be applied to these three representative shipments for each six-month time period beginning in January of 1974. The average of the smelting costs for these three representative shipments at a point in time is then used as standard smelting costs for the following six-month period. Only one standard smelting cost is used for all grades of tin concentrate. If a higher standard smelting cost was used for low-grade concentrate while the same standard mining cost was used for all shipments, then this would present an incentive to the mining sector to ship low-grade concentrates. This would not only result in economically inefficient practices but would be directly counter to expressed government policy of raising the grade of mineral concentrate exported from Bolivia.

Rather than using only COMIBOL'S smelting contracts to determine standard smelting costs, an average of COMIBOL'S contracts and of medium mine or international contracts could be used. However, this procedure would increase administrative difficulties of maintaining a consistent cost series through time, as the private sector would then have an incentive to give misleading information. If instead only COMIBOL'S smelting contracts are used the private mining sector has an incentive to attempt to negotiate more favorable contracts. In this way the efficiency of the private part of the mining sector would be improved, which is a goal of the government's mining policy.

A *representative shipment* has to be defined for each of the minerals for which standard smelting costs are required. COMIBOL'S smelting contract for the mineral is then applied to this representative shipment to determine standard smelting costs for the mineral at any point in time.

Smelting costs generally change through time either because the discount for grade changes, or because the energy and the pollution charges are altered. Therefore, by maintaining the same representative shipment or shipments for a mineral over time the changes in these costs will be measured accurately. This will not be the case if the type of shipment on which smelting costs are estimated is allowed to change from period to period.

STANDARD TRANSPORTATION COSTS

Standard Transportation Costs are estimated as an average cost (pesos) per unit of refined metal for transportation from the collection points in Bolivia to the foreign smelter. This involves estimating both the transportation costs for a mineral from the Bolivian collection points to the ports for ocean transport and from the South American ports to the foreign smelters.

To carry out this estimation the representative shipments, which were developed to estimate smelting costs through time, are again utilized. For these representative shipments the transportation costs from Bolivia to the port for ocean transportation are calculated as an average from the various collection points in Bolivia. A distinction should be made between those shipments that are transported through Chile and those through Peru. These overland transportation costs will be primarily determined by railway rates, but trucking costs could also be included in the average costs of transport. There will be two standard transportation costs for overland transport for each metal depending on whether or not the concentrate is shipped via Chile or Peru. (See Appendix A for an example of the transportation costs calculated for mines in Peru.)

Ocean transportation costs should be calculated as an average transportation cost per unit of metal using the information available through ocean transportation brokers or agents. COMIBOL regularly compiles these rates. As in the case of overland transport, these ocean transport rates are applied to the same representative shipments each time period to estimate the level and movement of ocean transportation costs.

The total Standard Transportation Costs including both overland and ocean transport costs can thus be calculated for each metal with a differentiation made between the country of the ocean port (Chile or Peru) through which the concentrate is shipped.

STANDARD OTHER COSTS

Using the representative shipments previously defined, other costs (including legal fees and port charges) would be calculated every six months for each metal and included in the total for Standard Realization Costs.

Appendix G

Revenue Estimates for the Present Tax Structure: Data, Assumptions and Methodology

Table G-1. Revenue Projections for Identifiable Taxes Paid by Mining Sector, Assuming No Changes in Basic Tax Structure or Tax Rates, Millions of Pesos (nominal terms)

	1974 (actuals)	*1975 (est.)*	*1976 (projected)*	*1977*	*1978*	*1979*	*1980*
I. *National government* (A + B)	1,486.0	842.6	1,286.5	1,515.7	1,637.2	1,724.0	1,867.5
A. Customs	1,473.5	828.5	1,271.2	1,499.2	1,619.3	1,704.6	1,846.5
1. *Regalias* (central government share only, inclusive of 10% share reimbursed to COMIBOL)[a]	881.8	449.8[c]	740.7[a]	888.9[a]	959.6[a]	1,014.3[a]	1,069.0[a]
2. Export taxes	481.4	290.4[c]	435.6[b]	508.3[b]	550.0[b]	608.5[b]	650.9[b]
3. Other customs (import duties, DIVISA tax on mining)	110.4	88.3[d]	94.9[d]	102.0[d]	109.7[d]	117.8[d]	126.6[d]
B. *Internal Revenue*	12.5	14.1	15.3	16.5	17.9	19.4	21.0
1. Landrents[e]	1.1	1.2	1.3	1.4	1.5	1.6	1.7
2. Income tax on commercializadores[f]	4.0	4.3	4.6	4.9	5.2	5.6	6.0
3. Gold and platinum[f]	5.4	5.8	6.2	6.6	7.1	7.6	8.1
4. Others[g]	2.5	2.8	3.2	3.6	4.1	4.6	5.2
II. *Earmarked taxes*	192.2	123.6	223.8	246.4	268.7	286.5	297.2
A. Subnational govt. share of royalties[h]	144.7	93.4	176.8	195.1	210.6	222.7	234.7
B. Educational institutions[i]	7.8	8.3	8.8	9.4	10.0	10.6	11.3
C. Dept. dev. corp.[j]	29.7	14.2	26.9	29.7	32.1	33.9	35.7
D. Govt. agencies (sulphur)[k]	0.9	1.0	1.1	1.2	1.3	1.4	1.5
E. Mining and Metallurgical Institute (IIM)[l]	9.1	6.7	10.2	11.0	11.8	13.0	14.0
III. *Earmarked taxes* (Internal Revenue)[m]	1.2	1.3	1.4	1.5	1.6	1.7	1.8
IV. *ENAF*[n]	29.5	21.0	29.1	35.6	38.4	40.4	42.4
Grand Total, taxes on mining sector (identifiable)	1,708.9	988.5	1,540.8	1,800.2	1,943.0	2,047.7	2,208.9

Notes:

a) (i) Price and quantity assumptions for 1976-1980 are as follows:

Mineral	*Quantity** (*in thousands of tons)*					*Price*** *per* fine pound *unless otherwise noted* *(US $)*				
	1976	*1977*	*1978*	*1979*	*1980*	*1976*	*1977*	*1978*	*1979*	*1980*
Tin	29.70	29.84	29.98	30.11	30.25	3.20	3.67	3.96	4.29	4.62
Copper	8.42	8.77	9.13	9.48	9.84	0.75	0.90	1.00	1.10	1.20
Zinc	51.34	53.17	54.99	56.8	58.69	0.37	0.40	0.43	0.47	0.50
Lead	21.02	22.12	23.21	24.30	25.39	0.27	0.27	0.28	0.30	0.33
Wolfram	3.21	3.38	3.55	3.72	3.89	80.00[1]	86.40[1]	93.31[1]	100.78[1]	108.24[1]
Antimony	14.44	14.93	18.42	15.91	16.42	18.29[1]	19.75[1]	21.33[1]	23.04[1]	24.88[1]
Silver	0.22	0.24	0.26	0.27	0.29	4.54[2]	4.90[2]	5.30[2]	5.72[2]	6.18[2]
Bismuth	0.77	0.80	0.84	0.89	0.93	5.06	5.46	5.90	6.37	6.88

*Authors' estimates of quantities, assuming some lags in implementation of certain key projects (e.g. copper, zinc). Metric tons unless otherwise noted.

**Authors' estimates, based largely but not wholly on IBRD forecasts.

[1] Per "unidad larga" (10.16 kilos) (wolfram, antimony).

[2] Per troy ounce (silver).

(ii) Although this table assumes that no significant *structural* changes will be made in taxes on the mining sector through 1980, the estimates do contain a built-in allowance for increases in presumptive costs over the period. [Note the significant increase for tin in 1976 and the more moderate increases thereafter.] These estimates, together with the actual presumptive costs for 1975, are as follows, for all minerals having a presumptive cost:

	Probable Presumptive Costs, 1976-1980–US $					
	Tin[a]	*Copper*[a]	*Zinc*[a]	*Lead*[a]	*Wolfram*[b]	*Antimony*[b]
1975						
Medium Mining	1.50	0.59	0.19	0.15	30.00	7.50
COMIBOL	1.55	0.60	0.195	0.155	31.00	8.00
Small Mining	1.60	0.61	0.20	0.16	32.00	8.50
1976						
Medium Mining	2.00	0.65	0.21	0.17	34.00	8.50
COMIBOL	2.05	0.66	0.215	0.175	35.00	9.00
Small Mining	2.10	0.67	0.22	0.18	36.00	9.50
1977						
Medium Mining	2.20	0.72	0.23	0.19	38.00	9.50
COMIBOL	2.25	0.73	0.235	0.195	39.00	10.00
Small Mining	2.30	0.74	0.24	0.20	40.00	10.50
1978						
Medium Mining	2.40	0.79	0.25	0.21	42.00	10.50
COMIBOL	2.45	0.80	0.255	0.215	43.00	11.00
Small Mining	2.50	0.81	0.26	0.22	44.00	11.50

	Probable Presumptive Costs, 1976-1980–US $					
	Tin[a]	*Copper*[a]	*Zinc*[a]	*Lead*[a]	*Wolfram*[b]	*Antimony*[b]
1979						
Medium Mining	2.70	0.87	0.28	0.23	46.00	11.50
COMIBOL	2.75	0.88	0.285	0.235	47.00	12.00
Small Mining	2.80	0.89	0.29	0.24	48.00	12.50
1980						
Medium Mining	3.00	0.96	0.31	0.25	50.00	12.50
COMIBOL	3.05	0.97	0.315	0.255	51.00	13.00
Small Mining	3.10	0.98	0.32	0.26	52.00	13.50

[a]Presumptive costs per fine pound.
[b]Presumptive costs per "unidad larga" (10.16 kilos).

a. (iii) The breakdown of projected royalties paid to the central government as between COMIBOL, the Medium Mining Sector and the Small Mining Sector is as follows:

	(in millions of Bolivian pesos)						
	1974 (actuals)	*1975 (est.)*	*1976*	*1977*	*1978*	*1979*	*1980*
COMIBOL	531.9	223.0	448.0	532.3	571.4	601.3	623.4
Medium Mines	228.2	166.8	216.2	264.4	288.8	307.5	331.6
Small Mines[1]	129.7	60.0	76.5	92.2	99.4	105.5	114.0
Totals	881.8	449.8	740.7	888.9	959.6	1014.3	1069.0

[1]Slower increase for small mining sector after 1975 reflects assumption that total production of metal content of various minerals in sector will grow at a much slower rate for most metals.

b) (i) For the export tax, export values are projected on the same basis as for royalties and are as follows:

	Projected Value of Exports (US $ millions)			
	COMIBOL	*Medium Mines*	*Small Mines*	*Total*
1976	$237.0	$ 90.2	$35.8	$363.0
1977	275.9	106.2	41.5	423.6
1978	302.9	117.9	45.3	466.1
1979	343.6	130.9	50.1	524.6
1980	371.5	144.4	55.1	571.0

b) (ii) Effective rates of export tax (collections as a percent of export values) were 6.00% in 1973 and 6.23% in 1974. Effective rates for 1976-80 are assumed to decline slightly from 1975 onward (assuming no discretionary changes in rates) owing to the declining tendency in recorded grade of concentrates that is expected (note that the export tax on tin is geared to the grade of the concentrate). Thus projected effective rates of export tax on gross export values are: 1976–6.20%; 1977–6.10%; 1978–5.90%; 1979–5.80%; 1980–5.70%.

[c]Ministry of Finance 1975 estimates (as of July 1975) for royalties and export taxes. Estimates assume that International Tin Council Quotas will remain in effect throughout 1975.

[d]Based on the following assumptions: For 1975, dutiable imports and purchase of exchange by the mining sector will fall by 20%, owing to slowdown in mining activity after March 1975. For the period 1976-1980, dutiable imports and purchase of exchange are both assumed to grow at 7.5% per annum in nominal dollar terms.

[e]Assumes that growth in patentes revenues over the period 1975-1980 will match the historical growth of 7.5% (compound) per annum.

[f]Assumes that growth in this revenue source over the period 1975-1980 will match the compound rate of growth projected for export values for the mining sector (6.9%) for the *six*-year period.

[g]Revenue growth from this source over the period is assumed to match historical (1965-74) patterns (12.5% compound growth).

[h]The subnational government share is calculated as 18% of gross projected royalties, projected on the same basis as in note (a) above.

[i]Revenues from this source (6 per mill tax on mining exports plus assorted other taxes) assumed to grow at same rate as projected export growth (mining sector only).

[j]Departmental development corporations: earmarked revenues to these entities depend on royalties. Revenues from this source after 1975 projected to grow at the same rate as projected mineral exports.

[k]Revenues from this source projected to grow at same rate as projected increase in value of sulphur exports (7%).

[l]Earmarked revenues accruing to IIM come from a 1.1% tax on a share of medium mining sector exports. Revenues from this source projected to grow at same rate as projected export taxes.

[m]This category includes a variety of earmarked taxes collected by *renta interna* from the mining sector (earmarked for certain educational institutes, certain municipalities etc.). Revenues from this source are assumed to grow at the same rate as projected mining sector exports over the 1975-1980 period.

[n]ENAFBOL receives the central government share of royalties on the export of 1,200 fine metric tons of tin each year. Note: for 1974 the central government share of royalties per metric ton of tin are projected as follows for the period 1975-1980:

1975–US $ 926 (assumes total 1975 production to be held to 24,300 tons due to ITC quotas)
1976–US $ 1,356
1977–US $ 1,490
1978–US $ 1,600
1979–US $ 1,684
1980–US $ 1,767

Appendix H

Details of Revenue Estimates of Interim Tax Reform Measures

I. Lower Revenue Target: Presumptive Tax Rate Schedule (A) and a 3% Production Royalty on Standard Export Value.

A. *Presumptive Income Taxes* (millions of pesos)

Mineral	*1976*	*1977*	*1978*	*1979*[a]	*1980*[a]
1. *Tin*					
Gross revenues	699.3	950.6	1,055.3	1,160.5	1,276.0
Central govt. share	573.4	779.5	865.1	951.6	1,046.3
Subnational share	125.9	171.1	790.2	208.9	229.7
2. *Antimony*					
Gross revenues	150.6	163.0	187.5	206.2	226.0
Central govt. share	123.5	133.7	153.6	169.1	195.3
Subnational share	27.1	29.3	33.9	37.1	40.7
3. *Wolfram*					
Gross Revenues	166.2	191.8	220.2	242.0	266.2
Central govt. share	136.3	157.3	180.6	198.4	218.3
Subnational share	29.9	34.5	39.6	43.6	7.9
4. *Copper*					
Gross revenues	5.0	21.6	26.5	29.8	33.5
Central govt. share	4.1	17.7	21.7	24.5	27.5
Subnational share	0.9	3.9	4.8	5.3	6.0
5. *Zinc*					
Gross revenues	66.7	79.3	97.8	112.4	123.7
Central govt. share	54.7	65.0	80.2	92.2	101.4
Subnational share	12.0	14.3	17.6	10.2	22.3

Mineral	*1976*	*1977*	*1978*	*1979*[a]	*1980*[a]
6. *Lead*					
Gross revenues	19.4	12.8	12.7	13.0	13.8
Central govt. share	15.9	10.5	10.4	10.7	11.3
Subnational share	3.5	2.3	2.3	2.3	2.5
7. *Bismuth*					
Gross revenues	6.8	7.7	9.0	9.8	11.2
Central govt. share	5.6	6.3	7.4	8.0	9.3
Subnational share	1.2	1.4	1.6	1.8	1.9
8. *Silver*					
Gross revenues	35.0	45.7	47.8	61.9	63.6
Central govt. share	28.7	37.5	39.2	50.8	52.1
Subnational share	6.3	8.2	8.6	11.1	11.5
Totals: Gross	1,149.0	1,472.5	1,656.8	1,835.6	2,014.0
Central	942.2	1,207.5	1,358.6	1,505.2	1,651.5
Subnational	206.8	265.4	298.2	330.4	362.5

[a]1979 and 1980 estimates are extrapolations. Others are direct computations.

B. *Other Taxes* (Central Government) (millions of pesos)

Tax	*1976*	*1977*	*1978*	*1979*	*1980*
1. Production royalty @ 3% of Standard Export Value[a]	147.0	177.9	195.8	231.0	252.0
2. Patentes	2.5	3.0	3.8	4.9	6.6
3. Other central govt. taxes	142.8	152.7	164.5	176.0	188.3
4. Recoupment of losses from abolishing excess production incentive[b]	8.1	18.7	53.4	61.4	70.6
5. Minus revenue loss under bona fide presumptive tax from new tax credit for exploration and development	–5.0	–18.8	–30.0	–50.7	–65.9
Total, central govt. (this table only)	295.4	342.3	387.5	422.9	451.9

[a]Standard Export Values (gross export value minus Standard Realization Costs) are assumed at 70% of gross export value.

[b]Recoupment from gradual abolition: Revenue loss from this incentive was B $30.4 million in 1974. By 1975, given low minerals prices for that year (and thus lower royalties and export taxes per unit), the cost should not have exceeded B $35 million, in spite of a larger number of firms taking advantage of the incentive.

Index

About the Author

Malcolm Gillis is an Institute Fellow in the multidisciplinary Harvard Institute for International Development, and Lecturer in Economics in the Department of Economics, Harvard University. His broad interests are in public finance, natural resource economics and economic development. In directing this study, he has drawn upon past experience in dealing with taxation, resource and general development problems in Colombia, Chile, Ghana and Indonesia as well as Bolivia. Among his most recent publications are *Fiscal Reform for Colombia* with Professor Richard A. Musgrave and "Incidence of World Taxes on Natural Resources" with Charles E. McLure.

Meyer W. Bucovetsky is an Associate Professor of Economics at the Institute for Policy Analysis in the University of Toronto. Prof. Bucovetsky brings over a decade of experience in mineral taxes issues, having served as a consultant to the Carter Commission on Taxation in Canada in the mid-sixties. Among other titles, he is the author of *The Taxation of Mineral Extraction* and, with R.M. Bird, *Canadian Tax Reform and Private Philanthropy.*

Glenn P. Jenkins is an economist with extensive experience in taxation problems and in the theory and practice of project evaluation. He has recently served as special adviser to the Department of Regional Economic Expansion in Canada, and as a consultant on mining tax issues for the Government of Indonesia. Among his most recent publications is: "United States Taxation and the Incentive to

Develop Foreign Primary Energy Sources" in Ballinger's *Studies in Energy Tax Policy*. At present, Mr. Jenkins is an Institute Associate at the Harvard Institute of International Development and a Lecturer in Economics, Department of Economics, Harvard.

Ulrich Peterson is a geologist and mining engineer with broad experience in mining problems and geological studies in Latin America, Asia, Europe and North America. He is Professor of Mining Geology in the Department of Geological Sciences at Harvard University. Professor Petersen is recognized as one of the leading authorities on Andean Geology. Among his representative publications are "Regional Geology and Major Ore Deposits of Central Peru" and "Metallogenic Provinces in South America."

Louis T. Wells, Professor of Business Administration, Harvard Business School, is co-author of a recent Ballinger volume, *Negotiating Third World Mineral Agreements*, with David N. Smith. He has served as a consultant in concessions policies in over a dozen nations. Professor Wells's most recent research includes work on multinational enterprises, choice of production technology in developing countries and investments of firms in other developing nations. Together with Gillis, Jenkins and Petersen, he is presently engaged in a study of mining taxes in Indonesia.

Brian D. Wright is Assistant Professor of Economics at Yale University. He has done extensive research on energy tax policy and resource economics, and is co-author, with Glenn Jenkins, of "Taxation of Income of Multinational Corporations: The Case of the U.S. Petroleum Industry." Professor Wright teaches public finance and resource economics at Yale, and has served as a consultant on energy tax policy for government agencies and foundations.